Intelligent Transportation Systems (ITS)

Intelligent Transportation Systems (ITS)

Editor

Beatriz L. Boada

MDPI • Basel • Beijing • Wuhan • Barcelona • Belgrade • Manchester • Tokyo • Cluj • Tianjin

Editor
Beatriz L. Boada
Universidad Carlos III de Madrid
Spain

Editorial Office
MDPI
St. Alban-Anlage 66
4052 Basel, Switzerland

This is a reprint of articles from the Special Issue published online in the open access journal *Electronics* (ISSN 2079-9292) (available at: https://www.mdpi.com/journal/electronics/special_issues/itsi).

For citation purposes, cite each article independently as indicated on the article page online and as indicated below:

LastName, A.A.; LastName, B.B.; LastName, C.C. Article Title. *Journal Name* **Year**, *Volume Number*, Page Range.

ISBN 978-3-0365-0506-0 (Hbk)
ISBN 978-3-0365-0507-7 (PDF)

Contents

About the Editor

Beatriz Lopez Boada received the Industrial Engineering degree from University of Carlos III de Madrid (UC3M), Spain, in 1996. In 1997, she joined the Systems Engineering and Automation Department of UC3M with an FPI scholarship from the Spanish Government to complete her doctorate studies. In July of 1997, she made a predoctoral stay at the Helsinki University of Technology University (Finland). In July of 2002, she received the Ph.D. degree with the thesis entitled: "Construcci6n de un mapa topo-geométrico del entorno y localizaci6n de manera simultánea", with a distinction "cum laude" given unanimously. In 2002, she joined the Mechanical Engineering Department of UC3M as a Ph.D. Assistant Professor. In 2002, she was hired as a Doctor Contracted Professor, and she assumed the role of Professor in 2007. Since December 2019, she has been Full Professor. Since 2017, she has been co-responsible for the research group "Experimental Mechanics, Calculations and Transport (MECATRAN)". At the Mechanical Engineering Department, her research activity is focused on vehicle dynamics; designing vehicle controllers to improve their handling, stability, and comfort; designing vehicle state observers; mechatronics; and artificial intelligence such as neural networks and fuzzy logic. Her research has been conducted in the framework of different research projects. She has participated in a total of 14 research projects: 1 European project, 10 national projects, and 3 regional projects. She has been the primary investigator of two national projects and three regional projects. All of them have been related to vehicle safety. Additionally, she has participated in seven contracts belonging to article 83 and four MARCO agreements with different enterprises (Indra, FITSA, Michelin España Portugal S.A., Recreativos Franco S.A., Construcciones Espaciales y Dragados S.A., CAF, METRO de Madrid). Her research has generated more than 80 publications, 33 of which are articles in JCR-listed journals, 54 of which are conference papers, and 3 of which are international research book chapters. She has been Director of three PhD theses and, nowadays, she is supervising four Ph.D. theses. She has been an editorial board member of an international journal and has been on the technical committee of five international congresses. She is a referee for more than 20 international journals, the majority of them JCR-listed. She has participated in the organization of three international congresses and two national congresses. She has been assessor of the EMPLEA program and research projects for the Convocatoria de Ayudas de la Dirección General de Tráfico, de la ANEP para ayudas para contratos Torres Quevedo.

Preface to "Intelligent Transportation Systems (ITS)"

In recent years, the growth in the number of vehicles in cities has caused problems of mobility, environmental pollution, and road safety. In order to face these challenges, the intelligent transportation system (ITS) concept includes many advanced technologies, such as communication, sensing, and control, which are used for managing a large amount of information. The use of new technologies in communications such as Internet of Things (IoT) or 5G in ITS applications is becoming fundamental for the development of new advances in fleet management, traffic control, and connected vehicles.

The articles contained in this book cover a broad range of topics in the context of ITS applications. One of the topics covered is fleet management, as seen in "A Predictive Fleet Management Strategy for On-Demand Mobility Services: A Case Study in Munich" and "Management and Control System for Medium-Sized Cities Based in Intelligent Transportation Systems: From Review to Proposal in a City". The articles "Machine Learning-Based Driving Style Identification of Truck Drivers in Open-Pit Mines" and "A Vehicle Type Dependent Car-following Model Based on Naturalistic Driving Study" deal with the problem of driving behavior. The papers "The Multi-Station Based Variable Speed Limit Model for Realization on Urban Highway", "Floating Car Data Adaptive Traffic Signals: A Description of the First Real-Time Experiment with "Connected" Vehicles", and "A Traffic Flow Prediction Method Based on Road Crossing Vector Coding and a Bidirectional Recursive Neural Network" concern with traffic control research. The trajectory planning problem is treated in "Masivo: Parallel Simulation Model Based on OpenCL for Massive Public Transportation Systems' Routes" and "Deep, Consistent Behavioral Decision Making with Planning Features for Autonomous Vehicles". The articles "The Need for Cooperative Automated Driving" and "Decision-Making System for Lane Change Using Deep Reinforcement Learning in Connected and Automated Driving" are focused on connected vehicles. Finally, the paper "Efficiency Analysis and Improvement of an Intelligent Transportation System for the Application in Greenhouse" deals with energy consumption efficiency.

This book can be used as a reference for professionals and engineers in both academia and industry that would like to develop an understanding of ITS, as well as for students.

Beatriz L. Boada
Editor

Article

A Predictive Fleet Management Strategy for On-Demand Mobility Services: A Case Study in Munich

Michael Wittmann *, Lorenz Neuner and Markus Lienkamp

Chair of Automotive Technology, Technical University of Munich, 85748 Garching, Germany; lorenz.neuner@tum.de (L.N.); lienkamp@ftm.mw.tum.de (M.L.)

* Correspondence: michael.wittmann@tum.de

Received: 18 May 2020; Accepted: 16 June 2020; Published: 19 June 2020

Abstract: The global market for MoD services is in a state of rapid and challenging transformation, with new market entrants in Europe, such as Uber, MOIA, and CleverShuttle, competing with traditional taxi providers. Rapid developments in available algorithms, data sources, and real-time information systems offer new possibilities of maximizing the efficiency of MoD services. In particular, the use of demand predictions is expected to contribute to a reduction in operational costs and an increase in overall service quality. This paper examines the potential of predictive fleet management strategies applied to a large-scale real-world taxi dataset for the city of Munich. A combination of state-of-the art dispatching algorithms and a predictive RHC optimization for idle vehicle rebalancing was developed to determine the scale by which a fleet size can be reduced without affecting service quality. A simulation study was conducted over a one-week period in Munich, which showed that predictive fleet strategies clearly outperform the present strategy in terms of both service quality and costs. Furthermore, the results showed that current taxi fleets could be reduced to 70% of their original size without any decrease in performance. In addition, the results indicated that the reduced fleet size of the predictive strategy was still 20% larger compared to the theoretical optimum resulting from a bipartite matching approach.

Keywords: mobility-on-demand; fleet-management; taxi; predictive strategies; RHC

---

1. Introduction

The global mobility market is undergoing considerable changes, primarily due to developments in connectivity, autonomous driving, electrification, and shared mobility. Besides these four major technological advances, trends in population growth, shifting consumer behavior, and sustainability serve to promote these disruptive changes, especially in urban areas.

These developments are set to bring about multiple challenges in terms of sustainable traffic flow management and an increased demand for transport, which future megacities will have to face. Most notably, the demand for mobility as a service (MaaS) is expected to grow disproportionately in urban environments. Viewed in more detail, the global MaaS market capitalization is estimated to quadruple from a net present value of less than $2 trillion in 2017 to approximately $9 trillion in 2030 [1]. The rapid emergence of such new entrants as car or ride sharing companies, as well as software producers for mobility purposes serves to confirm this development.

Taxi drivers, in particular, are worried about losing their monopolistic position. This has resulted in frequent protests, especially in major European cities such as London, Paris, and Barcelona. Demonstrations against online ride hailing operators such as Uber, CleverShuttle, or MOIA have also taken place in some German cities, such as in Berlin in February 2019 [2]. In the same context, the great competitive pressure among new entrants to the electric scooter market (Lime, Bird, TIER, etc.) demonstrates the increasing need for cost efficiency among on-demand operators. The predicted growth of the global fleet management market from $15.9 billion in 2019 to $31.5 billion in 2023 confirms this need [3].

Thanks to the availability of real-time data (e.g., GNSS positions, vehicle occupancy, or customer demand), supply and demand predictions can be used to optimize the routing, customer matching, and scheduling of entire fleets. In particular, it is anticipated that the use of demand predictions will lead to a reduction in empty fleet mileages and, in turn, to a decrease in variable costs, such as fuel and wear and tear. It is therefore quite possible that predictive fleet management strategies will increase the operational efficiency of operators, thus giving this industry a competitive advantage.

2. Outline

The objective of this paper is to demonstrate the benefits of a predictive fleet management strategy on operational efficiency, with the example of a taxi fleet in Munich. It is anticipated that a fleet strategy relying on demand predictions might both improve the service level measured in terms of customer waiting time and reduce the fleet size in terms of the number of vehicles. To test this hypothesis, two predictive fleet management strategies were developed and simulated and subsequently compared to the current strategy of a local taxi fleet.

The remainder of this paper is structured as follows: In Section 3, we present an overview of relevant publications in the field of passenger demand prediction, fleet sizing, dispatching, and rebalancing. Section 4 gives an overview of the underlying real-world dataset of a taxi fleet in Munich. Section 5 presents the implementation of the real-world reference strategy along with a profit-maximizing predictive fleet management strategy and an optimal fleet sizing approach. Section 6 describes the implementation of the strategies defined above and the chosen simulation parameters. Section 7 presents the results of a one-week simulation of the different fleet management strategies and their potential in optimizing overall fleet efficiency. We end with a discussion of our findings (Section 8) and a final conclusion in Section 9.

3. Related Work

The optimization of mobility-on-demand (MoD) services is of enormous importance as an emerging field of research, as real-world datasets become more and more accessible to the research community [4–6]. This paper focuses on the most relevant publications on the subjects of fleet sizing, dispatching, and rebalancing methods. It primarily highlights publications that utilize demand forecasts in combination with fleet control mechanisms to increase the efficiency of MoD services.

3.1. Passenger Demand Prediction

Methods of passenger demand prediction range from neural networks to simple probabilistic estimations. Oda and Joe-Wong [7] relied on a convolutional neural network to forecast pickup locations. Xu et al. [8] used two different neural networks to predict absolute demand and the destination distribution in an area. On the other hand, Miao et al. [9] used probability functions based on historic and real-time data to predict future demand. Tsao et al. [10] assumed a spatial and time variant Poisson process to predict the customer arrival rate over the following two hours with a short-term memory (LSTM) network. Dandl et al. [11] used a time-varying Poisson model, which was based on the historic average over a three-month-long period.

Further research employs a time-invariant Poisson process with a constant customer arrival rate to account for future demand, such as proposed by [10,12–14]. They rely on the same probabilistic process that does not consider potential fluctuations in customer demand. Existing ride-sharing approaches likewise utilize constant arrival rates [15,16] or historic probability distributions [17] that account for future demand without forecasting.

3.2. Dispatching and Rebalancing Strategies

Taxi fleet management techniques range from simple heuristic approaches to bipartite matching and multi-objective optimization models and on to negotiation-based and deep-learning methods. Maciejewski and Bischoff [18], Maciejewski et al. [19] applied a "nearest-idle-taxi" heuristic to large-scale taxi datasets from Berlin and showed the benefit of a combined heuristic relying on a "nearest-open-request" dispatch in the case of undersupply. Bipartite matching techniques minimize the total waiting times and arrival distances when matching a set of idle taxis and requests over a period of time [20,21]. Other mathematical optimization approaches typically aim to minimize a fleet's idle distances, supply-demand gap [9], and rebalancing flows [12] or else maximize driver reward [7]. These single- or multiple-objective optimization approaches are commonly formulated in the form of an integer linear program (ILP) or a mixed integer linear program (MILP). By solving two linear programs, Tsao et al. [10] used bipartite matching in conjunction with a rebalancing approach. Dandl et al. [11] combined a predictive rebalancing and matching approach in a single ILP.

Ruch et al. [22] presented a comparison of different fleet operating strategies [12,23,24] applied to real-world use cases in Chicago, San Francisco, and Zurich. To ensure comparability with real-world conditions, they developed an algorithm to estimate traffic flow based on taxi trip metrics.

Negotiation-based fleet management methods such as proposed by Seow et al. [25,26] allow for the exchange of information among several drivers in order to generate more efficient dispatching solutions. The authors were able to achieve a reduction in customer waiting time using this decentralized approach as compared with a centralized nearest-taxi heuristic. Deep-learning approaches were applied to manage the coordination of taxi fleets and commonly aspired to optimizing the behavior of an individual driver. Based on a set of different actions, the maximization of the driver reward function was shown in [7,27,28]. Oda and Joe-Wong [7] showed that a reinforcement-learning approach in the form of a deep Q network (DQN) could achieve lower waiting times and reject rates than a centralized multiple-objective optimization technique, but based on different time steps.

3.3. Fleet Sizing Strategies

The goal of fleet sizing is to estimate the number of vehicles required over a period of time, to minimize costs from the operator perspective. Bipartite matching approaches are commonly used to find the minimum fleet size, by integrating as many customer trips into the route of a vehicle [20,29]. A drawback of this method is that it assumes perfect knowledge of future customer demand over the respective period. This means that the origin and destination, as well as the start and end times of every customer ride are known over a period of time. By applying full demand knowledge, Vazifeh et al. [20] computed an average minimum fleet size for New York that was 40% smaller than its current taxi fleet. Similarly, Reference [29] showed that two thirds of New York's fleet would be sufficient for handling demand at most times of the day.

Zhu and Prabhakar [30] utilized a network flow model that discretized New York in time and space to approximate the minimum number of yellow cabs required. The authors found that 72% of the current fleet size could handle the rides occurring over a 12-hour period.

Further literature sources have investigated the impact of various fleets of constant sizes on such metrics as customer waiting time, rejection rate, or vehicle availability [11,20,31,32]. Vazifeh et al. [20] employed a batch-based matching procedure to maximize the number of clients served for fleet sizes that exceeded the optimal quantity. This method could serve over 90% of requests with a fleet 20% larger than the theoretical minimum, but only if a customer delay of 6 min was accepted. Hörl et al. [31] sized an autonomous fleet for Zurich such that the waiting time remained acceptable for a variety of fleet management strategies. Bischoff and Maciejewski [32] determined the autonomous taxi fleet size for Berlin by ensuring that utilization levels reached their limits only temporarily during peak hours and average waiting times stayed low. Spieser et al. [33] relied on a queueing network and a Poisson process to determine Singapore's fleet size, such that waiting time and supply availability were acceptable for two different time slots of the day.

3.4. Conclusion and Research Question

The literature analysis showed that only some studies relied on time-varying demand forecasts [7,8,10,11]. Oda and Joe-Wong [7] presented two predictive fleet management strategies, the first involving decentralized reinforcement learning and the second using an optimization approach based on centralized receding horizon control (RHC). Both strategies incorporated a local dispatching heuristic. This research issue was in line with the goal of this study: to improve the current taxi dispatch system by taking demand predictions into account. [21–24,31] presented various fleet management methods for large-scale real-world scenarios in San Francisco, Chicago, and Zurich. Since the way the taxi business works differs greatly from place to place, our study contributed a new real-world scenario of a representative German city to render the results globally comparable.

In our previous work [34], we mentioned that the current fleet size of the underlying dataset may not be optimal. Consequently, a second objective of this study was to compute an optimal fleet size and test the extent to which it could be attained using a predictive strategy. Existing predictive fleet strategies [7,8,11] assumed a constant fleet size over the simulation period, which did not, however, reflect real-world conditions. Some fleet-sizing approaches simply increase the number of simulation scenarios to determine an appropriate fleet size [11,20,31,32]. For instance, Dandl et al. [11] performed test simulations before selecting a sufficiently large fleet size, thereby allowing the predictive RHC optimizer to perform rebalancing actions. Vazifeh et al. [20] computed an optimal fleet size and upscaled it to test the performance of a reactive real-time dispatching method. However, the authors chose not to rely on demand predictions. To the best of our knowledge, there is no predictive strategy that applies a time-dependent fleet size to the taxi sector.

This led us to formulate the following two research questions:

- To what degree can the current fleet size be reduced with the aid of a predictive fleet strategy without decreasing the service quality?
- To what extent can a predictive fleet management strategy achieve the optimal fleet size with respect to a defined service level?

4. Reference Dataset

This study relies on a reference scenario based on real-world FCD provided by a local taxi agency in Munich. The total dataset currently comprises taxi rides from March 2015 to April 2020. This study focuses on the period from 1 January 2018 to 31 December 2018. The test fleet under consideration contained 490 vehicles in 2018 (14% of the total taxi fleet in Munich). An initial analysis of the dataset over a 19-week period in 2015 as already presented in [34].

Table 1 summarizes the key metrics of the dataset. In total, this dataset contained 1.87 million customer rides for 2018, with a total distance of 9.4 million km. This corresponded to an average customer trip length of 5.02 km. The entire intra-urban distance covered by the fleet totaled 16.97 million km in 2018. This implied that a vehicle traveled on average 55.39% of its total distances with a customer. Moreover, the mean speed of all customer rides was 22.31 km/h.

Table 1. Key metrics for reference dataset from 1 January 2018 to 31 December 2018.

Metric	Total	Occupied	Empty	Unit
Number of rides	4,642,962	1,873,599	2,769,363	-
Distance (sum)	16,965,948	9,397,954	7,567,994	km
Speed (mean)	18.75	22.31	16.34	km/h
Trip distance (mean)	3.65	5.02	2.73	km
Trip duration (mean)	12.73	13.88	11.96	min

Figure 1a shows a typical taxi demand curve in Munich for 2018. As found in [34], the demand was mainly affected by repetitive patterns on a weekly timescale. On regular weekdays, taxi demand was virtually identical from Monday to Thursday. The lowest number of requests was received in the early morning hours. The daily demand peaks in the morning and afternoon were about half the size of the weekly demand maximum. The weekly maximum was reached on Friday and Saturday evenings. Disregarding individual deviations during holidays or special events such as Oktoberfest, the taxi demand was regular, which resulted in a small standard deviation over 52 weeks. Figure 1b shows the average fleet utilization rate for 2018, aggregated at five-minute intervals. It can be seen that the mean utilization oscillated between 32% and 56%, with an overall mean of 41%. The highest measured utilization rate of 75% was observed at 1:55 a.m. on the night of New Years Eve 2018. This showed that the fleet spent on average 59% of its time in an idle state. These findings supported the assumption that a reduction in fleet size would offer huge leverage to reducing operational costs.

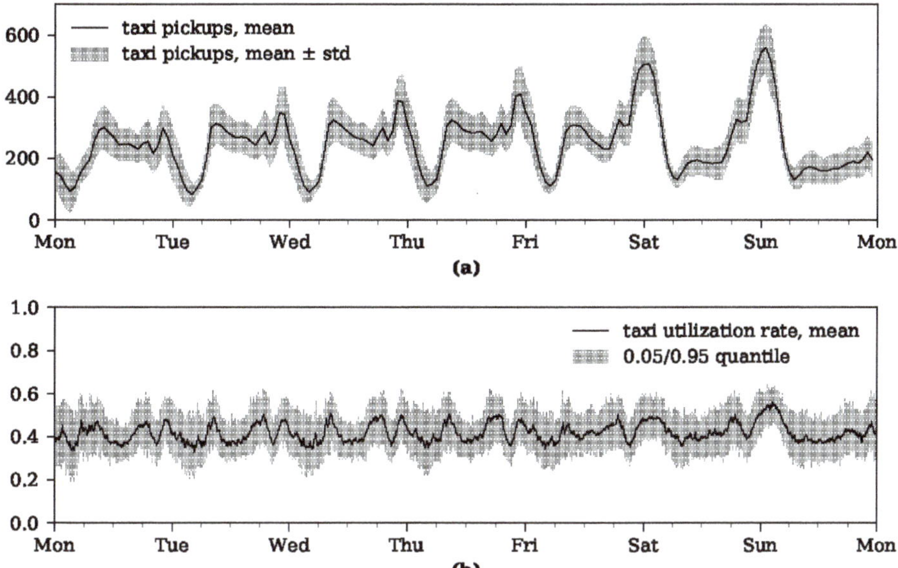

Figure 1. Demand and supply metrics for the reference dataset: (**a**) hourly aggregated taxi pickups in 2018; (**b**) taxi utilization rate in 2018 aggregated at 5-min intervals.

5. Fleet strategies for MoD Services

5.1. Munich Reference Strategy

In order to compare new fleet strategies, it is essential to build a reference scenario based on real-world circumstances. For this reason, the present strategy of a local taxi agency was modeled as a reference strategy. The strategy mainly resulted from the fleet management software in place. This software is an industry standard that is used in a multitude of European taxi agencies. Therefore, it was likely that similar results could be achieved in other major European cities.

In the case of Munich, the dispatching strategy relied on a heuristic based on defined sectors and ranks. For this purpose, the city was divided into 165 non-overlapping sectors and 167 taxi ranks, as displayed in Figure 2.

In general, the strategy resembled the nearest-idle-taxi dispatch approach, with fist come, first served (FCFS) sequencing on a regional scale. However, taxis awaiting customers at a rank had priority over taxis idling within the same zone. This dispatching strategy contained four major steps for assigning immediate requests to vehicles (cf. Figure 3):

1. The dispatching center first assigned an incoming request to a sector based on the start location. If at least one taxi was waiting at a rank within the same sector, the request was proposed to the first taxi in the queue.
2. If no taxis were present at a rank of the same sector, the ride was assigned to the closest idle taxi within the sector.
3. In case of unavailability, the dispatching unit looked for idle taxis at ranks of three predefined neighboring priority sectors. The dispatching center sequentially searched for available taxis at the ranks of these sectors.
4. If no taxis were present at the ranks of the neighboring sectors, open requests were offered to taxis idling in a neighboring sector.

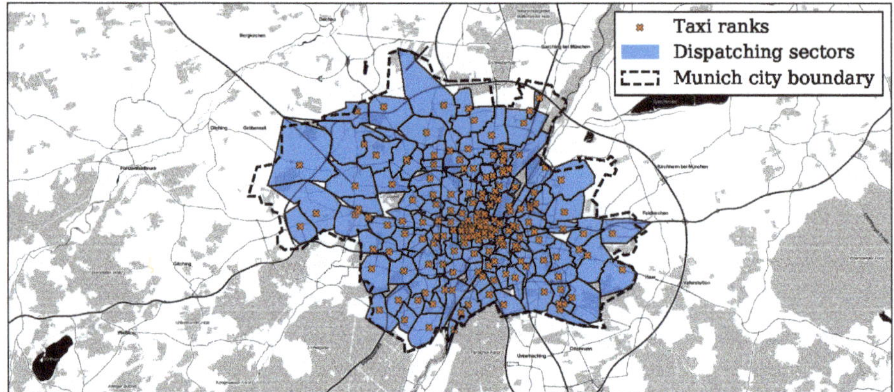

Figure 2. Overview of sectors and ranks used for a dispatching heuristic in Munich (map tiles by Stamen Design under CC BY 3.0; data by OSM [35] under ODbL).

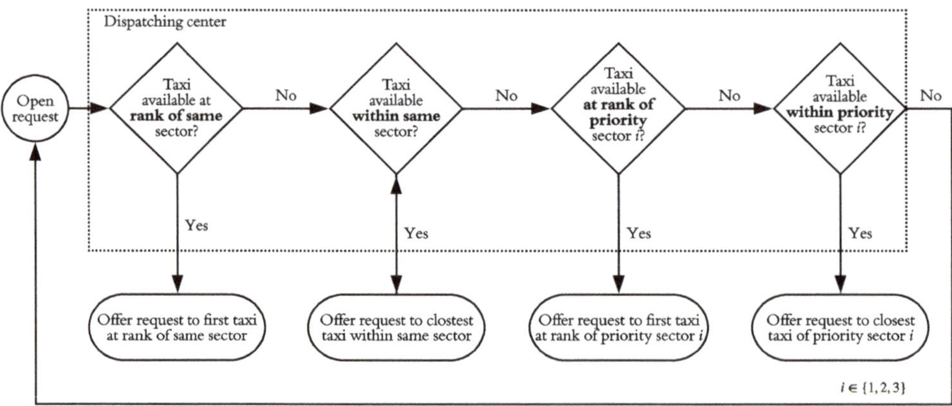

Figure 3. Reference dispatching strategy.

Advance bookings with at least 15 min of reservation time were only offered to taxis waiting at a rank. Hence, the current strategy did not consider idling taxis when allocating advance bookings to vehicles.

The present fleet strategy did not involve any rebalancing of vehicles. After a customer ride, taxi drivers usually returned to sectors/ranks where they assumed the best chance of obtaining a profitable ride. As this individual choice was hard to model with the underlying dataset, the model assumed that taxi drivers returned to the nearest (shortest distance) rank following a customer drop-off.

5.2. A Profit-Maximizing Predictive Fleet-Management Strategy

As evidenced by the literature analysis, heuristic and bipartite matching models are state-of-the-art dispatching methods. To conduct matching between taxis and clients, the proposed strategy relied either on a supply-demand balancing heuristic [18] or a bipartite matching model minimizing customer waiting time. Besides these two regional dispatching methods, a dual-objective RHC-based linear optimization model was used to relocate idle vehicles among regions of over-/under-supply, based on the predicted passenger demand. Both methods were subsequently compared to each other to determine the best predictive strategy.

5.2.1. Dispatching

This analysis compared two state-of-the-art dispatching methods (NTNR, BPM) matching idle vehicles to open requests.

Supply-Demand-Balancing Strategy: NTNR

The first dispatching method used the NTNR approach applied by Maciejewski and Bischoff [18], Maciejewski et al. [19,36]. This approach differentiates between periods of over- and under-supply. If the number of idle vehicles exceeds the number of open requests (oversupply), a client is assigned to the nearest taxi in FCFS order. In contrast, if the number of open requests is higher than the amount of idle taxis (undersupply), then a taxi drives to the closest customer. This matching procedure might lead to longer waiting times for individual customers in times of undersupply, since vehicles only pick up the nearest clients. On the other hand, it decreases average waiting times when demand is high, in comparison to a simple nearest-idle-taxi heuristic [18,19,36]. The implemented NTNR algorithm had a complexity of $\mathcal{O}(R \log T)$ for undersupply and respectively $\mathcal{O}(T \log R)$ for oversupply, with T as the number of taxis and R as the number of requests. To utilize the potential of demand forecasts fully, this strategy was applied at regular time intervals (e.g., 1 min) to the entire set of open requests and idle vehicles. Thus, the dispatching unit periodically collected incoming requests and idle taxi data over the entire area of Munich. By conducting matching over the whole area within Munich's borders, it was possible to increase the efficiency in terms of space. This excluded inefficient matchings, which occurred particularly at the borders of sectors such as performed by the currently practiced allocation strategy per sector. By neglecting partitioning in space, higher runtimes were accepted in favor of a better solution with respect to customer waiting time. This solution might only be practicable to a limited extent for real-time applications.

Maximum Bipartite Matching

The second dispatching strategy minimized the average customer waiting time. Here as well, idle vehicles and open requests were collected over a short period in time and matched in such a way that total waiting times were minimized. This optimization strategy is familiar from the literature as the minimum cost maximum bipartite matching technique [20,21,29]. Here, the set of incoming requests and idle vehicles was divided into two disjoint node sets. In this case, the edge cost between a vehicle and a request node represented the waiting time a client experienced when he or she was picked up by the respective taxi. A minimum cost maximum matching algorithm could be used to find the customer-vehicle allocation with the lowest total waiting time. A drawback of this approach is that several customers might suffer long waiting times in favor of the common good. From a holistic viewpoint, this approach was optimal with respect to the community. Since this study assumed constant hourly speeds, the minimized arrival times and arrival routes of taxis coincided. Thus, this technique helps a fleet operator to increase short-term profit, since the total arrival costs are minimal. However, if this results in customers who experience exceptionally long waiting times not returning as clients, this strategy can have a negative impact on an operator's long-term profits. Similarly, a maximum bipartite matching approach was applied to the entire area of Munich for a given interval period. As mentioned above, this may lead to an increase in runtime in favor of a better matching solution. The bipartite graph made up of customer requests and idle vehicles grew in space and time, i.e., the number of nodes increased linearly, while the number of edges increased quadratically with respect to the number of incoming requests and vehicles.

5.2.2. Rebalancing

Our approach was an extension of the RHC approach formulated by Oda and Joe-Wong [7]. To recapitulate, this linear optimization located the optimal set of rebalancing flows over several discrete time horizons by maximizing the total expected reward. However, this approach rejected all requests that could not be served within a time step. Therefore, this approach was extended by considering open requests for future optimization steps. In addition, Oda and Joe-Wong [7] assumed that vehicles departed a cell at the end of a time step and arrived at the target cell at the beginning of the following time step. Since this assumption did not reflect real-world circumstances, our adaption also considered a travel time of up to one rebalancing period Δt_{reb}. Both extensions were needed to assure realistic customer servicing and supply behavior.

The resulting optimization problem could be formulated as follows: The profit to be maximized was defined as the incremental income (additionally or previously served customers) minus the supplementary costs incurred when picking up these clients. Furthermore, the space was discretized into a set of two-dimensional grid cells G. We define the supply (number of idle vehicles) within a cell, g_i, at a given point in time t as s_i^t. Similarly, we define the demand (number of open requests) in g_i between t and $t + \Delta t_{\mathrm{reb}}$ as d_i^t. The optimization function is limited such that no additional income is generated if the supply exceeds the demand in a cell. Consequently, $\min(s_i^t - d_i^t, 0)$ up to the equilibrium of supply and demand.

By summing over all grid cells, this term counts the number of requests additionally or earlier served by rebalancing vehicles and is therefore proportional to the supplementary reward of the fleet operator. The costs of rebalancing movements result from the number of vehicles r_{ij}^t that are sent from g_i to g_j at a time t multiplied by variable costs, c_{ij}, based on the distance between g_i and g_j.

Since customers in undersupplied cells might be served at a later time, a weighting parameter λ is added to reflect the additional income from previously served requests. Hence, the weighting parameter can be interpreted as the value of time (VoT) of a customer and a driver. Furthermore, a second weighting parameter $\gamma(t)$ discounts the profit of future time steps in the objective function to account for prediction uncertainty. Since the customer demand in future time steps is more uncertain than the incoming demand of the next interval, near-term requests are assigned with a higher monetary value than long-term requests. In addition to maximizing the total profit over time, a set of constraints must be respected, as follows:

1. The number of rebalancing vehicles leaving cell g_i is limited by the number of available vehicles present in the respective cell. Equation 2 expresses this condition when inserting the supply definitions from below.
2. All rebalancing flows must be completed within a rebalancing period Δt_{reb}. This is ensured by setting Δt_{reb} such that the cell-to-cell distance can be covered on the basis of average hourly speed $v(t)$. In so doing, inter-cellular distances are approximated through street distances q_{ij} between cell centers g_i and g_j.

Finally, the optimization problem is formulated as:

$$\max \sum_{t=t_0}^{t_0+T} \gamma(t) \left(\sum_i \lambda \min(s_i^t - d_i^t, 0) - \sum_{i \neq j} c_{ij} r_{ij}^t \right) \tag{1}$$

restricted to:

$$s_i^t \geq 0 \quad \text{where} \quad t = t_0, ..., t_0 + (T-1), \quad \forall i \in G \tag{2}$$

and:

$$r_{ij}^t = 0 \quad \text{where} \quad \{i, j, t | d_{ij} \geq v(t)\Delta t_{\text{reb}}\}. \tag{3}$$

Describing the vehicle supply, we define:

$$s_i^t = x_i^t - \sum_{j=1}^{N} r_{ij}^t \quad \text{if} \quad t = t_0, \tag{4}$$

where x_i^t is the number of idle vehicles at time t in cell g_i and:

$$\begin{aligned} s_i^t = {} & \max(s_i^{t-1} - d_i^{t-1}, 0) + x_i^{t-1,t} + \sum_{j=1}^{N} (r_{ji}^{t-1} - r_{ij}^t) \\ & + \sum_{\tau=t_0}^{t-1} \sum_{j=1}^{N} 1\left(t = \tau + \left\lceil \frac{q_{ji}}{v(t)\Delta t_{\text{reb}}} \right\rceil\right) \mathbb{P}_\tau(i|j) \min(s_j^\tau, d_j^\tau) \quad \forall t > t_0, \end{aligned} \tag{5}$$

where:

- $\max(s_i^{t-1} - d_i^{t-1}, 0)$ is the number of idle vehicles remaining from the previous time step,
- $x_i^{t-1,t}$ is the number of occupied vehicles becoming empty between $t-1$ and t,
- r_{ji}^{t-1} is the number of rebalancing vehicles from the previous time step arriving at g_i, and r_{ij}^t is the number of vehicles, which are sent off from g_i at the beginning of the current interval.
- Furthermore, the number of currently idle vehicles predicted to pick up a customer in region g_j and drop off a customer in region g_i within Δt is estimated on the basis of the historic likelihood $\mathbb{P}_t(i|j)$ of possible origin-destination (OD) pairs in combination with the expected travel time of a customer ride from g_i to g_j depending on $v(t)$.

We define the customer demand over a period as:

$$d_i^t = \begin{cases} \hat{d}_i^t, & \text{if} \quad t = t_0 \\ \hat{d}_i^t + \max(d_i^{t-1} - s_i^{t-1}, 0), & \forall t \geq t_1 \end{cases}, \tag{6}$$

where $\hat{d}_i^t$ is the predicted number of requests for time step t, and $\max(d_i^{t-1} - s_i^{t-1}, 0)$ is the amount of remaining open requests from the previous interval.

5.3. An Optimal Fleet Sizing Approach

To test the extent to which this predictive strategy could achieve the optimal fleet size, we utilized a bipartite matching fleet sizing technique. As mentioned above, this approach relies on full demand knowledge over a period $\bar{H}$. It is assumed that a customer ride u_i can be uniquely described by the pickup and drop-off locations, $l_i^{\text{p}}, l_i^{\text{d}}$ and their timestamps, $t_i^{\text{p}}, t_i^{\text{d}}$, such that:

$$u_i := ((l_i^{\text{p}}, t_i^{\text{p}}), (l_i^{\text{d}}, t_i^{\text{d}})). \tag{7}$$

Minimizing the number of taxis for a set of customer rides corresponds to solving the minimum path cover problem on the corresponding directed graph $G(U, E)$ [20,29]. Thereby, the nodes of two trips u_i and $u_j \in U$ are only linked with an edge $e_{ij} \in E$ if ride u_j can be carried out after ride u_i by the same vehicle. This is only the case if the ride u_j starts after the drop-off of a previous trip u_i and if the driving distance θ_{ij} between d_i and p_j can be covered within that time. Furthermore, the idle time of taxis between trips is limited by a maximum empty trip time δ, as follows:

$$t_i^{\mathrm{d}} < t_j^{\mathrm{p}} \tag{8}$$

$$\theta_{ij} \leq (t_j^{\mathrm{p}} - t_i^{\mathrm{d}})v(t) \tag{9}$$

$$(t_j^{\mathrm{p}} - t_i^{\mathrm{d}}) \leq \delta \tag{10}$$

Zhan et al. [29] showed that solving the minimum path cover problem corresponds to solving a maximum bipartite matching procedure on the equivalent bipartite graph. Hence, $G(U,E)$ is transformed into a bipartite graph $G(U',E')$, such that the origin and destination vertexes represent two disjoint node sets ($U'^{\mathrm{p}} \cap U'^{\mathrm{d}} = \emptyset$). The maximum bipartite matching problem can then be solved with a maximum flow algorithm after adding a source and sink node to G' with respective connections to all destination and origin nodes. Although the problem is NP-hard, it can be solved in polynomial time, for instance, by applying the Hopcroft–Karp algorithm with a complexity of $\mathcal{O}(n^{5/2})$ where n represents the number of nodes of the bipartite graph [37].

6. Simulation Model and Parameters

6.1. Simulation Model

To compare the predictive BPM and NTNR strategies with Munich's state-of-the-art method, this approach made use of a modular simulation model that combined independent functions; see Figure 4.

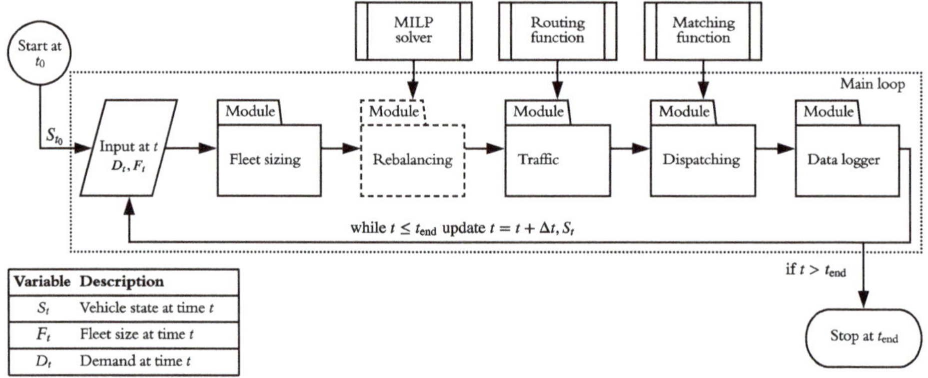

Figure 4. Structure of the simulation model.

First of all, the fleet sizing module adjusts the number of active vehicles as a function of the input fleet size F_t. If the current fleet size is lower than the input size, the module adds new vehicles to the simulation. If the number of active vehicles exceeds the desired fleet size, the module logs off idle taxis, until equilibrium is achieved. Then, the rebalancing module periodically directs idle vehicles to undersupplied cells by solving the MILP resulting from the linear program of Section 5.2.2 (only applies to predictive strategies). Finally, the traffic module routes the rebalancing and occupied vehicles throughout Munich's street network. The road network used relied on data from OSM [35]. All routes (arrival, customer, and rebalancing) resulted from Dijkstra's shortest path algorithm. Hence, the distance of every trip was minimized on the basis of the total edge length from the origin to the destination (speed differences between road classes were neglected). In the case of rebalancing vehicles, idle taxis were sent from their current location to the center node of the destination cell in accordance with the current optimal solution r_{ij}^{0}.

The dispatching module assigned customer requests to idle vehicles in accordance with one of the policies described in Section 5.2.1.

6.2. Simulation Parameters

For this study, a representative week in the year 2018 (11 May 2018 to 11 November 2018) was simulated. The simulated period comprised 35,602 customer requests, which reflected the average weekly demand in 2018. As the fleet of the reference dataset mainly operated in the city of Munich, only rides that started and ended within the city boundaries were considered. Furthermore, rides to and from the airport were neglected, as the reference dataset did not contain other taxi operators who mainly served this area. With a view to decreasing current costs, the current fleet size was either down- or up-scaled to the optimal fleet size. Since an analysis of the dataset showed that the real fleet size varied on a minute scale, the number of active vehicles was adjusted every 5 min so as to reflect real-world conditions.

The predicted customer demand was assumed to rely on perfect demand knowledge for computing the maximum achievable performance for the chosen algorithms as an upper bound.

The matching period not only influenced the choice of the dispatching technique, but also the dispatching behavior. The matching interval was therefore set to Δt = 1 min both to reduce computational effort and to represent near-realistic dispatch conditions. A customer request was rejected after 20 min corresponding to the 99th percentile of observed real-world customer waiting times. Empty, idle vehicles remained at their drop-off location if no rebalancing was applied.

The main parameters of the rebalancing module were grid shape, grid size, interval length Δt_{reb}, horizons T, and frequency f_{reb}. For reasons of simplicity, a square grid was chosen. The finer the grid, the higher the possible performance of an optimizer. However, the number of unknown variables increased quadratically with the number of cells, since the optimizer of Section 5.2.2 had N^2 T variables with N grid cells and T optimization horizons. Spatial analysis of the origin distribution of the dataset showed that demand was concentrated at the center of Munich and decreased towards the suburbs. Consequently, the grid covered the central area with a fine resolution of 1 km $\times$ 1 km and the suburban region with a coarser resolution of 2 km $\times$ 2 km. This hybrid approach resulted in a decrease from 375 cells to 143 cells, as well as a decrease in unknowns by a factor of 7, compared to a grid with a cell size of 1 km.

The rebalancing interval must allow vehicles to reach a neighboring cell within a certain period. On the other hand, in order to limit idle times, the period should not be significantly longer than the average rebalancing durations to neighboring regions, which varied between 2 min and 10 min depending on the travel speed. Thus, a period of Δt_{reb} = 10 min was selected to allow rebalancing vehicles to reach neighboring cells within an optimization period. To avoid over- and non-optimized periods, the rebalancing frequency was also set to 10 min.

The number of control horizons was fixed to T = 2, which corresponded to a prediction horizon of 30 min. To determine the weighting parameter λ, a range between 0.02 and 20 was analyzed, based on realistic combinations of customer and driver VoT and possible waiting time reductions. Based on test simulations, λ was set to 1, since this value achieved good waiting time reductions and reasonably limited the distance of a rebalancing trip to 5.39 km.

Additionally, the travel speed of a vehicle was assumed as the hourly mean speed for the dataset for all actions (idle and customer rides).

To calculate the optimal fleet size, the request collection period was set to $\bar{H}$ =1 h. This value represented a reasonable time span for the shift planning of a fleet operator. The maximum idle time of vehicles was limited by setting the inter-trip time to $\delta = 10$ min.

6.3. Validation

The reference strategy was validated by comparing fleet size, average customer waiting time, and distances (arrival, idling, and customer) between the simulation and the dataset. First of all, simulated and actual fleet size coincided precisely on a temporal interval of 5 min. The distributions of simulated and real occupied distances were rather similar, but 54% of the customer trips actually made were up to 1 km longer (taxis in the dataset covered 13% more distance with a customer). One reason for this might be that taxi drivers opt for slightly longer routes than the shortest path if they allow for a higher average speed. The mean simulated idle distance of 1.3 km fell below that of 2.6 km in the dataset, since the simulated vehicles returned to the nearest rank. However, this difference was considered as acceptable as long as the waiting time and arrival distances to the customer matched. This was indeed the case, since the distribution of arrival distances was similar, i.e., simulated taxis covered only 300 m more when picking up a customer. Additionally, the simulated customer waiting time of 5.4 min reflected the actual attendance time (5.5 min) for a matching interval of 1 min. To conclude, simulated vehicles covered slightly greater distances, but the service quality almost reflected the real-world circumstances.

7. Results

7.1. Performance Based on Current Fleet Size

To evaluate the performance of the predictive strategies, the current fleet size was linearly decreased, and customer waiting time, rejected requests, and profit increase for the operator were analyzed. The waiting time was calculated as the average difference between arrival and servicing of a request over the simulated week. The profit increase for the test fleet in this study was calculated as the sum of labor and distance-based savings over the week, assuming a labor cost of 11 €/h before social security contributions and fuel plus wear and tear cost of 18.55 cts/km (based on the assumptions in [38]).

Figure 5 shows the waiting times of predictive strategies in comparison to the reference strategy. Both predictive approaches, predictive maximum bipartite matching (PR-BPM) and predictive nearest-taxi/nearest-request (PR-NTNR), were able to reduce the fleet size by 30% while maintaining the current service level. Any further decrease in fleet size would aggravate the current situation irrespective of the strategy. For small fleet sizes, the reference strategy performed significantly worse than its predictive alternatives (because it distributed vehicles evenly across the space). Table 2 confirmed that the fleet size could be reduced to no more than 70%, since customers would otherwise have to be rejected. It likewise confirmed that the predictive strategies could serve more customers than the reference approach in the event of an extreme undersupply.

Besides upholding the current service level, a strategy was expected to maximize the profit of a fleet operator on the supply side. Apart from equal labor savings, the predictive strategies showed smaller arrival distances than their non-predictive counterpart, as reflected by the income differences shown in Table 3. A 70% fleet dispatched by the PR-BPM and PR-NTNR covered 234,228 km and 236,957 km respectively, whereas the fleet controlled by the reference strategy travels 253,427 km over the course of the simulated week (Figure 6). Similarly, it showed that the operator's incremental income increased for smaller fleets. Since fleet sizes of 60% and 50% were excluded due to the insufficient service level, an operator would opt for a PR-BPM strategy, which generated slightly more revenues than a PR-NTNR approach. The reason for this difference was that the PR-BPM minimized total arrival distances.

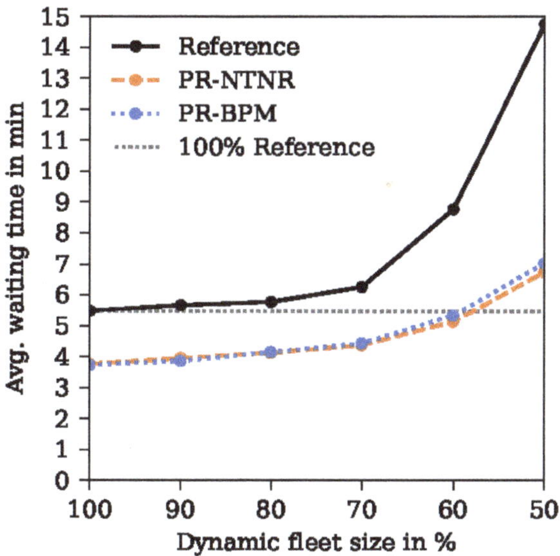

Figure 5. Customer waiting time with respect to the current fleet size.

Table 2. Denied requests.

Strategy/Fleet Size	70%	60%	50%
Reference	0	166	1605
PR-NTNR	0	9	271
PR-BPM	0	3	233

Table 3. Increase in variable profit in k€.

Strategy/Fleet Size	70%	60%	50%
Reference	117.1	151.0	163.4
PR-NTNR	120.2	158.5	192.6
PR-BPM	120.7	159.1	193.9

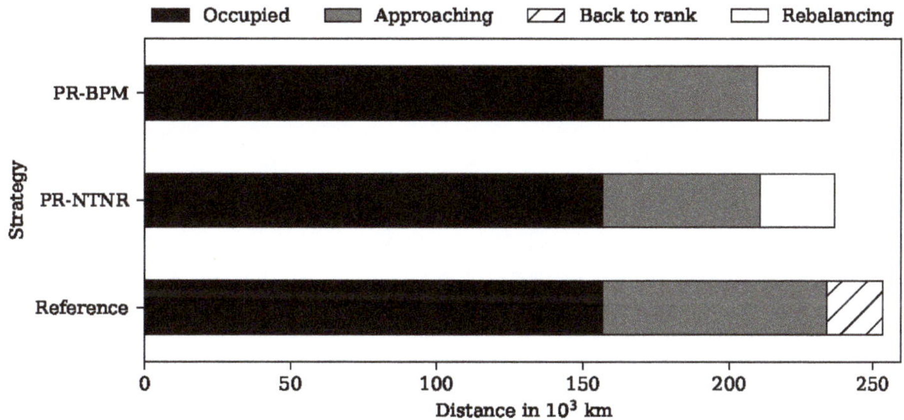

Figure 6. Cumulative distance by taxi status for different strategies at 70% fleet size.

7.2. Performance Based on Optimal Fleet Size

Before applying the proposed fleet management methods, the optimal fleet size in the week under consideration was computed as described in Section 5.3. This hourly optimum was then fed into the fleet sizing module to adjust the number of simulated vehicles. In detail, the optimal fleet supplied only 58% of the currently provided hours, i.e., the average number of active vehicles decreased from 177 to 104. Moreover, the hourly maximum number of taxis needed fell from 292 to 188.

Again, the performance of all three policies on the basis of optimal fleet size was evaluated. Table 4 shows that both approaches were able to uphold the current waiting time level of 5.5 min for the theoretical optimum, but were not able to fulfill all customer requests. Therefore, the optimal fleet size was increased by 10% and 20%, resulting in both strategies, PR-BPM and PR-NTNR, being able to assure the current service level at 120% of the optimal fleet size. All other combinations either exceeded the threshold or were not able to serve all customers. It is interesting to note that the reference strategy rejected fewer customers than the predictive approaches. This was because its FCFS dispatching method served requests in chronological order, whereas the other strategies rejected clients during a supply shortage on Wednesday evening.

To conclude, the predictive approaches were not able to attain the optimum fleet size without a deterioration in service level, since the degree of information accuracy provided to the controller was too imprecise in terms of space and time. Since the number of supplied hours for 120% of the optimal fleet was roughly equal to those of 70% of the current fleet and the waiting time was no lower, there was no advantage to be had by upscaling the theoretical optimum in this case. Hence, the optimal fleet sizing approach did not increase the performance by supplying the right amount of hours at the right point in time. This may also have been because a maximum customer waiting time of 20 min was relatively high and provided for the possibility of compensating for short periods of undersupply.

Table 4. Waiting time in minutes (rejected requests) for the percentages of the optimal fleet size.

Strategy/Fleet Size	100%	110%	120%
Reference	7.12 (0)	6.32 (0)	6.07 (0)
PR-NTNR	5.12 (13)	4.64 (1)	4.42 (0)
PR-BPM	5.02 (5)	4.62 (1)	4.40 (0)

8. Discussion

The aim of this study was to test the potential of a predictive strategy under a realistic MoD scenario. The performance of the proposed strategy depended first of all on the input data and parameters of the optimizer. It also relied on forecasts of pickup points; however, we assumed that the OD distribution resembled historic travel behavior (Section 5.2.2). Future models might also incorporate predictions of drop-off distributions. Furthermore, rebalancing vehicles were sent to the center of a cell. Besides modifying the shape and size of the grid cells, a higher resolution of rebalancing destinations might yield another increase in performance, which would, however, entail greater computational effort. This could, for example, be realized by predictively directing empty vehicles to the centers of demand clusters. This approach did not account for the customers' shifting readiness to pay for a ride, e.g., clients' willingness to pay higher prices during nighttime. This could be improved in further studies by introducing a time-dependent reward parameter $\lambda(t)$.

Secondly, the reference model was not able to reproduce the idling behavior of taxi drivers accurately. To incorporate this, we would need to take into account the individual decision making process of a single driver. Nevertheless, we always compared the results with the simulated reference model rather than actual values of the real-world dataset, in order to exclude side effects and possible model inaccuracies. Furthermore, we relied on a batch matching technique, with a period of $\Delta t = 1$ min, although this could be reduced further to simulate real-time dispatching.

Finally, we found that there was no advantage of upscaling the perfect fleet size by reducing the current amount of vehicles in the week investigated. This result might vary if the reject period of 20 min changes. The importance of a predictive strategy and accurate vehicle positioning might increase if the customers' willingness to wait for a taxi decreases. Besides this simple percentage-based modification, we aim to test fleet scaling methods that integrated demand forecasts.

The present simulation model lacked computational efficiency. In order to compute simulations over a long-term period (one year), we plan to integrate our algorithms into a more efficient and scalable framework such as amodeus [23].

The presented fleet strategy assumed that fleet operators were able to control actions taxi drivers usually decide as individuals. Even if this strategy offered cost-reductions and operational benefits in a global perspective, not all drivers would profit from this approach. For example, drivers who had high turnovers, because they anticipated demand or adapted quickly to dynamic changes, were now forced to act in the interest of the global optimum. Fleet operators may therefore rethink their salary models to compensate such effects.

This study did not reflect on the dynamic reassignment of taxis. It has to be investigated to what extent dynamic reassignment of customers is beneficial to further reduce waiting times and operational costs.

9. Conclusions

A literature analysis of predictive fleet management strategies, including fleet sizing, was conducted for MoD scenarios. None of the existing predictive approaches were tested in conjunction with a dynamic, time-varying fleet size. Our contribution was to improve an existing model and apply it to different time-dependent fleet sizes, as well as to the optimal fleet size, of a large-scale taxi dataset.

As a result, it was found that the current fleet size could be reduced by 30% without aggravating the current service level. By applying a PR-BPM model, a fleet operator could save up to 120,670€ per week and reduce the fleet's total distance by 19,199 km. However, the simulation results also showed that the the optimum fleet size could only be attained to a limited extent using the proposed strategies. Both predictive strategies in conjunction with the theoretical fleet minimum upscaled by 20% were able to serve all customers without increasing the present waiting time threshold.

We further intend to show that this reduction potential also holds for other periods of the year, for instance by performing simulations of low-demand (e.g., Christmas) and high-demand periods (e.g., Oktoberfest). Additionally, we aim to test the controller in combination with demand forecasts of an LSTM neural network to evaluate the extent to which prediction accuracy influences the optimizer's performance. This work represents a first step towards testing the potential of predictive strategies for dynamic fleet sizes based on a real-world dataset.

Author Contributions: M.W. as the first author primarily developed the research question and made an essential contribution to the overall structure of the software, the methodology of the algorithms, and the validation of the data. Conceptualization, M.W. and L.N.; data curation, M.W.; formal analysis, M.W. and L.N.; investigation, M.W. and L.N.; methodology, M.W.; resources, M.W.; software, M.W. and L.N.; supervision, M.W. and M.L.; validation, M.W. and L.N.; visualization, M.W.; writing, original draft, M.W. and L.N.; writing, review and editing, M.W., L.N., and M.L. All authors have read and agreed to the published version of the manuscript.

Funding: The Chair of Automotive Technology at the Technical University of Munich funded the project independently.

Acknowledgments: The authors would like to thank IsarFunk Taxizentrale GmbH & Co. KG for providing useful taxi data.

Conflicts of Interest: The authors declare no conflict of interest.

Abbreviations

MoD	mobility-on-demand
MaaS	mobility as a service
GNSS	Global Navigation Satellite System
RHC	receding horizon control
DQN	deep Q network
ILP	integer linear program
MILP	mixed integer linear program
FCD	floating car data
LSTM	long short-term memory
FCFS	fist come, first served
VoT	value of time
OD	origin-destination
BPM	maximum bipartite matching
NTNR	nearest-taxi/nearest-request
PR-BPM	predictive maximum bipartite matching
PR-NTNR	predictive nearest-taxi/nearest-request
OSM	Open Street Map

Nomenclature

G	Set of grid cells
N	Number of grid cells
Δt	Matching period
Δt_{reb}	Rebalancing period
f_{reb}	Rebalancing frequency
T	Number of control horizons
s_i^t	Total number of available vehicles in cell i at t
d_i^t	Number of open requests in cell i between t and $t + \Delta t_{\text{reb}}$
$\hat{d}_i^t$	Number of predicted requests in cell i between t and $t + \Delta t_{\text{reb}}$
r_{ij}^t	Number of rebalancing vehicles sent from cell i to j at t
x_i^t	Number of already idle vehicles in cell i at t
$x_i^{t-1,t}$	Number of occupied vehicles turning idle in cell i between $t-1$ and t
c_{ij}	Variable cost for traveling from cell i to j
λ	Weighting parameter
$\mathbb{P}_t(i \mid j)$	Probability distribution of the destination cell j based on starting cell i at time t
q_{ij}	Street distance between cell centers g_i and g_j
γ	Discount factor
$v(t)$	Vehicle speed at time t
l_i^{p}	Pickup location of ride i
l_i^{d}	Drop-off location of ride i
t_i^{p}	Pickup time of ride i
t_i^{d}	Drop-off time of ride i
$\tilde{H}$	Optimization horizon
θ_{ij}	Street distance between the drop-off location of ride i and the pickup location of ride j
δ	Maximum trip connection time

References

1. Keeney, T. Mobility-As-a-Service: Why Self-Driving Cars Could Change Everything. Available online: https://research.ark-invest.com/self-driving-cars-white-paper (accessed on 16 April 2020).
2. Willmroth, J. Freie Fahrt für Freie Taxler. Available online: https://www.sueddeutsche.de/auto/fahrdienste-taxi-uber-1.4336689 (accessed on 19 February 2020).
3. MarketsandMarkets. Fleet Management Market. Available online: https://www.marketsandmarkets.com/updatedreport/fleet-management-market-1020.asp (accessed on 22 February 2020).

4. NYC Taxi & Limousine Commission. TLC Trip Record Data. Available online: https://www1.nyc.gov/site/tlc/about/tlc-trip-record-data.page (accessed on 16 April 2020).
5. Piorkowski, M.; Sarafijanovic-Djukic, N.; Grossglauser, M. CRAWDAD Dataset Epfl/Mobility. 2009. Available online: https://crawdad.org/epfl/mobility/20090224 (accessed on 24 February 2019).
6. Data.world. Uber Pickups in NYC. Available online: https://data.world/data-society/uber-pickups-in-nyc (accessed on 16 April 2020).
7. Oda, T.; Joe-Wong, C. MOVI: A Model-Free Approach to Dynamic Fleet Management. In Proceedings of the IEEE INFOCOM 2018-IEEE Conference on Computer Communications, Honolulu, HI, USA, 16–19 April 2018; doi:10.1109/infocom.2018.8485988. [CrossRef]
8. Xu, J.; Rahmatizadeh, R.; Boloni, L.; Turgut, D. Taxi Dispatch Planning via Demand and Destination Modeling. In Proceedings of the 2018 IEEE 43rd Conference on Local Computer Networks (LCN), Chicago, IL, USA, 1–4 October 2018; doi:10.1109/lcn.2018.8638038. [CrossRef]
9. Miao, F.; Han, S.; Lin, S.; Stankovic, J.A.; Zhang, D.; Munir, S.; Huang, H.; He, T.; Pappas, G.J. Taxi Dispatch With Real-Time Sensing Data in Metropolitan Areas: A Receding Horizon Control Approach. *IEEE Trans. Autom. Sci. Eng.* **2016**, *13*, 463–478, doi:10.1109/tase.2016.2529580. [CrossRef]
10. Tsao, M.; Iglesias, R.; Pavone, M. Stochastic Model Predictive Control for Autonomous Mobility on Demand. In Proceedings of the 2018 21st International Conference on Intelligent Transportation Systems (ITSC), Maui, HI, USA, 4–7 November 2018; doi:10.1109/itsc.2018.8569459. [CrossRef]
11. Dandl, F.; Hyland, M.; Bogenberger, K.; Mahmassani, H.S. Evaluating the impact of spatio-temporal demand forecast aggregation on the operational performance of shared autonomous mobility fleets. *Transportation* **2019**, *46*, 1975–1996, doi:10.1007/s11116-019-10007-9. [CrossRef]
12. Pavone, M.; Smith, S.; Frazzoli, E.; Rus, D. Load balancing for mobility-on-demand systems. In *Robotics: Science and Systems VII*; The MIT Press: Cambridge, MA, USA, 2012; pp. 249–256.
13. Zhang, R.; Pavone, M. Control of robotic mobility-on-demand systems: A queueing-theoretical perspective. *Int. J. Robot. Res.* **2015**, *35*, 186–203, doi:10.1177/0278364915581863. [CrossRef]
14. Iglesias, R.; Rossi, F.; Wang, K.; Hallac, D.; Leskovec, J.; Pavone, M. Data-Driven Model Predictive Control of Autonomous Mobility-on-Demand Systems. In Proceedings of the 2018 IEEE International Conference on Robotics and Automation (ICRA), Brisbane, Australia, 21–25 May 2018; doi:10.1109/icra.2018.8460966. [CrossRef]
15. Tsao, M.; Milojevic, D.; Ruch, C.; Salazar, M.; Frazzoli, E.; Pavone, M. Model Predictive Control of Ride-sharing Autonomous Mobility-on-Demand Systems. In Proceedings of the 2019 International Conference on Robotics and Automation (ICRA), Montreal, QC, Canada, 20–24 May 2019; doi:10.1109/icra.2019.8794194. [CrossRef]
16. Miller, J.; How, J.P. Predictive positioning and quality of service ridesharing for campus mobility on demand systems. In Proceedings of the 2017 IEEE International Conference on Robotics and Automation (ICRA), Singapore, 29 May–3 June 2017; doi:10.1109/icra.2017.7989167. [CrossRef]
17. Alonso-Mora, J.; Wallar, A.; Rus, D. Predictive routing for autonomous mobility-on-demand systems with ride-sharing. In Proceedings of the 2017 IEEE/RSJ International Conference on Intelligent Robots and Systems (IROS), Vancouver, BC, Canada, 24–28 September 2017; doi:10.1109/iros.2017.8206203. [CrossRef]
18. Maciejewski, M.; Bischoff, J. Large-scale Microscopic Simulation of Taxi Services. *Procedia Comput. Sci.* **2015**, *52*, 358–364, doi:10.1016/j.procs.2015.05.107. [CrossRef]
19. Maciejewski, M.; Bischoff, J.; Nagel, K. An Assignment-Based Approach to Efficient Real-Time City-Scale Taxi Dispatching. *IEEE Intell. Syst.* **2016**, *31*, 68–77, doi:10.1109/mis.2016.2. [CrossRef]
20. Vazifeh, M.M.; Santi, P.; Resta, G.; Strogatz, S.H.; Ratti, C. Addressing the minimum fleet problem in on-demand urban mobility. *Nature* **2018**, *557*, 534–538, doi:10.1038/s41586-018-0095-1. [CrossRef] [PubMed]
21. Hörl, S.; Ruch, C.; Becker, F.; Frazzoli, E.; Axhausen, K. Fleet operational policies for automated mobility: A simulation assessment for Zurich. *TRansp. Res. Part C Emerg. Technol.* **2019**, *102*, 20–31, doi:10.1016/j.trc.2019.02.020. [CrossRef]
22. Ruch, C.; Hörl, S.; Hakenberg, J.; Frazzoli, E. The Impact of Fleet Coordination on Taxi Operations. 2019. Available online: https://www.research-collection.ethz.ch/handle/20.500.11850/379519 (accessed on 18 June 2020).

23. Ruch, C.; Horl, S.; Frazzoli, E. AMoDeus, a Simulation-Based Testbed for Autonomous Mobility-on-Demand Systems. In Proceedings of the 2018 21st International Conference on Intelligent Transportation Systems (ITSC), Maui, HI, USA, 4–7 November 2018; doi:10.1109/itsc.2018.8569961. [CrossRef]
24. Ruch, C.; Gächter, J.; Hakenberg, J.; Frazzoli, E. The +1 Method Model-Free Adaptive Repositioning Policies for Robotic Multi-Agent Systems. 2019. Available online: https://www.research-collection.ethz.ch/handle/20.500.11850/322945 (accessed on 18 June 2020).
25. Seow, K.T.; Dang, N.H.; Lee, D.H. Towards An Automated Multiagent Taxi-Dispatch System. In Proceedings of the 2007 IEEE International Conference on Automation Science and Engineering, Scottsdale, AZ, USA, 22–25 September 2007; doi:10.1109/coase.2007.4341673. [CrossRef]
26. Seow, K.T.; Dang, N.H.; Lee, D.H. A Collaborative Multiagent Taxi-Dispatch System. *IEEE Trans. Autom. Sci. Eng.* **2010**, *7*, 607–616, doi:10.1109/tase.2009.2028577. [CrossRef]
27. Gao, Y.; Jiang, D.; Xu, Y. Optimize taxi driving strategies based on reinforcement learning. *Int. J. Geogr. Inf. Sci.* **2018**, *32*, 1677–1696, doi:10.1080/13658816.2018.1458984. [CrossRef]
28. Xu, Z.; Li, Z.; Guan, Q.; Zhang, D.; Li, Q.; Nan, J.; Liu, C.; Bian, W.; Ye, J. Large-Scale Order Dispatch in On-Demand Ride-Hailing Platforms. In Proceedings of the 24th ACM SIGKDD International Conference on Knowledge Discovery & Data Mining, London, UK, 19–23 August 2018; doi:10.1145/3219819.3219824. [CrossRef]
29. Zhan, X.; Qian, X.; Ukkusuri, S.V. A Graph-Based Approach to Measuring the Efficiency of an Urban Taxi Service System. *IEEE Trans. Intell. Transp. Syst.* **2016**, *17*, 2479–2489, doi:10.1109/tits.2016.2521862. [CrossRef]
30. Zhu, C.; Prabhakar, B. Reducing inefficiencies in taxi systems. In Proceedings of the 2017 IEEE 56th Annual Conference on Decision and Control (CDC), Melbourne, Australia, 12–15 December 2017; doi:10.1109/cdc.2017.8264609. [CrossRef]
31. Hörl, S.; Ruch, C.; Becker, F.; Frazzoli, E.; Axhausen, K.W. Fleet control algorithms for automated mobility: A simulation assessment for Zurich. In Proceedings of the Transportation Research Board 97th Annual Meeting, Washington, DC, USA, 7–11 January 2018, doi:10.3929/ethz-b-000251651. [CrossRef]
32. Bischoff, J.; Maciejewski, M. Simulation of City-wide Replacement of Private Cars with Autonomous Taxis in Berlin. *Procedia Comput. Sci.* **2016**, *83*, 237–244, doi:10.1016/j.procs.2016.04.121. [CrossRef]
33. Spieser, K.; Treleaven, K.; Zhang, R.; Frazzoli, E.; Morton, D.; Pavone, M. Toward a Systematic Approach to the Design and Evaluation of Automated Mobility-on-Demand Systems: A Case Study in Singapore. In *Road Vehicle Automation*; Springer International Publishing: Berlin, Germany, 2014; pp. 229–245; doi:10.1007/978-3-319-05990-7_20. [CrossRef]
34. Jäger, B.; Wittmann, M.; Lienkamp, M. Analyzing and Modeling a City's Spatiotemporal Taxi Supply and Demand: A Case Study for Munich. *J. Traffic Logist. Eng.* **2016**, *4*, 147–153, doi:10.18178/jtle.4.2.147–153. [CrossRef]
35. OpenStreetMap Contributors. Map Data. 2020. Available online: https://download.geofabrik.de/europe/germany/bayern.html (accessed on 16 April 2020).
36. Maciejewski, M.; Salanova, J.M.; Bischoff, J.; Estrada, M. Large-scale microscopic simulation of taxi services. Berlin and Barcelona case studies. *J. Ambient. Intell. Humaniz. Comput.* **2016**, *7*, 385–393, doi:10.1007/s12652-016-0366-3. [CrossRef]
37. Hopcroft, J.E.; Karp, R.M. An $n^{5/2}$ algorithm for maximum matchings in bipartite graphs. In Proceedings of the 12th Annual Symposium on Switching and Automata Theory (swat 1971), East Lansing, MI, USA, 13–15 October 1971; doi:10.1109/swat.1971.1. [CrossRef]
38. Bundesverband Taxi und Mietwagen e.V. *BZP Geschäftsbericht 2017/2018*; Technical Report; Bundesverband Taxi und Mietwagen e.V.: Frankfurt am Main, Germany, October 2019.

Review

Fleet Management and Control System for Medium-Sized Cities Based in Intelligent Transportation Systems: From Review to Proposal in a City

Beimar Rojas [1], Cristhian Bolaños [1], Ricardo Salazar-Cabrera [1,*], Gustavo Ramírez-González [1], Álvaro Pachón de la Cruz [2] and Juan Manuel Madrid Molina [2]

1. Telematics Engineering Research Group (GIT), Telematics Department, Universidad del Cauca, Popayán 190001, Colombia; beimaro@unicauca.edu.co (B.R.); crissebas@unicauca.edu.co (C.B.); gramirez@unicauca.edu.co (G.R.-G.)
2. Information Technology and Telecommunications Research Group (I2T), ICT Department, Universidad Icesi, Cali 760001, Colombia; alvaro@icesi.edu.co (Á.P.d.l.C.); jmadrid@icesi.edu.co (J.M.M.M.)

* Correspondence: ricardosalazarc@unicauca.edu.co; Tel.: +57-313-586-0304

Received: 31 July 2020; Accepted: 24 August 2020; Published: 27 August 2020

Abstract: In medium-sized cities in developing countries, transit services without dedicated lanes have issues related to route compliance, schedules, speed control, and safety. An efficient way for dealing with this issue is the use of Information and Communication Technologies (ICT), to implement a Fleet Management and Control Systems (FMCS). Such implementation can be performed using Intelligent Transportation Systems (ITSs), which allow integration of services and adequate standardization. This article features: (a) a literature review, related to FMCS based on ITS and enabling technologies, (b) design of the ITS architecture of an FMCS, and (c) some advances in the development of the proposed FMCS in a Colombian city (Popayán). The results of the literature review allowed identifying the most important requirements of FMCS in order to design the ITS architecture and build a prototype featuring the suggested technologies. Finally, some experiments were performed to evaluate the operation of the developed prototype. The results showed evidence of adequate operation in sending and receiving messages from and to four prototypes developed for the vehicles, also complying with the established requirements of location, tracking, exchanged data, and security. This allows continuing the development of the proposed FMCS, with some adjustments.

Keywords: Intelligent Transportation Systems; ITS architecture; fleet management; transit vehicles; ITS enabling technologies

1. Introduction

Cities in developing countries such as Colombia face mobility problems, such as high traffic congestion on roads, high traffic accident rates, and high levels of air pollution and noise. Many of these problems are caused by public transportation [1,2].

Therefore, a system seeking to improve control and monitoring of transit vehicles for cities of Colombia and other Latin American countries could indirectly contribute to improving the mobility problems of such cities. Medium-sized cities (with populations between 100,000 and 1,000,000 inhabitants) use "collective" service as the principal transit service. "Collective" service is provided by small buses or vans with a capacity between 12 and 20 passengers, through public roads with other types of vehicles (e.g., private vehicles). The service has specific stops for passenger pick up/drop off, but a considerable number of users and drivers do not respect this, stopping anywhere along their route. The "collective" service also has problems related to speeding, noncompliance with regulations, vehicle routes, schedules, and frequencies [3,4].

Information and Communication Technologies (ICT) emerge as the best option for the implementation of an efficient transport system because they allow centralized traffic control, navigation, security, access to travel information, and real-time assistance, among other services [5]. Such systems belong in this work to the context of the Intelligent Transportation Systems (ITSs) and Fleet Management and Control Systems (FMCSs) [6,7].

The Fleet Management Control Systems (FMCS) are responsible for controlling vehicle operation and evaluating compliance with scheduled services. An FMCS monitors vehicles in real-time and generates alerts when events happen. In general, Fleet Management Systems (FMS) reduce risks, increase the quality of service, and improve operational efficiency at a minimal cost [8].

The increase of quality in transit service implies providing a better service to users. The main need for users of "collective" public transport is to know the current location of the vehicle they are waiting for, to be able to roughly calculate the approximate time of arrival at a stop [9]. To determine the current location of the transit vehicle, it is necessary to continuously track its location (with the use of ICT); then, by factoring variables as actual traffic level and historic route time history, an approximate time of arrival at a certain stop can be calculated. Uncertainty about the timely arrival of the vehicle is one of the main reasons why the level of use of this means of transport is relatively low in times of low demand [7]. Hence, continuous monitoring of the transit vehicle is a very important requirement to consider in an FMCS, to improve the quality and usability of the transit service.

FMCS can improve operational efficiency in aspects such as fuel consumption, which is a considerable percentage of operational cost. Fuel consumption generates CO_2 emissions, which considerably affect the air quality of a city [10,11]. Fuel consumption is related to the way the vehicle is driven [12], so monitoring the speed of a vehicle on the route can help to identify inappropriate driving behaviors, e.g., running a red light [13]. Speed control is also an important factor for reducing risks, considering that one of the main causes of road accidents in the world is speeding. Therefore, speed control of the vehicle along its route is an important requirement of an FMCS, to reduce risks and decrease fuel consumption.

FMCS developed recently commonly use cellular data transfer (GSM, GPRS, or LTE) for communication between transit vehicles and control centers, which causes issues related to high operating costs, and difficulty in maintaining the system [14]. Some emerging communication technologies such as low-power wide-area networks (LPWANs) can improve upon some factors of cellular communication technologies, such as range, availability, frequency band, and cost reduction both in acquisition and operation [15]. The use of adequate communication technology between transit vehicles and the control centers is an important requirement for an FMCS to improve operational efficiency at a minimal cost.

It is also important that an FMCS is not complex in its handling and operation because this could mean more delays in processes, which could make the system slow and inefficient. That is why it is convenient to properly distribute the workload between components and operating personnel [16]. The simplicity of handling and division of work does not imply a lack of security because it is also necessary to include security components to guarantee confidentiality, integrity, privacy, and availability of the managed information [17,18].

Finally, it is important that an FMCS must also consider the safety of passengers because it is necessary to protect passengers' well-being, thus building trust in the system [19]. Speed control is a factor that contributes to increasing safety in the vehicles that are part of the FMCS. Proper vehicle tracking can also identify some events that may pose a safety risk for passengers, such as unscheduled routes or stops at prohibited sites.

Considering ITS developments, the significant disadvantage of most of the ITS services developed at the international and Colombian level is that they are not based on adequate architectures, which brings integration, interoperation, and reusability issues [20]. Many models of FMCS have been proposed, but most of them are implemented with services not based on an appropriate architecture.

Although there are several approaches to Colombian ITS architectures, the proposed modeling is very general and not suitable for cities with specific characteristics [21].

To address this situation with transit service in the mentioned context, this research focuses on:

- A literature review (with scientific mapping and systematic review) of FMCS.
- Proposal of an ITS architecture for the FMCS in medium-sized cities in the Latin American context, by previously identifying suitable requirements for an FMCS in such context. In addition to the ITS architecture, suggested technologies and a prototype were defined.
- Experiments for validating usage of the suggested technology and prototype in a case study (Popayán medium-sized Colombian city).

The article structure is as follows: Section 2 presents the methods and materials used in this work and the criteria for the selection of a communication technology; Section 3 presents the literature review with SciMAT tool [22], and Preferred Reporting Items for Systematic Reviews and Meta-Analyses (PRISMA) methodology [23]; Section 4 presents the identification of suitable requirements of an FMCS, a study of ITS reference architectures, a proposal for an ITS architecture for the FMCS, and a proposal for an FMCS prototype; Section 5 contains experiments performed to validate the proposal presented in the above section; Section 6 presents a discussion of the obtained results; finally, Section 7 presents the conclusions.

2. Materials and Methods

This section presents the methodologies, methods, processes, and materials used in this work.

2.1. Literature Review

Initially, a scientific mapping analysis was performed with the help of the tool called SciMAT [22], in order to analyze the social, conceptual, and intellectual evolution of the interest area. In this way, relevant aspects and topics were identified and included in the literature review. Then, a systematic review of the literature was carried out in some databases related to the research topic. The review sought to identify different ITS services and architectures used in FMCS, as well as the methods used for their development, the communication technologies, and the obtained results.

Several tools for bibliometric and scientometric studies are available. SciMAT was selected because it allows the analysis of scientific mapping within a longitudinal framework and provides different modules that help the analyst perform all the steps of the scientific mapping workflow.

The procedure for scientific mapping included the following steps:

- General Flow Settings. In this step, the configuration for mapping was determined. A query string was determined and used with the SCOPUS database (due to its compatibility with SciMAT and high worldwide recognition). In the definition of time segments to allow a longitudinal analysis and discover the conceptual evolution of the field of study, three consecutive periods were considered: 2014–2015, 2016–2017, 2018–2020.
- Analysis of Results. Strategic diagrams were initially generated to measure the performance, quality of the subject, and subject areas detected. Subsequently, a review of the evolution was performed, considering the number of keywords and the number of shared keywords in the different subperiods, and the thematic evolution of the research.
- Mapping conclusions. Relevant topics and other important aspects were identified to perform the systematic review.

To address some limitations found in the scientific mapping (related to the type of product and license used) with respect to the number of available databases and the impossibility of detailed identification of documents to find some comparison parameters, a systematic review was performed using the Preferred Reporting Items for Systematic Reviews and Meta-Analyses (PRISMA) methodology. The phases of such review are detailed below:

- Identification phase. A bibliographic search was performed on the following topics: Intelligent Transportation System, communications, security, and Fleet Management and Control System (FMCS). The three selected databases were: IEEE Xplore Library, eBook Academic Collection (EBSCO), and ScienceDirect. The search was limited to documents published over the last five years. This search was concluded in March 2020. These databases were chosen in addition to SCOPUS (which was used in the scientific mapping) to achieve two objectives: (1) expand the scope of the review, thus enriching the model through different sources of information, and (2) to check the relationship between scientific mapping and systematic review.
- Screening phase. All the documents from the different sources were joined and duplicates were eliminated. Then, the abstract of each document was reviewed; those documents not showing a direct relationship to the purpose of the review were discarded.
- Eligibility phase. The remaining documents were fully reviewed and filtered according to in-depth criteria defined in the previous phases of the systematic review and the results obtained from the scientific mapping.
- Included phase. The documents were divided into four different groups in this phase. Subsequently, a qualitative synthesis was performed for all groups, and according to the parameters defined for each one, quantitative synthesis was made for the same four groups.

Finally, the scientific mapping and systematic review allowed identifying features essential to define the requirements of an FMCS in the context of a "collective" service for a medium-sized city.

2.2. Proposal of ITS Architecture for FMCS and Prototype

Initially, the requirements of an FMCS for the specific context were compiled from the literature review and analysis of the environment. Next, in order to find an adequate reference ITS architecture, the most widely used architectures worldwide were analyzed, identifying the essential aspects of each one. This allowed the creation of a proposal of ITS architecture for the FMCS in the specified context.

From the ITS architecture, a scope was determined for the development of the FMCS prototype. Once the scope of the prototype was defined, the technologies to be used in this prototype were selected. According to the results of the literature review, such technologies are related to the following topics:

- Transit vehicle location technology (considering speed measurement).
- Communication technology between the transit vehicles and Transit Management Center.
- Technology to guarantee the security of the information transmitted between vehicles and the Transit Management Center.
- Technology to guarantee the security between other modules of the system.

The systematic review showed that almost all works featuring vehicle tracking used Global Navigation Satellite System (GNSS) services for positioning, particularly the Global Positioning System (GPS) technology. Some other works recommend the use of Short-Range Wireless technologies (SRWT), but this option needs a large infrastructure and does not allow continuous tracking, leaving GNSS as the best option. GNSS technology also allows speed measurement, a key aspect for an FMCS.

The communication technology to be used between the transit vehicles and the Transit Management Center (TMC) is one of the decisions that should be discussed. Reviewing [24], an analysis of different communication technologies for FMCS was performed, including cellular network technology, Ubiquiti, radio trunking, low-power wide-area network (LPWAN), and SRWT such as Bluetooth, ZigBee, and WiFi. The most important parameters in a communication system such as coverage, costs (of implementation and operation), scalability (appropriate for the growth of the network infrastructure), low latency, and frequency were considered, yielding LPWAN as the best for an FMCS system. In addition, Reference [25,26] analyze LPWAN technologies such as SigFox, narrowband Internet of Things (NB IoT), DA SH7 alliance protocol (D7A), Ingenu, Telensa, random phase multiple access (RPMA), and long range (LoRa). This review deemed LoRa [27] as the most appropriate for communication between vehicles and TMC, due to its high level of coverage; low implementation and operation costs; use of frequencies in the industrial, scientific, and medical (ISM) radio bands, and low latency.

LoRa is suitable for an FMCS between the mentioned modules because of the following features:

- Range (approximately 5 km in urban areas with line of sight).
- Transmission speed (37.5 kbps).
- Use in "free" frequencies (ISM), which do not require a spectrum use license.
- Its ability to send unlimited messages. Here it is important to highlight that the packet size of this technology is relatively small (255 bytes), however, its ability to send unlimited messages counters this weakness.
- Its status in the global market, achieving rapid growth for IoT communications, rivaling NB IoT as the best in the LPWAN field [28].
- The level of security. LoRa implements a handy protocol called "Long Range Wide-Area Network" (LoRaWAN), designed to wirelessly connect "things" to the internet with a bidirectional communication featuring end-to-end security [27].
- Data packets can be sent simultaneously, thus eliminating collision problems.

Regarding a technology to guarantee the security of the information transmitted between vehicles and the Transit Management Center, the LoRaWAN protocol is a good option. This protocol also allows point-to-point encryption of messages, ensuring that the information is not altered or eavesdropped by possible intruders. Finally, it guarantees that only information from registered devices is received, thus guaranteeing the integrity and confidentiality of the information [29].

The proposed architecture has different modules that must communicate through the network, so all communication must be done through the secure hypertext transfer protocol (HTTPS) which assures the information reaches only the intended recipient, and its headers send an authentication value with a key. In the same way, for the transmission of data, the secure file transfer protocol (FTPS) is used, and to guarantee the security of access to information, a hierarchy of users with limited permissions to obtain data is proposed.

Once the technologies were selected, the components of the prototype to be developed were defined. The module that was determined suitable for transit vehicles has the following principal components: (1) Heltec WiFi LoRa 32 (v2) device (which has a LoRa SX1276 chip, an OLED 0.96-inch display, and an ESP32 microcontroller unit or MCU); (2) Ublox 6M GPS module, to obtain vehicle location. These two components were integrated, achieving a low-level prototype. The selected LoRaWAN gateway was the Dragino LPS8 and the platform used was The Things Network (TTN).

2.3. Prototype Experiments for Validation

The proposed prototype was tested in a city used as a case study, the city of Popayán. Popayán is a medium-sized city in the southwest of Colombia (Latin America), it has an extension of 512 km^2 (including rural and urban areas) and a population of approximately 300,000 inhabitants. Considering the choice of LoRa as the communication technology for the prototype, a coverage analysis of this technology was performed for the city, using the QGIS (Quantum Geographic Information System) tool. QGIS was used because it has a plugin that allows the estimation of radio coverage taking into account demographic data and city buildings [30]. The study was performed in order to find an adequate location for the gateways and the total number of these devices that would be necessary to cover the city.

Once the coverage analysis was performed, a suitable location in the city was determined to develop the experiments. The objective of the experiments was to validate sending and receiving messages from and to the modules located in the transit vehicles (using four of these devices), determining parameters such as signal level, distance reached, percentage of packets received, and operation of the data security layer.

Section 5 of this document explains in detail the design and results of experiments.

3. Literature Review Results

3.1. Scientific Mapping

The scientific mapping yielded the following results:

- General Flow Settings. The initial configuration for mapping yielded 1729 results, three of them were deleted due to format issues or replication. Once filtered by time period, 434 documents were included in the 2014–2015 category, 543 in the 2016–2017 category, and 749 in the 2018–2020 category.
- Analysis of Results. The strategic diagrams show that in all the studied subperiods, the emerging (wireless-telecommunication-systems, sustainable-development, Internet of Things) and driving (security, intelligent systems, ITS, ad hoc networks) topics achieved the highest citations scores and impacts. It was also possible to determine a series of trends in the characteristics and technologies used for the systems under investigation, including security and Intelligent Transport Systems (ITSs). Figure 1 presents the strategic diagrams of the three time periods considered, based on the number of published documents. Through the revision of the evolution, it was evident that three themes (security, ITS, wireless communication systems) found in the strategic diagrams are identified as relevant, confirming their importance for the systematic review.
- Mapping conclusions. The topics related to security, ITS, transit systems, wireless communication technologies, and IoT, should be deemed as relevant in the systematic review.

Only a few details about the process and the results of the scientific mapping are presented. Nevertheless, the authors can be consulted for any additional information or the generated diagrams.

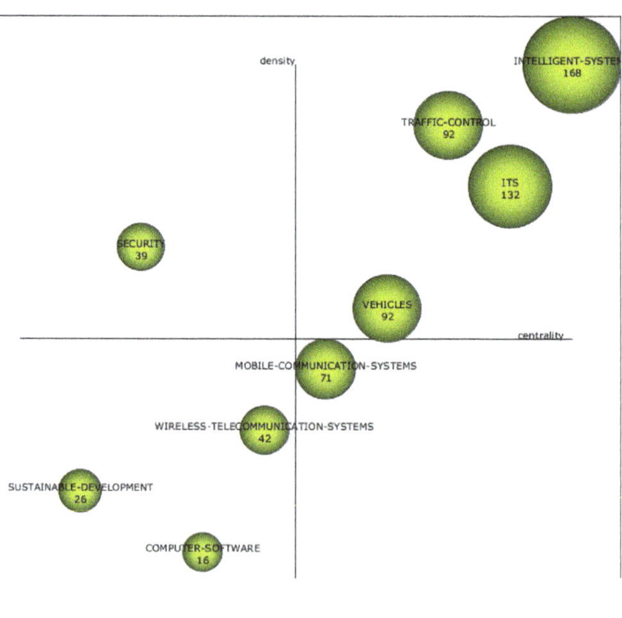

(a)

Figure 1. *Cont.*

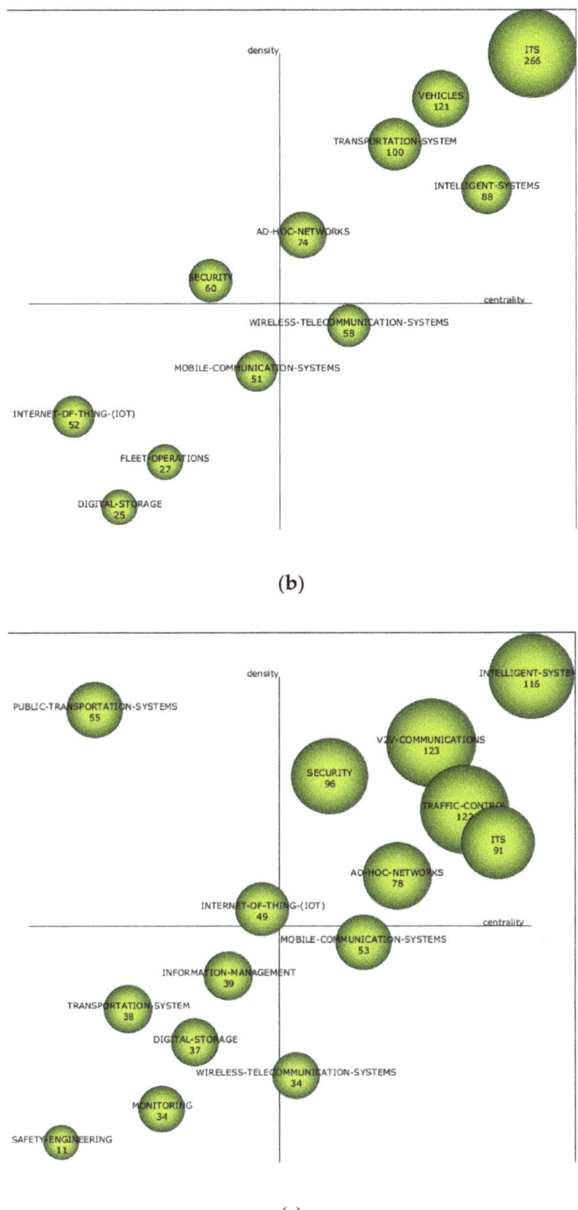

Figure 1. Strategic diagrams based on the number of documents published: (**a**) 2014–2015 time period; (**b**) 2016–2017 time period; (**c**) 2018–2020 time period.

3.2. Systematic Review

The systematic review was done using the Preferred Reporting Items for Systematic Reviews and Meta-Analyses (PRISMA) methodology. The searches were performed in the three databases mentioned in Section 2.1 (IEEE Xplore Library, EBSCO, and ScienceDirect). The documents found in such databases were identified using some search strings related to research terms, adding some recommended from external sources. The initial search yielded a total of 2139 documents. These documents were evaluated and filtered according to the PRISMA methodology. In the final phase, 56 documents of interest were selected, and quantitative and qualitative analysis was performed for such documents.

Figure 2 shows the exclusion and inclusion results obtained in each of the phases, and the percentage of distribution of documents in each group.

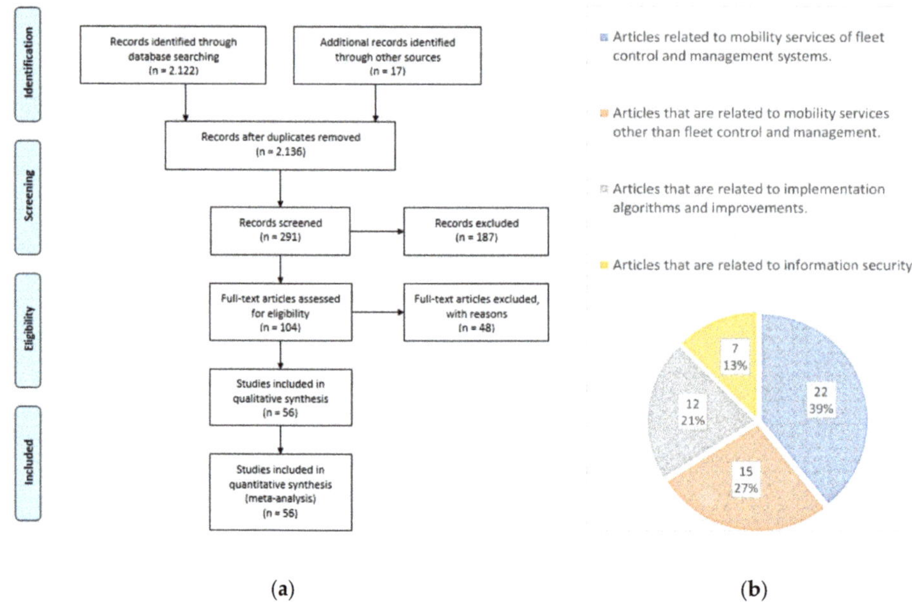

Figure 2. Results of the systematic review: (**a**) Preferred Reporting Items for Systematic Reviews and Meta-Analyses (PRISMA) Flow Diagram Results; (**b**) Classification by groups.

Groups identified in the included phase were: (1) articles related to mobility services of fleet control and management systems, (2) articles that are related to mobility services other than fleet control and management, (3) articles that are related to implementation algorithms and improvements, and (4) articles that are related to information security.

The following section presents only the results obtained in the last phase of the systematic review (quantitative and/or qualitative synthesis for each one of the four groups of documents).

3.2.1. Mobility Services of FMCS

Details of the documents reviewed in this group are shown in Table 1. From the total of articles in this group (56), 39% (22) are related to FMCSs. The quantitative study of this group was based on some pre-established parameters, such as the type of technology used for communications, the application environment, the use of security, the foundations of ITS, and the type of implementation. Accordingly, it was determined that none of the proposals made a full implementation of the defined parameters because many do not make use of ITS architectures or services (59%), little security implementation is seen (13%) and only a few are intended for implementation in Colombian or Latin American medium-sized cities (26%).

Table 1. Documents of mobility services of Fleet Management and Control Systems (FMCS).

Article	Communication Technology	Based on ITS	Implementation Type	Applied Environment	Use of Security
Architecture based on open-source hardware and software for designing a real-time vehicle tracking device. (English) [31].	Cellular network (GSM)	No	Real	Medium-sized city for Latin American context	No
Performance analysis of advanced bus information system using LTE antenna [32].	Cellular network (LTE)	Yes	Not specified	Not specified	No
IoT-based predictive maintenance for fleet management [33].	WiFi and cellular network	Yes	Real	Not specified	No
A Decision Support System based on smartphone probes as a tool to promote public transport [34].	Cellular network (GSM, GPRS and UMTS)	Yes	Simulation	Small town for European context	No
TransMilenio: A High Capacity—Low-Cost Bus Rapid Transit System Developed for Bogotá, Colombia [35].	Cellular network	No	Does not apply	Big city for Latin American context	No
Study on Real-time Bus Arrival Information System Based on Bluetooth [36].	Bluetooth and cellular network	Yes	Not specified	Not specified	No
An Internet-of-Things-Enabled Connected Navigation System for Urban Bus Riders [37].	WiFi and cellular network	Yes	Real	Big city for European context	No
Fleet Management and Control System from Intelligent Transportation Systems perspective [20].	LoRa	Yes	Not specified	Medium-sized city for Latin American context	No
Data analysis and information security of an Internet of Things (IoT) intelligent transit system [38].	WiFi	No	Real	Small town for American context	Yes
A low-cost M2M architecture for intelligent public transit [39].	ZigBee and cellular network	No	Real	Not specified	No
Real-time vehicle fleet management and security system [40].	Cellular network (GSM)	No	Real	Not specified	Yes
Monitoring System for Intelligent Transportation System Based in ZigBee [41].	ZigBee	No	Simulation	Medium-sized city for Latin American context	No
IoT enabled intelligent bus transportation system [42].	RFID and cellular network	No	Real (laboratory)	Not specified	No
Smart Bus Station-Passenger Information System [43].	WiFi and cellular network	No	Real	Cities in a European context	No
GPS based Public Transport Arrival Time Prediction [44].	Cellular network (GSM)	No	Real	Big city for Asian context	No
A New Framework of Intelligent Public Transportation System Based on the Internet of Things [45].	ZigBee, Bluetooth, and cellular network (GSM, GPRS and 4G)	Yes	Real	Big city for Asian context	No

Table 1. *Cont.*

Article	Communication Technology	Based on ITS	Implementation Type	Applied Environment	Use of Security
Public Transport Vehicle Tracking Service for Intermediate Cities of Developing Countries, based on ITS Architecture using Internet of Things (IoT) [46].	Cellular network and WiFi	Si	Real	Medium-sized city for Latin American context	Yes
A smart cost-effective public transportation system: An ingenious location tracking of public transit vehicles [47].	RFID and Ethernet	No	Simulation	Not specified	No
Technological web platform for integrated public transport system (SITP) of the West Center Metropolitan Area in Colombia [48].	Does not apply	No	Real	Medium-sized city for Latin American context	No
Integration of Smartphone and IoT for development of Smart Public Transportation System [49].	Cellular network (3G)	No	Real	Not specified	No
Systematic Development of Intelligent Systems for Public Road Transport [50].	WiFi	Yes	Real	Not specified	No
Analysis of Radio Technologies for the Transmission of Information to the User of a Fleet Control System [24].	LoRa and Ethernet	No	Real	Medium-sized city for Latin American context	No

Regarding the type of communication used to connect vehicle devices with the cloud server, it was found that the cellular network (GSM, GPRS, 3G, LTE) is the most used (71%), followed by WiFi (28%), ZigBee (14%), Bluetooth (9%), LoRa (9%), Ethernet (9%) and RFID (9%). (in some cases, more than one type of technology is used, so the percentages add up to more than 100%).

Works that implemented technologies such as WiFi, ZigBee, Bluetooth, and RFID also located the vehicle when it passed through a checkpoint or bus stop. Although this allows for a low-cost location service, it does not allow vehicle tracking in real-time.

Some of the works in this group recommended that the FMCS managed certain types of information between the Transit Management Center and transit vehicles. It is recommended that the Transit Management Center (module in charge of managing all transit in a city) sends to transit vehicle information regarding assigned routes, traffic alerts, partial or total route changes, and the scheduled route performance for each of the stops. In turn, it is recommended that each vehicle informs the Transit Management Center: its location, current number of passengers, and compliance with schedules.

3.2.2. Mobility Services other than FMCS

This group features 15 documents (27% of the total). See Table 2 for more details. In this case, a more diverse use of alternative communication technologies is seen, with cellular communication dropping to 46% of participation and technologies such as ZigBee, Bluetooth, and WiFi increasing to 20%. LoRa only is featured in two documents (13%). Regarding the mobility services on which the documents focus, 53% correspond to traffic service in general, 20% to cargo truck fleets, 13.4% to emergency vehicles, and 14.6% to bicycles and school transport. It is interesting to note that different mobility services are involving technologies for the management and control of their systems, thus confirming the need to develop a scalable and interoperable system.

Table 2. Documents of mobility services other than FMCS.

Article	Communication Technology	Service	Based on ITS	Implementation Type	Applied Environment	Use of Security
Internet of Bikes: A DTN Protocol with Data Aggregation for Urban Data Collection [51].	WiFi	Bicycle tracking	Yes	Simulation	Medium-sized city for European context	No
Heterogeneous Intelligent Transportation System in Quito city under the paradigm of the SWE-SOS standard and IoT notifications [52].	Cellular network (GSM)	General traffic	Yes	Real	Medium-sized city for Latin American context	No
A vehicular network-based intelligent transport system for smart cities [53].	WiFi	General traffic	Yes	Simulation	Cities for Asian context	Yes
Cloudthink: a scalable secure platform for mirroring transportation systems in the cloud [54].	Cellular network (GPRS)	General traffic	Yes	Real	Not specified	Yes
A mobile platform system for driving assistance [55].	Cellular network, Bluetooth and WiFi	General traffic	Yes	Not specified	Not specified	Yes
Fleet Management System for Truck Platoons—Generating an Optimum Route in Terms of Fuel Consumption [56].	Cellular network	Cargo truck fleets	Yes	Not specified	Not specified	No
Fleet Management Cooperative Systems for Commercial Vehicles [57].	Cellular network	Cargo truck fleets	Yes	Not specified	Not specified	No
Smart fleet monitoring system using Internet of Things (IoT) [58].	Cellular network (GSM)	Truck fleets	No	Real	Not specified	No
Intelligent transportation system based on the principles of service-oriented architecture [59].	Cellular network (3G & 4G)	Emergency vehicles	No	Real	Large-size city in the Asian context	Yes
Analysis of handshake time for Bluetooth communications to be implemented in vehicular environments [60].	Bluetooth	General traffic	No	Real	Not specified	No
Field testing of Bluetooth and ZigBee technologies for vehicle-to-infrastructure applications [61].	Bluetooth and ZigBee	General traffic	No	Not specified	Not specified	No
The Application of ZigBee Technology to the Intelligent Bus Query System [62].	ZigBee	School transportation	No	Real	Not specified	No
A novel approach of using security-enabled ZigBee in vehicular communication [63].	ZigBee Area Network	General traffic	Yes	Simulation	Not specified	Yes

Table 2. *Cont.*

Article	Communication Technology	Service	Based on ITS	Implementation Type	Applied Environment	Use of Security
i-Car System: A LoRa-based Low-Power Wide-Area Networks Vehicle Diagnostic System for Driving Safety [64].	LoRa	General traffic	No	Simulation	Not specified	No
Traffic Light Mobility Management Prototype for Ambulances, Under the Smart City Concept [65].	LoRa	Ambulance system	No	Real	Medium-sized city for Latin American context	No

Regarding the use of security, there is also an increase in participation, reaching up to 33%, and greater use of ITS services or architectures (53%), although there is a reduction in the implementation of the proposals in real environments (63%).

3.2.3. Implementation of Algorithms and Improvements Related to FMCS

This group features 12 documents (21% of the total). There is a significant increase in the use of WiFi technology (50%) and a reduction to 25% of the use of cellular networks. Technologies like LoRa, ZigBee, TETRA, and Ultra-Wideband have a usage percentage of 12% for each, thus evidencing the growth of particular technologies for IoT.

Similarly, there is an increase in the use of ITS architectures or services (75%) and a considerable reduction in the level of application in real environments (37%). WiFi technology stands out in this group (because simulations are commonly performed with algorithms).

The types of implementation for which the algorithms are designed do not differ much in their participation because of 12 documents found in this group, those that focus on communications, ITS architectures, transmission capacity and generation of alerts have the same number of occurrences (corresponding to 17%); those related to routing and data management account for 8% of the total; and the one with the most participation is "data collection", which has 25%. Sending alerts is an aspect to consider regarding the application of the algorithms, because as identified in the first group of this systematic review, it is important to send alerts from the Transit Management Center to transit vehicles so that they adjust their routes and are able to comply with the established schedules.

The high percentage of documents related to "data collection", and the science mapping analysis (specifically the strategic diagrams for the period 2018–2020) show a close relationship to digital storage and monitoring, two emerging issues within the field of research. Such issues are highly relevant to FMCS, so they should be considered for the proposed system [21,66–70].

3.2.4. Security Applications Related to FMCS

This last group features seven documents. In this group, the context of security application stands out, which is mainly on communications (43%), and it also has a high use of architectures or services (ITS) and a low implementation for real environments (one document).

Based on these data and those of previous groups, security is an important issue within the context of the study, a conclusion that ratifies the results of the science mapping analysis and verifies the need to implement this topic on the development of the proposed system. Similarly, the relationship between ITS architectures and security is observed in 71% of the documents, which stresses the importance of this topic in correctly defined systems that achieve interoperability and scalability [71–73].

3.3. Literature Review Conclusions

Based on the presented information, the predominant technology to geographically locate transit vehicles continues to be GNSS because it allows continuous tracking. Some other technologies such as ZigBee, WiFi, RFID, and Bluetooth can be used to locate vehicles, but only when it passes very close (a few meters) to a certain point where the reader device is located, for example at stops to pick up/drop off passengers.

Regarding the communication technology between the vehicles and the Transit Management Center, cellular communication technologies still prevail; however, some important emerging technologies were identified. LPWAN technologies have been evaluated in some documents, identifying LoRa as a promising technology (comparing it with technologies such as Sigfox or NBIoT), for its already mentioned features.

A gradual increase in the use of applications with security was also observed in the proposals for various fields of systems (communications, authentication, and information security), Although, in most cases, appropriate tests are not conducted to validate the proposals, this increase infers the need to implement security mechanisms in mobility services systems.

The security applications in the analyzed documents focus on various aspects. Authentication of vehicles is one of them; various strategies are proposed to reduce time and improve the efficiency of this function. Other proposals focus on communication security, proposing improvement algorithms on the security protocols such as LTE, emphasizing the need to improve security requirements such as confidentiality, integrity, privacy, availability, and nonrepudiation. Compliance with these requirements is important for the exchange of data between the vehicles since, otherwise, the information could be vulnerable to external attacks and inappropriate handling. It is also important to highlight security in access to the system, guaranteeing that unauthorized parties are not allowed to access the system. The issue of anonymity must also be considered in the security component. Travelers, for example, do not need to identify into the system to obtain certain information; while other actors, such as vehicles and operators, need to identify themselves in order to monitor their actions.

4. ITS Architecture Proposed for the FMCS

4.1. Identification of FMCS Requirements

According to the literature review, the following aspects were identified in a development validation of an FMCS in the aforementioned contexts:

- The location of the transit vehicle in an FMCS is mainly done through GNSS technology. Although there is evidence of other technologies used for location (Bluetooth, WiFi, RFID), they do not allow continuous tracking, only detecting the arrival of the vehicle at stops, due to the limited range they have.
- Controlling the speed of the transit vehicle is a relevant factor in identifying accident risks, complying with traffic regulations, and identifying inappropriate driving behaviors. The use of GNSS technology for location has the additional advantage of measuring the speed of the vehicle on the route.
- Communication technologies for mobility services such as fleet management, applied in the context of this work, must also involve low costs (due to budget constraints these types of cities have). Aspects such as: wide range operation, sufficient data rate, sufficient bandwidth, licensing of the frequency spectrum used, and suitable packet size must be taken into account.
- Most recent cases use the LPWAN LoRa technology as a means for communication between vehicles and Transit Management Center.
- Few cases consider information security. However, this feature should be taken into account in the implementation of these systems, for guaranteeing trust and integrity. The information security in the system must be evaluated from different fronts, one is integrity and authentication between the vehicles and the Transit Management Center; another is user authentication for system managers; finally, exchange of information between modules is a key issue because there is no manual authentication process, nevertheless, the use of REST APIs mitigates this problem by means of a transport layer security (TLS) encryption. The HTTPS protocol also allows the integration of

authentication headers. There are different authentication methods, the one recommended in this proposal is based on an API key, which also allows configuring the authorization to the information (providing limited access to resources), also offering the possibility of performing analysis and supervision of who uses the API, when and how.

- Most of the implemented security schemes are related to vehicle communications and authentication. Regarding vehicle communications, some proposals improve upon aspects such as reliability, operational range, transmission rates, and attack prevention. In authentication, the focus is towards the identification and access of system users, which in turn enable validity in the information they transmit and receive.
- The exchanged data between vehicles and the system control modules in an FMCS must allow the vehicle to send data such as its location, arrival at assigned stops, and in some cases, the number of passengers. In turn, the vehicle must receive the information about the assigned route, the changes in this route, alerts, and assigned schedules.
- In an FMCS the vehicle must receive information sent from the Traffic Management Center, so the communication technology must implement "downlink" messages to facilitate this process.
- The use of suitable ITS architectures and their related services facilitate scalability and interoperability with other services. In most of the studied cases, these principles are taken into account; however, a large percentage of such cases do not correspond to architectures or services that are appropriate to a particular context.

Summarizing, the FMCS should comply with the requirements presented in Table 3.

Table 3. Requirements determined for the FMCS.

Requirement	Related to	Proposal	Observation
Location technology of transit vehicles must allow continuous and efficient monitoring.	Transit vehicle location	Use of GNSS for location.	Although there are some alternative location technologies, these would not allow continuous vehicle tracking.
FMCS must allow speed control of the transit vehicle	Transit vehicle speed	Use of GNSS to measure speed, use of speed alerts and detection of inappropriate driving behaviors	Through speed alerts, the risks of accidents can be reduced. By detecting inappropriate driving behaviors, steps can be taken to reduce fuel consumption.
FMCS must allow interoperability and integration (with other systems or services) in the context where it is implemented.	ITS framework	Use of adequate ITS architecture for FMCS in the specific context.	An appropriate ITS architecture should be proposed, reviewing relevant references, and considering the context.
FMCS must use an adequate communication technology for the application context	Communication technology	Use of LoRa technology (using the network protocol LoRaWAN)	LoRa offers some notable advantages over other technologies for the specific service. LoRaWAN protocol can be a viable option that improves communication security and efficiency.
The exchange of data between the FMCS modules must be secure.	Information security	Secure information exchange (through a network protocol such as LoRaWAN, HTTPS, TCP). Access to the data with SHA passwords and user hierarchy.	LoRaWAN protocol will be of great help by providing the possibility of encoding messages and validation of devices in vehicles. Regarding the information that circulates through the software network, the TCP and HTTPS protocols will guarantee the secure exchange of information. Finally, each user will be given a password to access the system and a user hierarchy will be implemented.

Table 3. *Cont.*

Requirement	Related to	Proposal	Observation
Exchanged data between the vehicles and the control modules must allow permanent updating of the information in both directions, to provide a better service.	Exchanged data between modules	Transit vehicles send their location, arrival at assigned stops, and the number of passengers. The vehicles receive the assigned route, the changes in this route, alerts, and assigned schedule.	An FMCS is more than a tracking service. All required information regarding schedules, routes, changes, and alerts should be considered.

4.2. *Reference ITS Architectures Review*

An ITS architecture offers a vision of what an Intelligent Transport System will look like from a system design perspective. There are three pioneering world reference architectures: ISO 14813 [74], ARC-IT [75], and Framework Architecture Made for Europe (FRAME) [76], which establish certain groups of transport services. An ITS architecture can be presented at several detail levels, depending on the scope of the system. A review of the FMCS-related services proposed each reference architecture follows.

4.2.1. ARC-IT (Architecture Reference for Cooperative and Intelligent Transportation)

This American architecture establishes 12 areas of transportation services [77]. The two ARC-IT services of interest for this work are Public Transportation Transit Vehicle Tracking, and Transit Fleet Management (abbreviated as PT01 and PT06). The first one focuses on monitoring the current location of the transit vehicle using an Automated Vehicle Location System, while PT06 allows scheduling and automatic traffic maintenance monitoring.

Both architectures keep certain similarities in their design because they share several of the physical objects that compose them, such as the Transit Management Center (TMC), in charge of managing the fleet of transit vehicles, and providing operations, maintenance, customer information, planning, and management functions for transit property in coordination with other modes and transport services.

4.2.2. Framework Architecture Made for Europe (FRAME)

FRAME establishes 10 areas of transport services. The Manage Public Transport Operations area is related to the FMCS service and includes seven high-level functions. Each function encompasses several specific fleet management services, including general programming of services and generation of information that can be made available to travelers [78]. Among these functions, four were identified as appropriate for the context of this study: Monitor PT Fleet, Plan PT Service, Control PT Fleet, and Provide On-Demand Services.

4.2.3. ITS Architectures Initiatives in Colombia

In Colombia, there is no established ITS architecture that serves as a model for the subsequent development and interoperability of ITS services, although some ITS services have been established in the past. Work is currently underway to adapt the services established solely by the ISO 14813 standard [79], for achieving consistency in the national ITS plan.

In 2010, the Colombian government developed a national ITS architecture based on the ARC-IT architecture, to be implemented in different cities in Colombia; however, it has never been implemented [80].

A proposal related to the design of an FMCS from the perspective of ITS was also revised [20]. Such a proposal is based on American architecture (ARC-IT). Due to its relationship with the current work, the architecture proposed in the following section took into account the architecture presented in that work.

4.3. Proposed ITS Architecture

According to the obtained and analyzed information, the ARC-IT American architecture was taken as a reference, due to the level of detail it presents for each service, which facilitates their development. Additionally, when the service areas of the different architectures are compared, ARC-IT is the one with the largest covered areas, which gives it a higher scope factor. However, this architecture must include the context of medium-sized cities in developing countries. The ITS architecture must also take into account the aspects identified in the systematic review as important requirements in this type of system (FMCS). One of the most critical requirements is security for communications, reliability, authenticity, among others. Another relevant requirement is the measurement of vehicle parameters and the choice of reliable, affordable communication technologies.

The proposed architecture design is presented in Figure 3. The architecture is presented using ARC-IT notation. In addition, Table 4 briefly explains the flow of information in the proposed architecture.

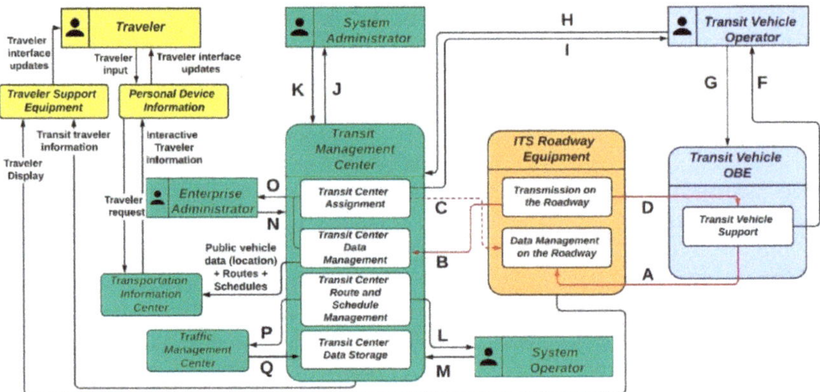

Figure 3. Intelligent Transportation System (ITS) architecture proposal for FMCS.

This architecture features a series of submodules (functional objects) for the Transit Management Center TMC and Transit Vehicle OBE physical objects. The ITS Roadway Equipment, with functionalities for the equipment on the road, is also included. Finally, two new actors (System Administrator and Enterprise Administrator) and new communication information were added. Thus, the ARC-IT base is preserved, keeping the Transit Vehicle OBE and Transit Management Center objects, and the Public Vehicle Operator actor together with most of the initial functions. It also includes several of the high-level functions of FRAME's Manage Public Transport Operations area, such as operational functions, continuous monitoring, continuous data collection, routes, schedule management, and provision of information for travelers, among others. Additionally, the on-demand service provisioning function of FRAME's Provide On-demand Services area was added.

To keep the architecture fully functional and scalable, communication messages with the other modules are defined so that information can be shared between them. To comply with already mentioned security requirements, the messages that go through the "Transit Vehicle OBE" and "TMC" modules will need layers of security to guarantee confidentiality, integrity, and availability of the information. Messages requiring security are marked with red lines in Figure 3. Likewise, authentication methods are applied for all information communication services between actors and modules.

It is important to emphasize that this architecture is generalizable to other medium-sized cities that handle the same context of the public transportation system ("collective" service). The use of an adequate architecture will allow the city to allow interoperability with other systems, in the same way, it facilitates the integration of new services that can be implemented to the system through modules and messages.

Table 4. Information flow of the ITS architecture proposed.

Flow	Name	Description
A	Location data + Vehicle information + Public vehicle schedule performance	Measured parameters in vehicles.
B	Location data + Vehicle information + Public vehicle schedule performance	The same information collected by the OBE is verified and passed to the Transit Center Data Management.
C	Public schedule info + Public operator info + Route + Alerts + Addressee	Assignment information to Transit Vehicle Support, this includes to identify the recipient.
D	Public schedule info + Public operator info + Route + Alerts	Information on schedules and routes assigned to the corresponding vehicle that requests it.
F	Public vehicle operator display	Information on the vehicle operator's work parameters, such as the assigned route number, schedules for that route, and alerts.
G	Public vehicle operator control	Manual interaction to operate the OBE device.
H	Public vehicle operator input	Data the vehicle operator sends to the system.
I	Route assignment + Vehicle assignment + Change alerts	Detailed information on the vehicle operator's work parameters, also includes detailed alerts on work parameter changes.
J	System information	Information generated each time the administrator has made a request to the system.
K	System administrator input	Access parameters of the system administrator and information request.
L	Public operation status + Traffic alerts + Suggestions	Information related to traffic alerts generated and suggestions for changes in routes and schedules.
M	System Operator input	Input or access data for the system operator.
N	Enterprise administrator input	Access parameters and requests information for the company administrator.
O	Enterprise information + change request	Responses to requests made by the company administrator.
P	Public vehicle data (speed and location) + Public vehicle conditions + Service demand quantity	Information related to different parameters measured and processed of the vehicles.
Q	Traffic images + Road conditions + Incidents	Predefined messages sent by "Traffic Management Center" to detect different traffic conditions.

4.4. *FMCS Prototype Network Diagram*

After establishing the ITS architecture proposal for the FMCS, the scope of the prototype to be developed must be defined. A network diagram, with the modules selected for the prototype and the proposed communication technologies for the architecture, is presented in Figure 4. The prototype focuses on the three main modules of the architecture presented in Figure 3: TMC, ITS Roadway Equipment, and Transit Vehicle OBE.

Figure 4 shows the path followed by the information between the different modules, and also shows module location. The OBE is located in each vehicle, which communicates with the ITS Roadway Equipment located in the city buildings, in which a gateway is located.

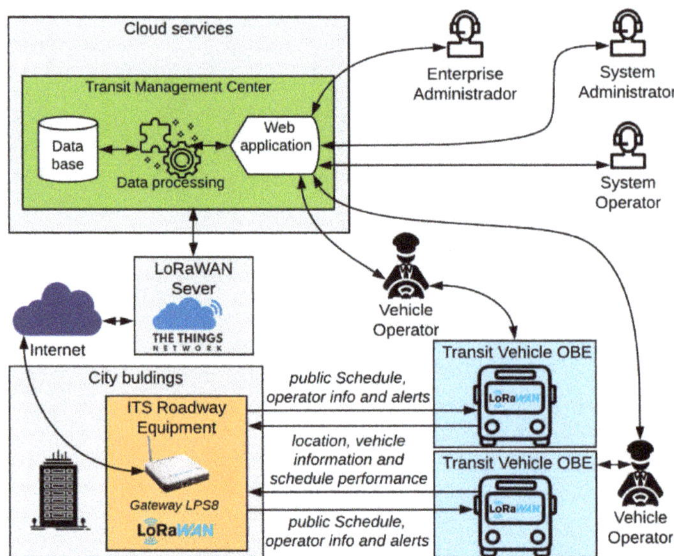

Figure 4. Network diagram for the proposed FMCS prototype.

The selected communication technology between Transit Vehicle OBE and ITS Roadway Equipment was LoRa, using the network protocol LoRaWAN. This was preferred over LoRa messages for two reasons. The first is that LoRaWAN provides an adequate security layer, the information is encrypted and the devices must join the network using the Over-the-Air Activation (OTAA) method in order to start sending messages. The join operation has several authentication parameters. The second reason is to avoid packet collision at the gateway (problem identified in the systematic review [14,81–83]) as LoRaWAN network protocol facilitates the dynamic change of LoRa communication parameters to minimize the probability of packet collision.

The use of the LoRaWAN network protocol requires the use of a LoRaWAN server, which is in charge of all the upper layers of communication; the proposed prototype uses The Things Network (TTN) platform [84]. TTN provides the possibility of sending the information coming from the LoRaWAN devices to servers in the cloud via the HTTPS protocol; this functionality was used to store the data received from the OBEs in a database. The code used for the Heltec cards (which includes integration with the GNSS module, the Ublox 6M GPS), the configuration of the Gateway Dragino LPS8 (mainly the setting of the adequate channels) and the configuration of the TTN platform (including the codification and decodification of data, the integration with a cloud server for sending data and downlink messages) is available upon request.

The next section describes analyses and experiments performed with the communication technology used between the OBE and the ITS Roadway Equipment, in order to determine the feasibility of the FMCS prototype.

5. Design and Results of Prototype Experiments

The coverage area analysis determined that a total of 13 gateways was necessary to cover the whole city of Popayán.

The experiments were developed to evaluate the proper functioning of the developed OBE devices, the selected gateways, the LoRa wireless communication technology, and the security of LoRaWAN protocol. The experiment used four OBE modules and one gateway. The location of the gateway was fixed, and the OBEs were able to move between 10 and 100 m of the gateway, all of them transmitting at the same time. To verify operation at greater distances, the OBEs were subsequently located at distances between 100 and 400 m.

The obtained results were satisfactory, managing to receive 100% of the packets transmitted simultaneously in all the tests performed with Line of Sight (LoS) or minor obstacles. It was possible to visualize the online data sent from the OBEs to the gateway thanks to the TTN platform.

Problems arose with the presence of large obstacles, where the RSSI decreased significantly, thus receiving only 10% of the packets at a distance of 80 m. In this type of test with large obstacles, the maximum distance between devices was approximately 50 m to receive 90% of the packets. This is the reason why the gateway devices should be located in an area with few obstacles and at an elevated location. It is estimated that if 80% of packets are received correctly, the system would operate correctly.

The security of the LoRaWAN protocol was also verified. The joining process that the devices must execute to start sending messages with payload has the adequate security parameters (which are configured both in the code of the microcontroller card, and in the LoRaWAN server platform, TTN). Security tests were performed, trying to communicate a correctly configured gateway, with devices that were not correctly configured (with the LoRaWAN security parameters in OTAA mode), in these cases, the gateway detects the request, but does not allow the joining of the device, therefore it does not continue to receive packets with data. A security test was also executed, trying to connect a gateway that was not properly configured, in that case, it is not possible to join the devices (they had the correct security settings) and packets with data are not received from the devices.

The proposed FMCS requires duplex communication to send messages to and from the modules located in the transit vehicles, so tests of sending downlink messages were performed. In these cases, the packets are received at certain specific times by the device in the vehicle because the program that runs sequentially sends the sensed data at a certain point and receives the receipt confirmation (ACK). Once the data has been sent, it verifies the reception of "downlink" type messages. These delays detected in the reception of messages in the vehicle module could be considered thanks to the performed tests.

Finally, the tests related to the consumption of services through the TTN platform were successful. The server correctly stored the information received by the gateway. It is important to mention that this platform limits access only to authorized users.

With the results obtained in the experiments, it was possible to show that the requirements of the FMCS related to location, tracking, exchanged data, and security are met for the proposed design. However, it is important to clarify that the experiments were performed with a prototype of the system, which has a limited scope of FMCS.

6. Discussion

The literature review allowed identifying an overview of the approach for research on transport services, where ITS architectures or services and issues related to security are increasingly taken into account. This review allowed the identification of specific parameters in the FMCS, according to a considerable number of references from different databases in the last five years. The need to implement communication technologies alternative to cellular service was established (reviewing the identified parameters), which implies lower costs and probably better operation. The use of an ITS architecture was also identified as relevant because although many documents related to FMCS and other mobility services and/or systems mention the use of ITS or its services, almost none use an ITS architecture as the basis for their developments, which considerably makes difficult to standardize and integrate services. Another critical aspect identified in the literature review was the low number of proposed mobility services (especially FMCS), which had validation tests in real settings. This is very important because key aspects of operation can be identified and improved in application environments.

In addition, it was possible to identify the suitable requirements of an FMCS in the context of a "collective" transit service in a medium-sized city. These requirements can be considered relevant in the development of this type of system for Latin American cities that have similar characteristics.

In the case of cities with other characteristics, the process performed in this work to identify relevant aspects in FMCS can be applied in order to obtain results for the new context under study.

To propose the architecture of the FMCS, the principal reference architectures that exist worldwide were taken into account, specifically in proposals on public transportation. In addition, the requirements identified in the review for the context of the study and other related previous initiatives were considered.

The inclusion of the LoRa technology and its LoRaWAN network protocol was considered suitable for communications in the designed architecture, due to its particular characteristics that reduce operation and implementation costs. It also features a wide range of coverage with low power consumption and can be deployed easily.

There are some limitations regarding the proposed idea, specifically in the type of communication technology (LoRa) between OBEs and gateways. One of the most significant issues is the range limitation in places where there is no Line of Sight (LoS) and large obstacles between the different devices, reducing the number of packets received. Therefore, the location of the gateway in an elevated place is a relevant aspect. Another issue is the LoRa message size, which is small (255 bytes), however, this is not critical because the service works well with a low data rate, and additionally, the information sent can be encrypted, minimizing security risks. Regarding the maximum number of devices, although theoretically, a LoRa gateway can support up to one million nodes, it is important to perform scalability tests to validate this because the frequency of sending packets in an FMCS can affect this aspect.

Sending alerts to the vehicles is another important aspect in terms of limitations. The alerts must be handled in a relatively short time, so if the number of devices increases greatly, it will be difficult to handle this type of request due to the large amount of broadcasted information.

Rain can also significantly increase the level of losses in transmission levels, therefore it is recommended to factor this type of loss when defining the coverage area. Finally, the operating temperatures of the electronic elements used in OBEs are important aspects to consider as a limitation.

The proposed architecture has been designed to be implemented in medium-sized cities (100,000 to 1,000,000 inhabitants) that have a public transport system similar to the "collective" (buses of 12 to 20 passengers) and road infrastructure shared with private vehicles. It is important to consider these design aspects and the limitations (or restrictions) of the system, if this proposed architecture is going to be implemented in a similar city. For example, the geographical characteristics of the city must allow the correct location of the gateway devices to provide an optimal power level for transmission and an adequate LoS.

Regarding GNSS, tests showed that this type of technology is adequate to continuously determine the location of the vehicle with an error between 1 and 5 m, however, due to high power consumption and a long time to connect to satellites, a study of the different GNSS devices should be conducted to find the most suitable one.

The results obtained with the performed experiments show that the developed prototype meets certain requirements set out for the FMCS, however, its scope in functionality must be increased and tests with a greater number of nodes in an urban environment as close as possible to the real environment must be conducted for a more complete validation.

Finally, this proposal could be taken into account by local and national authorities related to the Strategic Public Transport Systems (SPTS) that are currently being implemented in eight cities in Colombia. The majority of these projects have not yet developed their technological components (FMCS among them), therefore, new requirements could still be included in the analysis of this system, related to requirements identified for FMCS, the ITS architecture, and the developed prototype. Considering these aspects, important benefits could be obtained in terms of system efficiency, standardization, and savings in implementation and operation costs.

7. Conclusions

With the performed literature review, the principal requirements of an FMCS applied in the context of medium-sized cities (for the "collective" transit service, which is common in many countries of Latin America) were identified, such as interoperability and integration with other transport services, the use of adequate location, tracking, and communication technologies, and the measurement of suitable parameters for analysis and implementation of security protocols that guarantee the correct flow of information.

The design of the proposed architecture was done through a decentralized work approach, by dividing the operations of the system between different modules and actors, yielding a better distribution of processes and a decrease in the workload.

LoRa technology, with its network protocol LoRaWAN, was a good option as a communication technology between modules for the designed ITS architecture because it meets the requirements of low cost, good performance, and an adequate level of security, identified as necessary for the implementation of such architecture. Tests performed with LoRa technology (only LoRa packets) showed problems when two or more devices transmitted simultaneously, so the implementation of the LoRaWAN protocol was necessary, thus achieving effective uplink and downlink communication.

With the results obtained, it is feasible to continue with the development of the FMCS, by making some adjustments to the designed, developed and tested prototype to improve its operation, increase its functionality and verify its operation by scaling some parameters. An important aspect to improve the operation is to evaluate the possibility of changing the employed GNSS module, to improve the connection time to the satellite network and the energy consumption. To increase functionality, it is necessary to advance in the development of operations performed by all types of actors in the system. Finally, to validate the operation by scaling some aspects, tests have to be performed at a larger scale (with more gateways, more nodes, and larger distances), in order to guarantee the correct operation of the proposed architecture in real work environments (using vehicles on the roadways).

Funding: This research received no external funding.

Acknowledgments: Authors wish to thank Universidad del Cauca (Telematics Department) and Universidad Icesi (ICT Department) for supporting this research.

Conflicts of Interest: The authors declare no conflict of interest.

References

1. INRIX. INRIX 2018 Global Traffic Scorecard. Available online: http://inrix.com/press-releases/scorecard-2017/ (accessed on 17 August 2019).
2. Moreno, S.; Cifuentes, S.; Bohórquez, G.; Forero, C.; Hernández, H.; Insuasty, J.; Rodríguez, J.; Romero, J.; Solano, C.; Tello, J.; et al. Behavior of Deaths and Injuries by Transportation Accidents 2018. In *Foreins, Datos Para la Vida*, 2018th ed.; National Institute of Legal Medicine and Forensic Sciences: Bogotá, Colombia, 2018; pp. 295–336.
3. Departamento Nacional de Planeación (DNP). Colombia-Baseline-SETP Armenia-General information. 2012. Available online: https://anda.dnp.gov.co/index.php/catalog/58/overview (accessed on 13 August 2020).
4. Yepes, T.; Junca, J.C.; Aguilar, J. The Integration of Urban Transport Systems in Colombia, A Reform in Transition. Available online: https://repository.fedesarrollo.org.co/bitstream/handle/11445/175/La%20integracion%20de%20los%20sistemas%20de%20transporte%20urbano%20en%20Colombia%20-%20Findeter.pdf?sequence=1&isAllowed=y (accessed on 13 August 2020).
5. Gössling, S. ICT and transport behavior: A conceptual review. *Int. J. Sustain. Transp.* **2018**, *12*, 153–164. [CrossRef]
6. Rueda, R.Z.; Díaz, D.F.C.; Zambrano, C.O. Use of Information and Communication Technologies in Urban Mobility in Intelligent Cities, from a Systematic Review. *Espacios* **2018**, *39*.
7. Seguí Pons, J.M.; Martínez Reynés, M.R. Intelligent Transport Systems and Their Effects on Urban and Interurban Mobility. In *Scripta Nova*; Electronic Journal of Geography and Social Sciences: Barcelona, Spain, 2004.
8. Billhardt, H.; Fernández, A.; Lemus, L.; Lujak, M.; Osman, N.; Ossowski, S.; Sierra, C. Dynamic Coordination in Fleet Management Systems: Toward Smart Cyber Fleets. *IEEE Intell. Syst.* **2014**, *29*, 70–76. [CrossRef]

9. Billhardt, H.; Fernández, A.; Lemus, L.; Lujak, M.; Osman, N.; Ossowski, S.; Sierra, C. ITS development in Colombia: Challenges and opportunities. In *2018 ICAI Workshops (ICAIW)*; IEEE: New York, NY, USA, 2018; pp. i–vi. [CrossRef]
10. Casanova, O.L.; Flores, R.G.; Ortiz, J.M. Implementation of an Efficient Transportation Fleet Management System for Economic Sustainability in a Transport Company. Ph.D. Thesis, Universidad Autónoma de Tamaulipas México, Ciudad Victoria, Mexico, 2012.
11. Marín, C. *Método Para la Gestión Eficiente del Combustible en Flotas de Vehículos Con Rutas Fijas. Aplicación a Una Empresa de Construcción*; Universidad de Sevilla: Sevilla, Spain, 2012.
12. Álvarez León, J.C.; Calle Erráez, D.F. Determination of the Operation Costs for the Transportation of Passengers in the Bus-Type, in the Urban Sector of the City of Cuenca, Based on the New Integrated Transportation System. Bachelor's Thesis, Universidad del Cauca, Popayán, Colombia, 2014.
13. World Health Organization (WHO). Traffic Accidents. Available online: https://www.who.int/es/news-room/fact-sheets/detail/road-traffic-injuries (accessed on 13 August 2020).
14. Boshita, T.; Suzuki, H.; Matsumoto, Y. IoT-based Bus Location System Using LoRaWAN. In Proceedings of the 2018 21st International Conference on Intelligent Transportation Systems (ITSC), Maui, HI, USA, 4–7 November 2018; IEEE: New York, NY, USA, 2018; p. 6.
15. Chaudhari, B.S.; Zennaro, M. *LPWAN Technologies for IoT and M2M Applications*; Elsevier: Amsterdam, The Netherlands, 2020.
16. Wang, C.-S.; Wang, W.-H.A.; Chang, T.-R.; Lin, M.-C. Integrated design structure system for modular design in products development. In Proceedings of the 2009 IEEE International Conference on Industrial Engineering and Engineering Management, Hong Kong, China, 8 December 2009; pp. 1121–1125. [CrossRef]
17. Dellios, K.; Papanikas, D.; Polemi, D. Information Security Compliance over Intelligent Transport Systems: Is It Possible? *IEEE Secur. Priv.* **2015**, *13*, 9–15. [CrossRef]
18. Medina Iriarte, J. Standards for Information Security with Information Technologies. Bachelor's Thesis, Universidad de Chile, Santiago, Chile, 2014.
19. Villaveces, A. Defense of Safe and Healthy Public Transport. Available online: https://iris.paho.org/bitstream/handle/10665.2/28274/9789275331408_spa.pdf?sequence=1&isAllowed=y (accessed on 13 February 2020).
20. Salazar, R.; Cruz, A.; Molina, J. Fleet management and control system from intelligent transportation systems perspective. In Proceedings of the 2019 2nd Latin American Conference on Intelligent Transportation Systems (ITS LATAM), Bogota, Colombia, 19–20 March 2019; IEEE: New York, NY, USA, 2019; pp. 1–7. [CrossRef]
21. Salazar, R.; Pachón, Á. Methodology for design of an intelligent transport system (ITS) architecture for intermediate colombian city. In *Ingeniería y Competitividad*; Universidad del Valle: Valle del Cauca, Colombia, 2019; Volume 21, pp. 49–62. [CrossRef]
22. Cobo, M.J.; Herrera-Viedma, E.; Herrera, F.; López-Herrera, A. SciMAT: A new science mapping analysis software tool. *J. Am. Soc. Inf. Sci. Technol.* **2012**, *63*, 1609–1630. [CrossRef]
23. Moher, D.; Liberati, A.; Tetzlaff, J.; Altman, U.G. Preferred Reporting Items for Systematic Reviews and Meta-Analyses: The PRISMA Statement. *PLoS Med.* **2009**, *6*, e1000097. [CrossRef]
24. Palacios Ibarra, M.A.; Zúñiga Pérez, S.L. Analysis of Radio Technologies for the Transmission of Information to the User of a Fleet Control System. Bachelor's Thesis, Universidad del Cauca, Popayán, Colombia, 2019.
25. Qadir, Q.M.; Rashid, T.A.; Al-Salihi, N.K.; Ismael, B.; Kist, A.A.; Zhang, Z. Low Power Wide Area Networks: A Survey of Enabling Technologies, Applications and Interoperability Needs. *IEEE Access* **2018**, *6*, 77454–77473. [CrossRef]
26. Raza, U.; Kulkarni, P.; Sooriyabandara, M. Low Power Wide Area Networks: An Overview. *IEEE Commun. Surv. Tutor.* **2017**, *19*, 855–873. [CrossRef]
27. About LoRaWAN®|LoRa Alliance®. Available online: https://lora-alliance.org/about-lorawan (accessed on 22 July 2020).
28. LPWAN: The Fastest Growing IoT Communication Technology. IoT Now—How to Run an IoT Enabled Business. 29 October 2018. Available online: https://www.iot-now.com/2018/10/29/89895-lpwan-fastest-growing-iot-communication-technology/ (accessed on 9 August 2020).
29. Adelantado, F.; Vilajosana, X.; Melia-Segui, J.; Watteyne, T.; Tuset-Peiro, P.; Martinez, B. Understanding the Limits of LoRaWAN. *IEEE Commun. Mag.* **2017**, *55*, 34–40. [CrossRef]
30. Navarro Astudillo, E.O.; Quintero Flórez, V.M. Radio Electric Coverage Estimation Software Prototype Tool for Planning an IoT Network. Bachelor's Thesis, Universidad del Cauca, Popayán, Colombia, 2019.

31. Rodríguez, D.R.C.; LaGuardia, A.S.M.; Abreu, A.G. Architecture based in open source hardware and software for designing a real time vehicle tracking device. *Sist. Telemática* **2018**, *16*, 49–61. [CrossRef]
32. Shin, D.-K.; Jung, H.; Chung, K.-Y.; Park, R.C. Performance analysis of advanced bus information system using LTE antenna. *Multimed. Tools Appl.* **2013**, *74*, 9043–9054. [CrossRef]
33. Killeen, P.; Ding, B.; Kiringa, I.; Yeap, T. IoT-based predictive maintenance for fleet management. *Procedia Comput. Sci.* **2019**, *151*, 607–613. [CrossRef]
34. Vitale, A.; Festa, D.C.; Guido, G.; Rogano, D. A Decision Support System based on Smartphone Probes as a Tool to Promote Public Transport. *Procedia Soc. Behav. Sci.* **2014**, *111*, 224–231. [CrossRef]
35. Sandoval, E.E.; Hidalgo, D. TransMilenio: A High Capacity—Low Cost Bus Rapid Transit System Developed for Bogotá, Colombia. *Urban Public Transp. Syst.* **2004**, 37–49. [CrossRef]
36. Wang, Q.; Wang, Q. Study on real-time bus arrival information system based on Bluetooth. In Proceedings of the 2013 IEEE Third International Conference on Information Science and Technology (ICIST), Yangzhou, China, 23–25 March 2013; p. 70. [CrossRef]
37. Handte, M.; Foell, S.; Wagner, S.; Kortuem, G.; Marron, P.J. An Internet-of-Things Enabled Connected Navigation System for Urban Bus Riders. *IEEE Internet Things J.* **2016**, *3*, 735–744. [CrossRef]
38. Data Analysis and Information Security of an Internet of Things (IoT) Intelligent Transit System—IEEE Conference Publication. Available online: https://ieeexplore.ieee.org/document/8374744 (accessed on 6 February 2020).
39. Magdum, N.; Patil, S.; Maldar, A.; Tamhankar, S. A low cost M2M architecture for intelligent public transit. In Proceedings of the 2015 International Conference on Pervasive Computing (ICPC), Pune, India, 8 January 2015; pp. 1–5. [CrossRef]
40. Gowda, V.R.C.; Gopalakrishna, K. Real time vehicle fleet management and security system. In Proceedings of the 2015 IEEE Recent Advances in Intelligent Computational Systems (RAICS), Kerala, India, 10–12 December 2015; IEEE: New York, NY, USA, 2015; pp. 417–421. [CrossRef]
41. Heredia, X.C.; Barriga, C.H.; Piedra, D.I.; Oleas, G.D.; Flor, A.C. Monitoring System for Intelligent Transportation System Based in ZigBee. In Proceedings of the 2019 UNSA International Symposium on Communications (UNSA ISCOMM), Arequipa, Peru, 28–29 March 2019; IEEE: New York, NY, USA, 2019; pp. 1–6. [CrossRef]
42. Geetha, S.; Cicilia, D. IoT enabled intelligent bus transportation system. In Proceedings of the 2017 2nd International Conference on Communication and Electronics Systems (ICCES), Coimbatore, India, 19–20 October 2017; IEEE: New York, NY, USA, 2017; pp. 7–11. [CrossRef]
43. Sungur, C.; Babaoglu, I.; Sungur, A. Smart Bus Station-Passenger Information System. In Proceedings of the 2015 2nd International Conference on Information Science and Control Engineering, Shanghai, China, 24–26 April 2015; pp. 921–925. [CrossRef]
44. Farooq, M.U.; Shakoor, A.; Siddique, A.B. GPS based Public Transport Arrival Time Prediction. In Proceedings of the 2017 International Conference on Frontiers of Information Technology (FIT), Islamabad, Pakistan, 18–20 December 2017; pp. 76–81. [CrossRef]
45. Luo, X.-G.; Zhang, H.-B.; Zhang, Z.; Yu, Y.; Li, K. A New Framework of Intelligent Public Transportation System Based on the Internet of Things. *IEEE Access* **2019**, *7*, 55290–55304. [CrossRef]
46. Public Transport Vehicle Tracking Service for Intermediate Cities of Developing Countries, Based on ITS Architecture Using Internet of Things (IoT)—IEEE Conference Publication. Available online: https://ieeexplore.ieee.org/document/8569906 (accessed on 8 February 2020).
47. Kumar, T.; Gupta, S.; Kushwaha, D.S. A smart cost effective public transporation system: An ingenious location tracking of public transit vehicles. In Proceedings of the 2017 5th International Symposium on Computational and Business Intelligence (ISCBI), Dubai, UAE, 11–14 August 2017; pp. 134–138. [CrossRef]
48. Mejía, J.A.S.; Orozco Gutiérrez, Á.Á.; Mejía, S.E. Technological web platform for integrated public transport system (SITP) of the West Center Metropolitan Area in Colombia. In Proceedings of the 2015 CHILEAN Conference on Electrical, Electronics Engineering, Information and Communication Technologies (CHILECON), Santiago, Chile, 28–30 October 2015; pp. 763–770. [CrossRef]
49. Sutar, S.; Koul, R.; Suryavanshi, R. Integration of smart phone and IOT for development of smart public transportation system. In Proceedings of the 2016 International Conference on Internet of Things and Applications (IOTA), Pune, India, 22–24 January 2016; pp. 73–78. [CrossRef]

50. García, C.R.; Quesada-Arencibia, A.; Cristóbal, T.; Padrón, G.; Alayón, F. Systematic Development of Intelligent Systems for Public Road Transport. *Sensors* **2016**, *16*, 1104. [CrossRef]
51. Zguira, Y.; Rivano, H.; Meddeb, A. Internet of Bikes: A DTN Protocol with Data Aggregation for Urban Data Collection. *Sensors* **2018**, *18*, 2819. [CrossRef]
52. Zambrano, A.; Calderón, X.; Ortiz, E.; Zambrano, O. Intelligent Heterogeneous Transportation System in Quito City Under the Paradigm SWE-SOS Standard and loT Notifications. In Proceedings of the 2019 14th Iberian Conference on Information Systems and Technologies (CISTI), Coimbra, Portugal, 19–22 June 2019.
53. Zaheer, T.; Malik, A.W.; Rahman, A.U.; Zahir, A.; Fraz, M.M. A vehicular network–based intelligent transport system for smart cities. *Int. J. Distrib. Sens. Netw.* **2019**, *15*. [CrossRef]
54. Wilhelm, E.; Siegel, J.E.; Mayer, S.; Sadamori, L.; Dsouza, S.; Chau, S.C.-K.; Sarma, S. Cloudthink: A scalable secure platform for mirroring transportation systems in the cloud. *Transport* **2015**, *30*, 320–329. [CrossRef]
55. Martín-Fernández, F.; Caballero-Gil, P.; Caballero-Gil, C. A Mobile Platform System to Driving Assistance. In Proceedings of the International Conference on Information Technologies, Libertad City, Ecuador, 10–12 January 2016; pp. 284–290.
56. Stancel, I.N.; Surugiu, M.C. Fleet Management System for Truck Platoons—Generating an Optimum Route in Terms of Fuel Consumption. *Procedia Eng.* **2017**, *181*, 861–867. [CrossRef]
57. Surugiu, M.C.; Stancel, I.N. Fleet Management Cooperative Systems for Commercial Vehicles. *Procedia Technol.* **2016**, *22*, 984–990. [CrossRef]
58. Penna, M.; Arjun, B.; Goutham, K.R.; Madhaw, L.N.; Sanjay, K.G. Smart fleet monitoring system using Internet of Things (IoT). In Proceedings of the 2017 2nd IEEE International Conference on Recent Trends in Electronics, Information Communication Technology (RTEICT), Bangalore, India, 19–20 May 2017; IEEE: New York, NY, USA, 2017; pp. 1232–1236. [CrossRef]
59. Mallegowda, M.; Nete, V.; Kanavalli, A. Intelligent transportation system based on the principles of service-oriented architecture. In Proceedings of the 2015 Twelfth International Conference on Wireless and Optical Communications Networks (WOCN), Bangalore, India, 9–11 September 2015; IEEE: New York, NY, USA, 2015; pp. 1–5. [CrossRef]
60. Gheorghiu, R.A.; Iordache, V.; Cormos, A.C. Analysis of handshake time for bluetooth communications to be implemented in vehicular environments. In Proceedings of the 2017 40th International Conference on Telecommunications and Signal Processing (TSP), Barcelona, Spain, 5–7 July 2017; pp. 144–147. [CrossRef]
61. Ordache, V.; Gheorghiu, R.A.; Minea, M.; Cormos, A.C. Field testing of Bluetooth and ZigBee technologies for vehicle-to-infrastructure applications. In Proceedings of the 2017 13th International Conference on Advanced Technologies, Systems and Services in Telecommunications (TELSIKS), Niš, Serbia, 18–20 October 2017; pp. 248–251. [CrossRef]
62. Cui, S.; Xing, Y.; Lu, B.; Wang, H. The Application of ZigBee Technology to the Intelligent Bus Query System. In Proceedings of the 2014 Sixth International Conference on Measuring Technology and Mechatronics Automation, Zhangjiajie, China, 10–11 January 2014; pp. 672–675. [CrossRef]
63. Shree, K.L.; Penubaku, L.; Nandihal, G. A novel approach of using security enabled Zigbee in vehicular communication. In Proceedings of the 2016 IEEE International Conference on Computational Intelligence and Computing Research (ICCIC), Chennai, India, 15–17 December 2016; pp. 1–5. [CrossRef]
64. Chou, Y.S.; Mo, Y.C.; Su, J.P.; Chang, W.J.; Chen, L.B.; Tang, J.J.; Yu, C.T. i-Car system: A LoRa-based low power wide area networks vehicle diagnostic system for driving safety. In Proceedings of the 2017 International Conference on Applied System Innovation (ICASI), Sapporo, Japan, 13–17 May 2017; pp. 789–791. [CrossRef]
65. Mompotes Pizo, D.R.; Muñoz Narvaez, J.A. Traffic Light Mobility Management Prototype for Ambulances, Under the Smart City Concept. Bachelor's Thesis, Universidad del Cauca, Popayán, Colombia, 2019.
66. Oh, H.; Ahn, S.-H. A Full-Duplex Relay Based Hybrid Transmission Mechanism for the MIMO-Capable Cooperative Intelligent Transport System. *Int. J. Distrib. Sens. Netw.* **2015**, *11*, 1–11. [CrossRef]
67. Cardoso, F.; Serrador, A.; Canas, T. Algorithms for Road Safety Based on GPS and Communications Systems WAVE. *Procedia Technol.* **2014**, *17*, 640–649. [CrossRef]
68. Polo, A.; Robol, F.; Nardin, C.; Marchesi, S.; Zorer, A.; Zappini, L.; Viani, F.; Massa, A. Decision support system for fleet management based on TETRA terminals geolocation. In Proceedings of the 8th European Conference on Antennas and Propagation (EuCAP 2014), The Hague, The Netherlands, 6–11 April 2014; IEEE: New York, NY, USA, 2014; pp. 1195–1198. [CrossRef]

69. Sadanandan, L.; Nithin, S. A smart transportation system facilitating on-demand bus and route allocation. In Proceedings of the 2017 International Conference on Advances in Computing, Communications and Informatics (ICACCI), Daegu, Korea, 19–20 May 2017; IEEE: New York, NY, USA, 2017; pp. 1000–1003. [CrossRef]
70. Wang, S.Y.; Chang, C.H. Supporting TCP-Based Remote Managements of LoRa/LoRaWAN Devices. In Proceedings of the 2019 IEEE 90th Vehicular Technology Conference (VTC2019-Fall), Honolulu, HI, USA, 22–25 September 2019; IEEE: New York, NY, USA, 2019; pp. 1–5. [CrossRef]
71. Siwach, A.; Verma, D. Analysis of Security Problems in Vehicular Ad-hoc Networks using LTE Cellular Network. *Int. J. Comput. Sci. Manag. Stud.* **2015**, *15*, 7–10.
72. Vasudev, H.; Das, D. Secure Lightweight Data Transmission Scheme for Vehicular Ad hoc Networks. In Proceedings of the 2018 IEEE International Conference on Advanced Networks and Telecommunications Systems (ANTS), Indore, India, 16–19 December 2018; IEEE: New York, NY, USA, 2018; pp. 1–6. [CrossRef]
73. Zhu, H.; Yuan, Y.; Chen, Y.; Zha, Y.; Xi, W.; Jia, B.; Xin, Y. A Secure and Efficient Data Integrity Verification Scheme for Cloud-IoT Based on Short Signature. *IEEE Access* **2019**, *7*, 90036–90044. [CrossRef]
74. International Organization for Standardization (ISO). ISO 14813-1:2015. 2015. Available online: https://www.iso.org/obp/ui/#iso:std:iso:14813:-1:ed-2:v1:en (accessed on 17 August 2019).
75. The National ITS Reference Architecture. *Architecture Reference for Cooperative and Intelligent Transportation*; Iteris: Santa Ana, CA, USA, 2019.
76. Relationship with the ITS Action Plan and ITS Directive|FRAME ARCHITECTURE. Available online: https://frame-online.eu/frame-architecture/detailed-information/relationship-with-the-its-action-plan-and-its-directive (accessed on 11 June 2020).
77. Service Packages. Available online: https://local.iteris.com/arc-it/html/servicepackages/servicepackages-areaspsort.html (accessed on 11 June 2020).
78. The Browsing Tool|FRAME ARCHITECTURE. Available online: https://frame-online.eu/frame-architecture/the-browsing-tool (accessed on 11 June 2020).
79. Andinatraffic. *8°va FERIA ANDINATRAFFIC y 1°er Congreso ITS LATAM*; Sofex Americas: Cundinamarca, Colombia, 2017; p. 136.
80. Arquitectura Nacional ITS de Colombia. Available online: http://www.consystec.com/colombia/web/ (accessed on 21 August 2019).
81. Bankov, D.; Khorov, E.; Lyakhov, A. On the Limits of LoRaWAN Channel Access. In Proceedings of the 2016 International Conference on Engineering and Telecommunication (EnT), Dolgoprudny, Russia, 29–30 November 2016; IEEE: New York, NY, USA, 2016; pp. 10–14. [CrossRef]
82. Sanchez-Iborra, R.; Sanchez-Gómez, J.; Ballesta-Viñas, J.; Cano, M.-D.; Gómez, A.F.S. Performance Evaluation of LoRa Considering Scenario Conditions. *Sensors* **2018**, *18*, 772. [CrossRef]
83. Georgiou, O.; Raza, U. Low Power Wide Area Network Analysis: Can LoRa Scale? *IEEE Wirel. Commun. Lett.* **2017**, *6*, 162–165. [CrossRef]
84. The Things Network. Available online: https://thethingsnetwork.org/ (accessed on 14 July 2020).

 MDPI

Article

Machine Learning-Based Driving Style Identification of Truck Drivers in Open-Pit Mines

Qun Wang [1], Ruixin Zhang [1,2], Yangting Wang [2,*] and Shuaikang Lv [1]

1 School of Energy and Mining Engineering, China University of Mining and Technology (Beijing), Beijing 100083, China; wangqun@student.cumtb.edu.cn (Q.W.); zhangrx@ncist.edu.cn (R.Z.); lvsk521@163.com (S.L.)

2 North China Institute of Science and Technology, Langfang 065201, China

* Correspondence: wangyting@ncist.edu.cn; Tel.: +86-187-3516-9600

Received: 26 November 2019; Accepted: 22 December 2019; Published: 24 December 2019

Abstract: The significance in constructing a driving style identification model for open-pit mine truck drivers is to reduce diesel consumption and improve training. First, we developed a driving behavior and mining truck condition monitoring system for an open-pit mine. Under heavy-load and no-load conditions of a mining truck, based on the same experimental truck and haulage road, the data of driving behavior and truck status of different drivers were collected. The driving style characteristic parameters of mining trucks under heavy-load and no-load conditions were constructed through Pearson correlation analysis. Using a k-means clustering algorithm, driving style can be divided into three types: normal type, soft type, and aggressive type, and we verified the validity of this driving style classification with a box plot. On this basis, the parameters of random forest, k-nearest neighbor, support vector machine, and neural network models were optimized and the accuracy was compared through a cross-validation grid search, and then a driving style identification model based on the random forest method was finally proposed. Driving style parameter weight values were obtained based on the Gini coefficient. Last, the fuel consumption characteristics of different driving styles were calculated. The results show that the driving style identification models based on random forest can effectively identify different driving styles when the mining truck is operating under heavy load and no load, and the overall accuracy of the model is 95.39% and 90.74% respectively. The fuel consumption of the aggressive driving style was the largest and was 10% higher than the average fuel consumption. The research results provide data support and new ideas for operation training and fuel-saving driving of mining trucks in open-pit mines.

Keywords: open-pit mine; driving style recognition; K-means clustering; random forest

1. Introduction

With the advantages of high mobility, strong climbing ability, and a short construction period, mining trucks have been widely used in open-pit mines all over the world. At the same time, there are also problems of high fuel consumption and transportation costs. We take the open-pit mine of Inner Mongolia Xilingol Baiyinhua Coal Power Co., Ltd. (hereinafter referred to as Baiyinhua No. 2 mine) as an example. From 2016 to 2018, the annual diesel consumption of 24 mining trucks was 9734 t, 8322 t, and 8589 t respectively, with an average annual diesel consumption of 370 t per mining truck. Driving style refers to the relatively stable and habitual internal driving behavior tendencies formed by a driver in the long-term driving process, which can be generally divided into aggressive type, normal type, and soft type [1,2]. Research showed that the difference of energy consumption between soft drivers and radical drivers was about 30% under the same route and vehicle type. Even under the condition of low requirements for driver operation, the difference of energy consumption was as high as 17% [3,4].

Similarly, researchers have shown that the driving style of mining truck drivers in open-pit mines has a direct impact on fuel consumption, when considering the influence factors of mining truck fuel consumption, driving style cannot be ignored as an important indicator [5,6]. John et al. [7] constructed a virtual 24/7 driving style model of open-pit mining trucks, dividing the style of driver into aggressive, normal, and soft types, and simulated the fuel consumption under different driving styles. The results showed that the fuel consumption of aggressive, normal, and soft driving styles were 330 L/h, 300 L/h, and 295 L/h respectively under heavy-load conditions. Under no-load conditions of mining trucks, the fuel consumptions of aggressive, normal, and soft driving styles were 210 L/h, 186 L/h, and 170 L/h respectively. In addition, among a series of fuel-saving technologies, in which drivers can make decisions, the best fuel-saving effects can be as high as 25%, by changing the driving style [8]. Driving skill training could bring about 2–5% fuel-saving effects [9–11], and summarized references about the influence of driving style on fuel consumption (as shown in Table 1).

Table 1. The influence of driving style on fuel consumption.

Number	Driving Style	Fuel Consumption
1	Comparison of fuel consumption between soft and aggressive driving styles [3,4]	30% less
2	Fuel consumption of aggressive type, normal type and soft type under the condition of heavy load of mining truck in open-pit mine [7]	Aggressive: 330 L/h Normal: 300 L/h Soft: 295 L/h
3	Fuel consumption of aggressive type, normal type and soft type under the condition of heavy no load of mining truck in open-pit mine [7]	Aggressive: 210 L/h Normal: 186 L/h Soft: 170 L/h
4	Optimizing driving style [8]	25% reduction
5	Driver skill training [9,11]	2–5% reduction

Marina Martinez et al. [12] pointed out that driving style played an important role in energy management and safe driving, and that using machine learning to identify driving styles and applying this to vehicle intelligent control has become the future development trend. Therefore, it is of great significance for driver training and fuel-saving driving to establish a high precision driving style identification model for open-pit mining truck drivers. At present, the research methods of driving style include questionnaire surveys and objective driving data analysis. Shinar et al. [13] showed that the identification effect of the questionnaire survey method was limited due to its strong subjectivity and lack of objective data support. Objective driving data analysis can be carried out on the basis of real vehicle roads or simulation test platforms, which use sensors to collect driving behavior data such as vehicle speed, throttle opening, and longitudinal acceleration [14,15]. This method is widely used due to the high data reliability and real reflection of driving style. With the development of data technology and modern communication technology, increasing amounts of driving behavior data can be collected, therefore, machine learning can be used to extract knowledge from the data.

Machine learning is divided into unsupervised learning and supervised learning, according to whether the data sample has labeled attributes [16]. Research based on machine learning applied to driving style classification can be roughly divided into the following three categories.

(1) Unsupervised machine learning algorithms: Researchers used the statistical values of speed and longitudinal acceleration as the driving style characteristic parameters, and extracted driving style characteristic parameters based on principal component analysis (PCA), and K-means, density-based spatial clustering of applications with noise (DBSCAN), and spectral cluster algorithms were used to cluster driving styles [14,15,17,18]. Han et al. [19] took speed and throttle opening as driving style characteristic parameters, and driving styles were classified by full Bayesian theory and kernel density estimation, compared with the traditional fuzzy logic (FL) method, and driving style identification accuracy was improved. Zhu et al. [20] proposed an unsupervised clustering method of driving style

based on Kullback–Leibler (KL) divergence, where Gaussian mixture model (GMM) was used to represent the statistical distribution of the drivers' real driving data, and driving style clustering was realized based on the similarity between different distributions.

(2) Supervised machine learning algorithms: Karginnova et al. [21] compared k-nearest neighbor (KNN), neural networks (NN), and random forests (RF) in terms of recognition performance, and found that random forest was suitable for short term real-time classification application scenarios. Bejani et al. [22] integrated KNN, support vector machines (SVM), multi-layer perceptron (MLP), and selected precision, recall, accuracy, f-scores as model evaluation indexes, and they were all greater than 92%. The overall effect was better than any classifier.

(3) Combined unsupervised and supervised algorithms: The authors in previous studies [23,24] adopted PCA to achieve a few comprehensive indicators through dimensionality reduction of driving behavior characteristic parameters, and then applied k-means or fuzzy c-means (FCM) to calibrate driving style data samples, and introduced supervised learning SVM algorithms to establish driver driving style identification models. The model identification accuracies were above 90%.

At present, research on driving behavior in open-pit mines focuses more on safety aspects such as truck collision prevention [25] and driver fatigue prevention [26]. Daily management of fuel consumption related to driving behavior and driving style is mainly restricted by administrative orders [27]. Supervise whether the driver carries out the operation or not is difficult, at the same time, the management basis is not sufficient, and the management effect is not obvious. We adopted objective driving data analysis, based on the same experimental truck and haulage road, sensors were used to collect the data of driving behavior and mining truck status, such as throttle opening, speed, longitudinal acceleration, position, etc., and we adopted k-means to classify the driving style for different drivers. On this basis, the parameters of random forest, k-nearest neighbor, support vector machine, and neural network models were optimized and the accuracy was compared through a cross-validation grid search, and the model with the best accuracy was selected and the driving style identification model was constructed for mining truck under the conditions of heavy load and no load. Finally, the fuel consumption characteristics under different driving styles were calculated.

2. Experimental Design and Data Preprocessing

The Baiyinhua No. 2 open-pit mine is in Baiyinhua energy and chemical park, Xilingol, Inner Mongolia. The approved production capacity is 17 million t/a, the rock stripping mainly adopts the shovel-truck discontinuous mining system, with four sets of excavators which are mainly used to extract rock from the blasting pile and load it into the mining truck. Meanwhile, there are 24 mining trucks, which are mainly used to transport the stripped rock.

2.1. Experimental Scene Design

To achieve the most realistic and reflective driving style data of mining truck drivers, it is necessary to plan the experimental truck, experimental time, experimental haulage road and other factors in advance, and try to reduce the interference of other factors on the driver's driving behavior.

2.1.1. Selection of Experimental Time

The driving behavior data of different drivers were collected during normal operation times with good weather and sight distances, from 15 June 2019 to 3 July 2019. The normal operation time was from 09:00 to 16:00.

2.1.2. Experimental Scene Design

To avoid the influence of different haulage roads and driving conditions on driving behavior, an experimental truck is selected, and its haulage road is fixed in the experiment. Under the condition of heavy load, the mining truck transports stripped rock from the loading point to the unloading point of the dump site. Instead, under the condition of no load, the mining truck returns to the original loading point from the unloading point of the dump site, completing the operation cycle of loading-point loading–heavy-load driving—dumping-site unloading–returning to the original loading point. Through the same haulage road, under the condition of no load, different drivers are replaced every day to carry out normal transportation operations. In the process, the data of driving behavior and mining truck status is collected. Additionally, the driver's salary is related to the number of mining truck operation cycles, under the premise of not exceeding the safe driving speed, drivers rely on their own experience and driving habits to carry out transportation operations at the maximum running speed.

2.1.3. Selection of Experimental Road

The Baiyinhua No. 2 open-pit mine has one inner dump haulage road and outer dump haulage road, respectively. Both are in the state of free-flow traffic, low speed, and low traffic flow. Compared with the inner dump haulage road, the outer dump haulage road has a longer transportation distance and higher lifting height, included more curved roads, straight roads, uphill roads, downhill roads, and so on. Therefore, the outer dump haulage road (as shown in Figure 1) was selected in this experiment, the transportation distance of the experimental haulage road was about 2.9 km, and the lifting height was about 105 m.

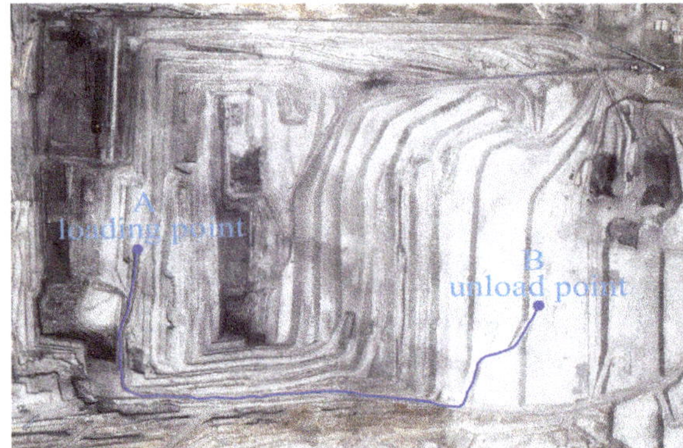

Figure 1. Sketch map of aerial photography of the experimental haulage road.

2.2. Design and Installation of Mining Truck Driving Behavior and Vehicle Condition Monitoring System

2.2.1. Selection of Experimental Truck

According to the maintenance history of mining trucks, the experiment selected a low failure rate mining truck, which was named No. 7, as the experimental truck. The main equipment parameters of this type of mining truck are shown in Table 2.

2.2.2. Driving Behavior and Mining Truck Condition Monitoring System Design

To collect the real-time data of the driver's driving behavior and vehicle status of the mining truck under heavy-load and no-load conditions, a system of driving behavior and vehicle status monitoring, which was based on an inertial navigation sensor and Advanced RISC Machine (ARM) micro-controller (Figure 2), was designed. Due to the sampling distance, time needed for the maximum pedal opening and heavy load, the data acquisition frequency was set as 2 Hz in the inertial navigation sensor uniformly, the driving behavior sensor collected data such as throttle opening and throttle pedal angular velocity, while the vehicle status sensor collected data such as speed, longitudinal acceleration, slope gradient, and position. The ARM micro-controller was applied to receive the data collected by the sensors and store it in the SD card, realizing the perception and storage of driving behavior and vehicle state of the mining truck in the open-pit mine. In addition, the sensor time was calibrated in real time through navigation satellite timing.

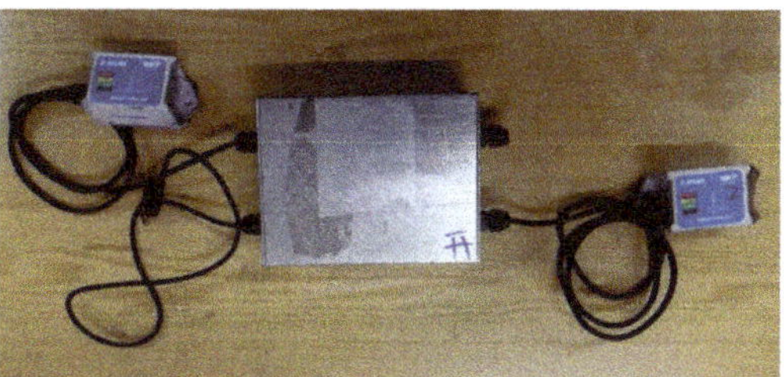

Figure 2. Photograph of the data acquisition system.

The system hardware mainly includes one ARM micro-controller (STM32f103), two ten-axis inertial navigation sensors (WTGAHRS2), one SD memory card, one protective shell, and the vehicle DC power supply. The ten-axis inertial navigation sensor integrates a high precision gyroscope, accelerometer, GPS, and other modules to form an integrated navigation unit named GPS-IMU. This unit has the advantages of high precision, low cost, low power consumption, small size, and so on. It can measure acceleration, angular velocity, angle, velocity, and other parameters accurately. The specific performance parameters of the ten-axis inertial navigation sensor are shown in Table 3.

Table 2. Main equipment parameters of the experimental truck.

Parameter Name	Parameter Value
Truck Type	BELAZ 75313
Truck Weight	107 t
Truck Load Rating	130 t
Boundary Dimension	11.5 m × 6.9 m × 5.7 m
Safe Driving Speed	Maximum speed of flat road: 30 km/h Maximum speed of no-load downhill: 35 km/h
Rated Capacity of Dump truck Bucket	70 m^3
Engine Power Rating	1194 kw
Traction Generator Power	1000 kw
Gears	Continuously variable speed

Table 3. Performance parameters of ten-axis inertial navigation sensor.

Parameter Name	Parameter Value
Volume	61.2 mm × 45.2 mm × 27.8 mm
Angular velocity precision	Static: 0.05°, Dynamic: 0.1°
Acceleration precision	0.01 g
Speed precision	0.05 m/s
GPS precision	±5 m

2.2.3. Driving Behavior and Vehicle Condition Monitoring System Installation

It was necessary to build a monitoring system of driving behavior and vehicle status in the cab of the mining truck before the experiment. The main work included the installation of the ten-axis inertial navigation sensor, the fixation of the protective shell and the external power supply, etc. During the installation, the No. 1 sensor was installed on the back of the throttle pedal along the X axis, while the No. 2 sensor was installed near the horizontal position in the cab along the Y axis. The installation diagram is shown in Figure 3. The installation of hardware did not cause safety hazards to the mining truck transportation operation, and the driver checked before each shift to ensure safe operation.

Drivers were not informed about the purpose of the installed monitoring equipment, and all data collected was anonymous and was not linked back to any individual operators.

Figure 3. Mining truck driving behavior and vehicle condition monitoring system installation. (**a**) Experimental truck. (**b**) Installation of sensor No. 1. (**c**) Installation of sensor No. 2.

2.3. *Tested Drivers Information*

The Baiyinhua No. 2 open-pit mine considers the factor of safe operation of mining truck transportation, and so drivers need 1–3 years of professional driving skills training before they can work skillfully, therefore, all drivers can skillfully drive the mining truck for transportation. According to the sequence of daily driver scheduling, 11 male drivers' natural driving data were collected in this experiment, the mental and physical state of the 11 drivers were normal. The essential information is shown in Table 4: the age range of drivers was between 30 to 50 years old, the average age was 38.2 years old, the driving experience range was between 3 to 10 years, the average driving experience was 8.6 years, and the education level was senior high school.

Table 4. Basic information of tested drivers.

Index	Range	Driver Number
Driver Age Distribution	30–40 years old	7
	40–50 years old	4
Driving Experience Distribution	3–5 years	3
	5–10 years	8
Education Level	Senior high school	11

2.4. Data Processing of Driving Behavior

2.4.1. Raw Data of Driving Behavior

Based on the same experimental truck and haulage road, 11 drivers covered about 650 km in actual transportation operation. The parameters of driving behavior and vehicle status are shown in Table 5.

Table 5. Sensor acquisition parameters.

No.	Name of Characteristic Parameter	Meaning of Characteristic Parameter	Unit
1	sensorID	Sensor number	-
2	time	Time	ms
3	rx3	Angle of throttle pedal to the horizontal	°
4	ry1	Angle of truck to the horizontal	°
5	wx3	Throttle pedal angular velocity	°/s
6	gpsv	Running velocity of truck	km/h
7	ax1	Longitudinal acceleration of truck	g
8	lon	Longitude	°
9	lat	Latitude	°

2.4.2. Driving Behavior Data Cleaning

Due to factors such as GPS signal occlusion and other electromagnetic interference, the sensor can output wrong and invalid driving behavior data, while data noise has a bad impact on the learning result of the machine learning algorithm. Therefore, driving behavior data should be processed before cluster analysis. The main processing contents include data extraction and data deletion, etc.

1. Data extraction
   The data collected by the No. 1 and No. 2 sensors was stored in the SD card separately, with the sensor number ID and time as the mark. The sensor data fusion program was developed using the Python programming language, which realized the splicing of data collected by two sensors at the same time, and provided complete driving behavior data for the subsequent data preprocessing.
2. Data deletion
   The data records with the speed of zero were eliminated, a speed of zero indicated that the mining truck was in a static state, while driving style recognition analysis was based on the dynamic transportation process. Records with a speed greater than 45 km/h were eliminated, the speed of mining truck has a set threshold, and the error caused by road bumps was considered. If the speed exceeded 45 km/h, this was considered to be abnormal data. Records with an acceleration greater than 0.55 m/s^2 were deleted, the total weight and load of mining truck was about 230 t. Due to the heavy weight and load of mining truck, the longitudinal acceleration of the mining truck was limited to 0.55 m/s^2.

2.4.3. Data Division of Driving Behavior Under Heavy-Load and No-Load Conditions

Under the conditions of heavy load and no load, there were great differences between throttle opening and vehicle speed (Figures 4 and 5). TPS means throttle pedal opening. Therefore, the data collected in the process of transporting stripped rock from each loading point to the unloading point of the dump site were taken as the heavy-load sample unit. On the contrary, the data collected in the process of returning to the original loading point through the same haulage road from the unloading point were regarded as the no-load sample unit. This paper divided the 11 drivers' natural driving behavior data into 111 heavy-load data samples and 108 no-load data samples.

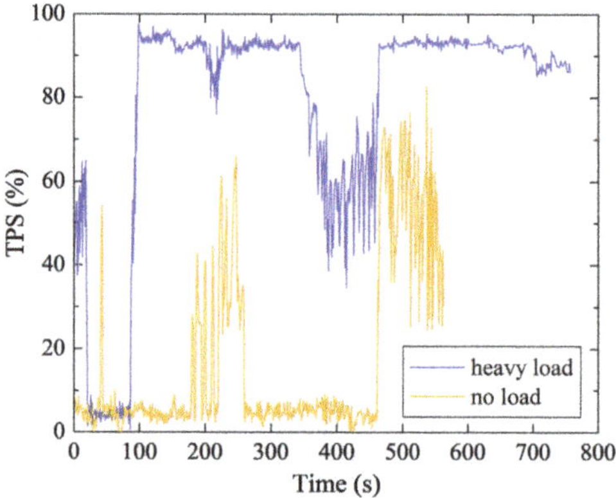

Figure 4. Throttle opening under heavy-load and no-load conditions.

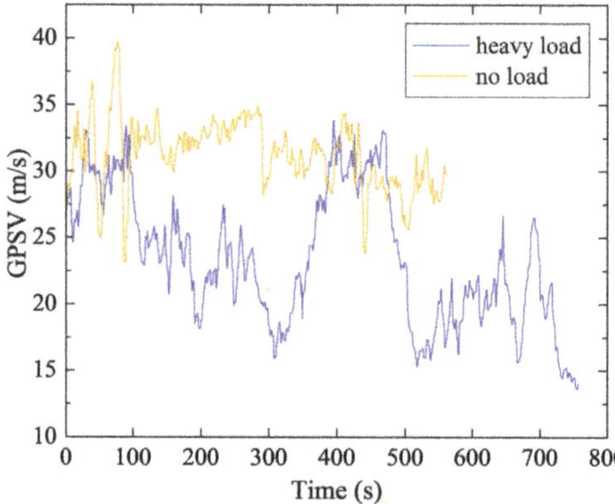

Figure 5. Running speed under heavy-load and no-load conditions.

3. Driving Style Classification and Recognition Based on Machine Learning

3.1. Selection of Driving Style Characteristic Parameters

3.1.1. Selection of Driving Style Characteristic Parameters for Mining Truck Drivers

To classify and identify the driving style of mining truck drivers in the open-pit mine, the characteristic parameters which can represent the driving style were determined first. Shi et al. [28] proposed that there was almost no difference in the size and range of brake pedal opening between soft and normal driving styles, while the size and range of throttle opening were significantly different in different driving styles, so the throttle opening was more suitable as a driving style characteristic parameter. Ma [29] summarized the driving style characteristic parameters that affect fuel consumption from the mechanism, mainly including speed, throttle opening, throttle opening change rate, etc. Some researchers took the statistical values (maximum, average, and standard deviation) of vehicle speed, throttle opening, and acceleration as characteristic parameters to classify driving styles [30,31].

In this research, the statistical values (maximum value, average value, and standard deviation) of throttle opening, throttle pedal angular velocity, speed, and longitudinal acceleration were selected to build driving style characteristic parameters. The main characteristic parameters are shown in Table 6:

Table 6. Driving behavior characteristic parameters.

No.	Name of Characteristic Parameter	Meaning of Characteristic Parameter	Unit
1	TPS_max	Maximum throttle pedal opening	%
2	TPS_mean	Average throttle pedal opening	%
3	TPS_std	Standard deviation of throttle pedal opening	%
4	wx3_max	Maximum angular velocity of throttle pedal	°/s
5	wx3_mean	Average angular velocity of throttle pedal	°/s
6	wx3_std	Standard deviation of angular velocity of throttle pedal	°/s
7	GPSV_max	Maximum speed	km/h
8	GPSV_mean	Average speed	km/h
9	GPSV_std	Standard deviation of speed	km/h
10	ax1_max	Longitudinal acceleration	m/s^2

3.1.2. Correlation Analysis of Driving Style Characteristic Parameters

When there is multicollinearity between driving style feature parameters, the weight of relevant feature parameters in the Euclidean distance calculation is higher, which affects the accuracy of clustering. The Pearson correlation coefficient is used to measure the linear correlation between two variables, X and Y. Its definition is shown in Formula (1). The redundant characteristic parameters can be eliminated using the Pearson correlation coefficient [32]. The Pearson correlation coefficient ranges from -1 to 1, the larger the absolute value is, the stronger the correlation will be; if the absolute value is close to 0, the weaker the correlation will be. Generally, a correlation coefficient between 0.6–0.8 is strongly correlated, and between 0.8–1.0 is an extremely strong correlation. Therefore, when the correlation coefficient is greater than 0.8, redundant feature parameters can be removed to improve the accuracy of clustering results.

$$r = \frac{\sum_{i=1}^{n}(x_i - \bar{x})(y - \bar{y})}{\sqrt{\sum_{i=1}^{n}(x_i - \bar{x}) \cdot \sum_{i=1}^{n}(y_i - \bar{y})^2}}. \tag{1}$$

The Pearson correlation analysis was used to calculate the correlation heat map between driving style characteristic parameters. The correlation of each characteristic parameter under heavy-load condition was less than 0.8 (Figure 6), which indicated that each driving behavior characteristic parameter has strong independence. Under the condition of no-load, the correlation between the average value of angular velocity (wx3_mean) and standard deviation of angular velocity (wx3_std) was 0.94 (Figure 7), which indicated that there was a strong positive correlation between the two variables, the redundant variable wx3_mean could be eliminated and wx3_std could be retained. Therefore, under the condition of heavy load and no load, the numbers of characteristic parameters of driving style for mining truck drivers were ten and nine, respectively. The specific characteristic parameters are shown in Table 7.

Table 7. Characteristic parameters under heavy and no-load conditions.

Operation State	Characteristic Parameter
Heavy-load feature parameters	TPS_max, TPS_mean, TPS_std, wx3_max, wx3_mean, wx3_std, GPSV_max, GPSV_mean, GPSV_std, ax1_max
No-load feature parameters	TPS_max, TPS_mean, TPS_std, wx3_max, wx3_std, GPSV_max, GPSV_mean, GPSV_std, ax1_max

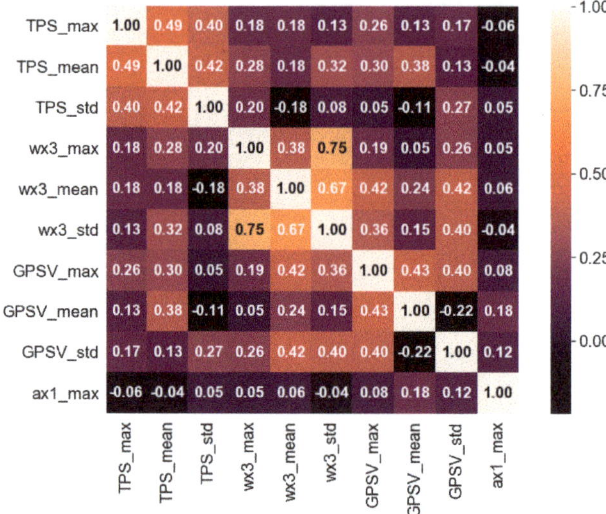

Figure 6. Heavy-load characteristic parameters correlation coefficient heat map.

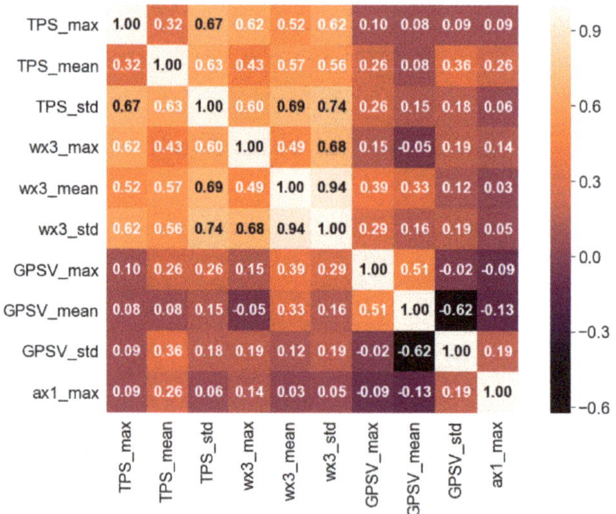

Figure 7. No-load characteristic parameters correlation coefficient heat map.

Based on Pearson correlation analysis, the driving behavior data samples of mining truck under heavy-load and no-load conditions are shown in Tables 8 and 9.

Table 8. Driving style data sample under heavy-load conditions.

No.	TPS_max	TPS_mean	TPS_std	wx3_max	...	wx3_std	GPSV_max	GPSV_std	ax1_max
1	97.08	74.05	24.93	61.22	...	3.77	33.21	4.71	0.18
2	100.00	73.53	25.34	37.72	...	2.85	35.15	4.73	0.16
3	96.23	71.50	23.38	59.02	...	3.37	32.93	4.77	0.16
4	97.16	72.70	26.70	50.05	...	2.98	34.15	4.96	0.15
5	96.88	72.31	25.20	44.98	...	2.87	33.22	5.16	0.13
6	95.70	71.47	23.23	50.66	...	3.14	35.60	4.55	0.17
7	100.00	77.05	28.37	68.42	...	3.50	32.98	5.92	0.12
8	100.00	75.12	27.30	67.26	...	3.80	34.77	5.48	0.17
9	98.20	70.13	28.31	54.26	...	3.90	32.58	5.65	0.16
10	89.96	70.91	20.20	28.57	...	2.64	32.78	5.02	0.13
11	98.75	73.29	20.29	33.33	...	2.43	35.02	5.37	0.16
12	94.02	72.19	20.39	29.54	...	2.78	33.21	5.17	0.15
...	...	...	...	...	...	...	...	...	...
99	96.57	77.45	23.84	41.38	...	5.08	35.58	6.09	0.15
100	99.84	75.12	25.60	37.17	...	4.51	39.61	6.51	0.17
101	97.16	77.70	22.77	61.65	...	5.57	36.04	6.10	0.15
102	97.49	76.69	22.20	46.81	...	5.30	35.78	5.89	0.20
103	97.49	78.38	21.82	56.64	...	6.00	36.43	6.08	0.17
104	92.02	75.21	22.68	48.71	...	5.35	35.89	5.95	0.14
105	99.10	79.25	24.83	81.57	...	6.01	34.19	5.79	0.15
106	97.90	74.76	22.55	36.01	...	3.27	34.52	6.43	0.18
107	95.08	70.18	22.58	81.97	...	4.25	36.61	7.60	0.19
108	100.00	72.51	19.81	75.99	...	4.55	37.41	5.99	0.18
109	99.78	69.70	24.61	32.35	...	2.59	33.45	5.75	0.17
110	97.47	68.06	23.95	28.81	...	2.42	31.56	5.96	0.17
111	97.72	68.30	23.22	33.26	...	2.32	33.08	6.07	0.17

Table 9. Driving style data sample under no-load conditions.

No.	TPS_max	TPS_mean	TPS_std	wx3_max	wx3_std	GPSV_max	GPSV_mean	GPSV_std	ax1_max
1	93.35	14.71	19.66	47.55	3.15	34.34	27.83	4.23	0.30
2	72.08	14.43	18.50	30.88	2.87	34.85	29.39	2.01	0.23
3	98.45	17.33	21.87	54.93	3.37	35.63	28.46	5.08	0.50
4	88.52	18.72	22.64	50.05	4.43	35.65	28.74	5.34	0.30
5	81.13	16.67	19.66	16.97	1.97	33.70	29.80	2.00	0.29
6	75.20	15.70	19.19	21.85	2.17	34.24	28.45	2.69	0.26
7	79.20	15.56	19.47	34.88	3.06	33.97	29.25	1.92	0.28
8	82.95	15.18	19.03	53.96	3.30	33.15	28.44	2.26	0.32
9	72.54	17.97	18.17	28.63	2.16	33.24	28.53	2.88	0.29
10	70.14	18.76	19.49	29.18	3.02	34.56	28.72	6.55	0.29
11	70.24	17.86	19.91	23.74	2.84	34.28	30.55	1.98	0.28
12	69.42	16.66	18.26	21.36	2.28	34.67	27.13	4.35	0.25
...	...	...	...	...	...	...	...		
96	66.96	20.97	20.62	38.70	3.83	33.58	28.72	2.44	0.26
97	72.02	23.01	20.16	34.00	2.95	34.84	23.22	5.89	0.30
98	82.78	22.10	23.34	46.05	6.85	36.52	28.77	5.93	0.28
99	82.72	20.08	21.53	38.45	6.50	36.50	28.69	7.76	0.27
100	84.40	21.31	22.06	57.25	6.39	38.95	32.14	2.00	0.31
101	84.93	22.28	23.12	44.62	6.91	37.73	29.16	8.91	0.31
102	76.64	21.76	21.79	58.65	7.01	40.37	32.00	2.06	0.30
103	82.41	21.84	21.87	41.14	5.97	37.06	30.46.	5.49	0.38
104	80.91	23.15	22.28	40.47	7.62	38.32	32.16	3.24	0.29
105	95.59	20.36	23.35	58.41	5.71	39.58	31.57	2.24	0.38
106	78.34	22.36	21.80	38.88	3.19	33.78	24.65	8.90	0.41
107	68.77	17.39	19.42	34.12	2.43	33.58	28.91	1.84	0.41
108	63.84	15.35	19.51	40.41	2.47	37.84	27.45	2.74	0.30

3.2. *Driving Style Classification*

Unsupervised clustering analysis can classify data into different clusters without pre-defined categories, so that the similarity of samples in each cluster is greater than that in other clusters. The clustering results can be transferred to regression or classification and other supervised machine learning models for further data analysis. The driver's driving style was unknown in actual transportation operations, and the clustering algorithm could be selected to classify the driving style.

This paper used the elbow rule to determine the number of driving style clustering centers, then the K-means clustering algorithm was applied to scientifically classify the driving style. A box plot was used to show the statistical distribution of different driving style characteristic parameters to verify the validity of the clustering results.

3.2.1. Clustering Algorithm

The commonly used clustering algorithms mainly include K-means, hierarchical clustering, fuzzy clustering, and DBSCN. K-means is superior to other algorithms in terms of operating efficiency and accuracy [33]. Therefore, this paper used K-means to classify driving styles.

The algorithm usually takes the sum of squares of errors in the cluster as the objective function for clustering, for the sample data of the same driving style, the error sum of squares within the cluster is small, which is highly similar and allocated to the same cluster. The sum of squares of errors in the clusters of different driving styles is large, and its similarity is low and allocated to different clusters. Samples in the driving behavior data set T can be represented by $Xi = x_{i1}, x_{i2} \ldots, x_{in}$, which takes m driving behavior samples grouped into k categories, and the clustering centers are $c_1, c_2 \ldots c_k$. The calculation Formulas (2) and (3) of the clustering center and error criterion function are as follows [16].

$$c_j = \frac{1}{n_j} \sum_{u \epsilon c_j} X_u \tag{2}$$

$$J = \sum_{j=1}^{k} \sum_{u \epsilon c_j} ||X_u - c_j||. \tag{3}$$

3.2.2. Selection of Driving Style Classification Number

In this research, the elbow rule was used to determine the number of driving styles. Under the condition of heavy load and no load, the sum of squares of errors in the cluster decreased significantly and then slowly when the numbers of clustering centers were three (Figure 8). Therefore, both the number of clustering centers of driving styles were three under the conditions of heavy load and no load.

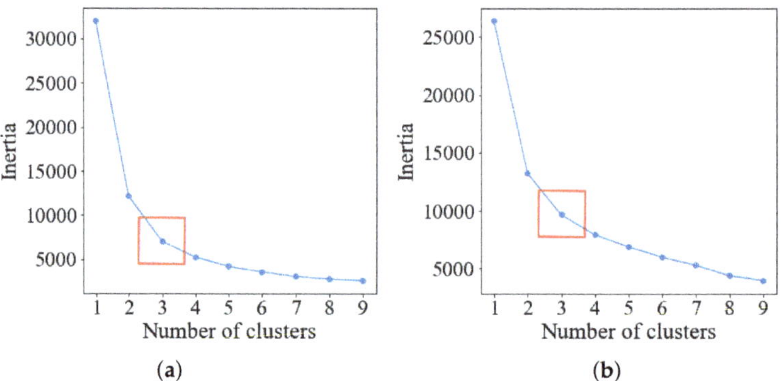

Figure 8. The elbow rule under the condition of heavy load and no load. (**a**) The elbow rule under heavy-load state. (**b**) The elbow rule under no-load state.

3.2.3. Analysis of Driving Style Clustering Results

Based on the K-means algorithm provided by Scikit-learn, the driving behavior data sets under heavy load and no load were fitted to realize the cluster analysis of driving styles, parameters such as the number of clustering centers three and maximum number of iterations 100 were set for unsupervised clustering analysis. The clustering results are shown in Tables 10 and 11.

Chen et al. [34] showed that the aggressive driving style has a larger throttle opening, throttle pedal change rate, and speed; the normal driving style has a moderate throttle opening, throttle pedal change rate, and speed; while the soft driving style has a smaller throttle opening, throttle pedal change rate, and speed. Under no-load conditions of the mining truck, the clustering centers of characteristic parameters related to throttle pedal opening, throttle pedal angular velocity, and speed in cluster2 were the largest, and the clustering centers of characteristic parameters related to throttle opening, throttle pedal angular velocity, and speed in cluster1 were the smallest. In this research, the characteristic parameters of different driving styles have similar distribution rules (as shown in Table 10). Therefore, under no-load conditions of the mining truck, driving style could be divided into three categories: normal (cluster0), soft (cluster1), and aggressive (cluster2). Similarly, under heavy-load conditions of the mining truck, the distribution rules of characteristic parameters related to throttle pedal angular velocity and speed were relatively obvious. According to the distribution rules of characteristic parameters related to throttle pedal angular velocity and speed (as shown in Table 11), driving style could also be divided into three categories: normal (cluster2), soft (cluster0), and aggressive (cluster1).

Table 10. Clustering results under no-load conditions.

Parameter	Implication	Cluster0 (Normal)	Cluster1 (Soft)	Cluster2 (Aggressive)
TPS_max	Maximum throttle opening	73.81	65.00	87.03
TPS_mean	Average throttle opening	17.62	16.29	19.02
TPS_std	Standard deviation of throttle opening	19.58	18.60	21.41
wx3_max	Maximum angular velocity of throttle pedal	36.15	24.33	53.25
wx3_std	Standard deviation of angular velocity	3.46	2.22	4.64
GPSV_max	Maximum speed	36.26	36.06	36.98
GPSV_mean	Average speed	30.06	29.81	29.92
GPSV_std	Standard deviation of speed	3.32	3.36	3.90
ax1_max	Maximum longitudinal acceleration	0.30	0.29	0.31
z	Classification number	42	51	15

Table 11. Clustering results under heavy-load conditions.

Parameter	Implication	Cluster2 (Normal)	Cluster0 (Soft)	Cluster1 (Aggressive)
TPS_max	Maximum throttle opening	98.23	97.13	97.96
TPS_mean	Average throttle opening	75.8	72.66	74.98
TPS_std	Standard deviation of throttle opening	24.87	23.46	24.52
wx3_max	Maximum angular velocity of throttle pedal	43.05	25.09	68.76
wx3_mean	Average angular velocity of throttle pedal	1.95	1.71	2.18
wx3_std	Standard deviation of angular velocity	3.24	2.41	4.34
GPSV_max	Maximum speed	34.75	34.01	34.91
GPSV_mean	Average speed	23.29	22.93	22.98
GPSV_std	Standard deviation of speed	5.53	5.3	5.77
ax1_max	Maximum longitudinal acceleration	0.17	0.17	0.17
z	Classification number	40	56	15

3.2.4. Verification of Driving Style Clustering Results

Liu et al. [23] identified the driving style of drivers based on the method of subjective evaluation by experts, and showed the statistical distribution rules of characteristic parameters of different driving styles using a box plot. For example, the median and upper quartile of throttle opening for the aggressive driving style were larger than those of normal and soft driving styles, while the throttle opening of the soft driving style was more distributed in the lower position. Therefore, a box plot can be used to show the statistical distribution of characteristic parameters of normal, soft, and aggressive driving styles, to verify the correctness of clustering results.

According to the box plot of driving style characteristic parameters under no-load conditions (Figure 9), it can be known that the median and upper quartile of characteristic parameters of throttle pedal opening and angular velocity of the aggressive driving style were greater than the other two driving styles, while the median and upper quartile of characteristic parameters of throttle pedal opening and angular velocity of the soft driving style were smaller than the other two driving styles.

Meanwhile, the standard deviation of velocity and maximum longitudinal acceleration also have similar distribution rules.

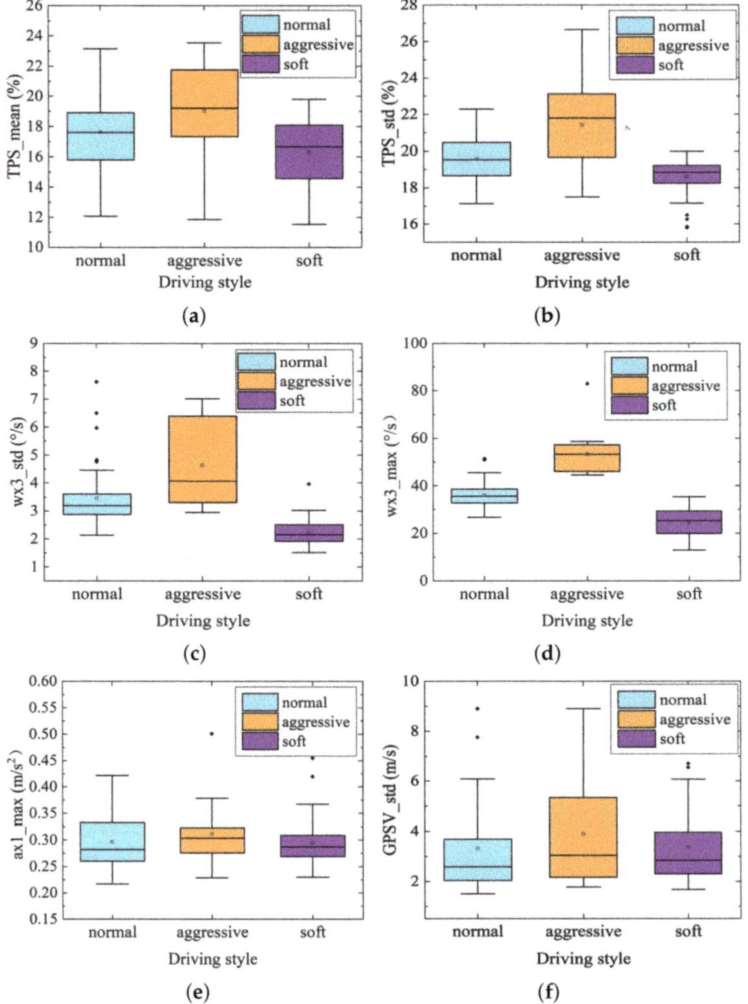

Figure 9. Box plot drawing of driving style characteristic parameters of mining truck under no-load state. (**a**) Average throttle opening. (**b**) Standard deviation of throttle opening. (**c**) Standard deviation of throttle pedal angular velocity. (**d**) Maximum throttle pedal angular velocity. (**e**) Maximum longitudinal acceleration. (**f**) Standard deviation of speed.

Similarly, from Figure 10, the median and upper quartile of characteristic parameters of throttle pedal angular velocity of the aggressive driving style were larger than the other two driving styles, and the median and upper quartile of characteristic parameters of throttle pedal angular velocity of the soft driving style were smaller than the other two driving styles. Meanwhile, the standard deviation of speed, maximum velocity, and maximum longitudinal acceleration also have similar distribution rules.

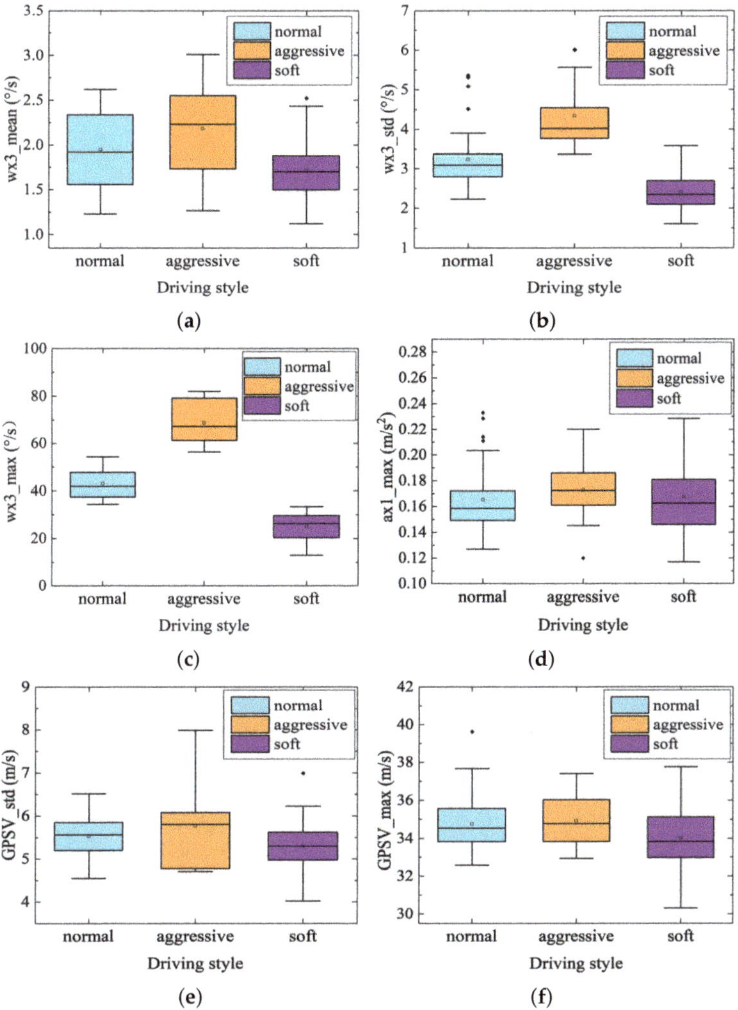

Figure 10. Box plot drawing of driving style characteristic parameters of mining truck under heavy-load state. (**a**) Average throttle pedal angular velocity. (**b**) Standard deviation of throttle pedal angular velocity. (**c**) Maximum throttle pedal angular velocity. (**d**) Maximum longitudinal acceleration. (**e**) Standard deviation of speed. (**f**) Maximum speed.

In addition, taking the mining truck driver as a statistical unit, the number of normal, soft, and aggressive driving styles of 11 drivers were counted respectively, (number means the number of different driving styles) (as shown in Figure 11): drivers have different driving styles in the same working day. Also, under the conditions of heavy load and no load, taking the third, fourth, fifth, sixth, and seventh drivers as example, the proportion of different driving styles of the same driver tended to be the same, which could also verify the effectiveness of the clustering results.

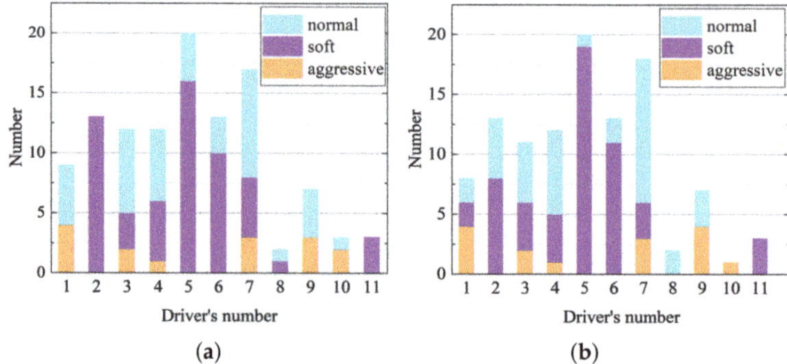

Figure 11. Clustering statistical results of mining truck under heavy-load and no-load conditions. (**a**) Statistics of clustering results of mining truck under heavy-load conditions. (**b**) Statistics of clustering results of mining truck under no-load conditions.

Under the conditions of heavy-load and low-speed operation of the mining truck, the distribution rules of characteristic parameters of different driving styles were in line with the research results of Liu et al. [23] and Chen et al. [34], which verified the validity of the clustering results. Moreover, under the no-load state, the distribution rules of characteristic parameters related to throttle pedal opening and angular velocity of different driving styles were evident. Also, under the condition of heavy load, the distribution rules of characteristic parameters of throttle pedal angular velocity of different driving styles were obvious.

3.3. *Driving Style Identification Based on Random Forest*

Based on clustering analysis of driving style of mining truck drivers, the random forest algorithm was introduced. The algorithm integrates multiple decision tree models to build a more effective driving style identification model, and the integrated model has a stronger generalization ability. It is suitable for processing unbalanced data sets and does not easily produce over-fitting. Simultaneously, the weights [35,36] of driving style parameters can also be obtained according to the Gini coefficient. In this research, the distribution of three driving styles was uneven under the conditions of heavy load and no load, which belongs to the imbalanced data set, it was suitable to use the random forest algorithm to build the driving style identification model and calculate characteristic parameter weights.

3.3.1. Random Forest Algorithm

Compared with a single decision tree, random forest is more tolerant to data noise and does not have to worry about the selection of hyper-parameters, only to focus on the number of decision trees needed to build a random forest. The progress of the implementation is as follows [16]:

1. Using the bootstrap sampling method, r samples are randomly selected from the training set, which are put back to train a decision tree. This can ensure that all trees in the random forest method are different.
2. According to the decision tree constructed in the first step, when the nodes are split, p features are randomly selected from the driving style feature parameters, and the Gini coefficient of all possible split methods for each feature are calculated respectively. Taking the minimum Gini coefficient as the objective function, the nodes are divided using the feature with the minimum Gini coefficient.

$$Gini(p) = 1 - \sum_{n=1}^{N} (\frac{C_n}{T})^2 \tag{4}$$

where n is the number of driving style categories, t is the sample set, and C_n is the sample subset of driving style category n.

3. Repeat step 1 and step 2 to get z decision trees, based on the majority vote. A random forest model for driving style identification of mining truck drivers is constructed.

In addition, the k-nearest neighbor (KNN), support vector machine (SVM), and neural network (NN) methods were applied to the classification of driving style, and the classification results were compared with the performance of the random forest algorithm. To obtain the optimal accuracy and generalization ability of driving style identification model, a ten-fold cross-validation grid search was used to obtain the optimal parameter combination of the above algorithm model. Thus, under the heavy-load and no-load conditions of the mining truck, the driving behavior data set was divided into a test set and training set respectively, with the proportion of 3:7. The parameters and accuracy of each driving style identification model are shown in Tables 12–15 as follows: under the heavy-load and no-load conditions, the performance of the random forest identification model was optimal and higher than that of the other algorithms, and the accuracy of the k-nearest neighbor identification model was the lowest.

The KNN algorithm mainly depends on the limited adjacent samples, which is simple and does not need many parameters to get good performance. However, due to the imbalance of the number of different driving styles, the accuracy of the KNN algorithm was the lowest. A kernel support vector machine (SVM) was used in this paper and can map data into higher dimensions to solve nonlinear separable problems. This algorithm performs well in various data sets and allows complex decision boundaries; however, it is difficult to understand and interpret the predicted results. Neural networks can acquire effective information from many data and build complex models, which requires a long training time and complex parameter adjustment.

The random forest model integrates multiple decision trees, which cannot only reduce over-fitting but also maintain the prediction ability of decision trees; this can also make up for the disadvantage of the poor generalization ability of decision trees. The model is easy to visualize and is not affected by data scaling. However, for sparse data with high dimensions, the performance of random forest is often poor.

Table 12. Random forest (RF) grid search parameter optimization.

Type	N-Estimator	Max_DEPTH	Accuracy
Load	20	7	95.49 %
No load	60	6	90.74%

Table 13. K-nearest neighbor (KNN) grid search parameter optimization.

Type	N_Neighbors	Weights	Accuracy
Load	16	uniform	77.48%
No load	4	uniform	73.15%

Table 14. Support vector machine (SVM) grid search parameter optimization.

Type	Kernel	C	Gamma	Accuracy
Load	rbf	100	0.01	82.88%
No load	rbf	10	0.01	87.96%

Table 15. Neural network (NN) grid search parameter optimization.

Type	Number of Layers	Number of Nodes	Max_Iter	Alpha	Accuracy
Load	2	100	1500	1	82.89%
No load	2	100	1500	0.5	82.41%

In summary, considering the performance, advantages, and disadvantages of each algorithm, we adopted the random forest algorithm to construct the driving style identification model for mining truck drivers in the open-pit mine.

3.3.2. Overall Evaluation of Driving Style Identification Model

This paper chooses the index of generalization ability and overall precision to evaluate the performance of the driving style identification model. The generalization ability of the identification model adopted the method of ten-fold cross-validation, which divided the data set into ten parts. In the process of ten iterations, nine sets of data were used for training in each iteration, and the remaining set was used as the test set. The average classification accuracy was calculated according to the results of each iteration. The overall accuracy could be calculated by a confusion matrix, which can show prediction results and actual results of the random forest classifier. The definition of the confusion matrix is shown in Table 16.

Table 16. Confusion matrix.

Actual Results	**Prediction Results**	**Prediction Results**
	Positive Samples	**Negative Samples**
Positive Samples	TP	FN
Negative Samples	FP	TN

Based on the Scikit-learn machine learning platform, using grid search optimal parameters combination as identification model parameters, we built driving style identification models for the mining truck under heavy-duty and no-load conditions. From Figure 12, under heavy-load conditions of the mining truck, the average cross-validation score of a driving style identification model was 97% and the overall accuracy was 95.49%; under no-load conditions of the mining truck, the average cross-validation score of a driving style identification model was 89% and the overall accuracy was 90.74%. As a result, the generalization ability and overall accuracy of driving style identification models based on random forest were excellent. It can effectively identify the mining truck driver's driving styles.

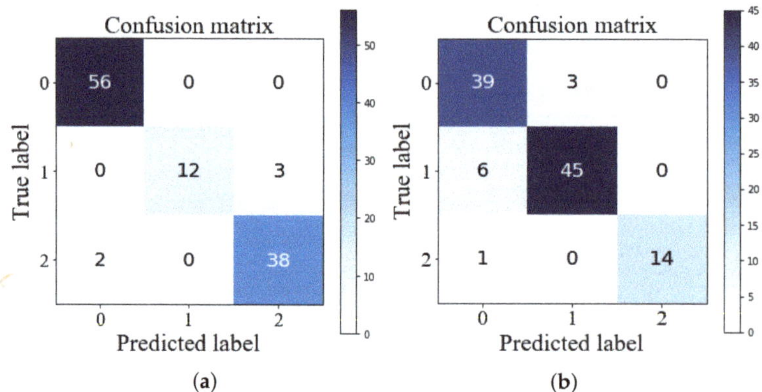

Figure 12. Confusion matrix of the mining truck under the conditions of heavy load and no load. (**a**) Confusion matrix of the mining truck under heavy-load conditions. (**b**) Confusion matrix of the mining truck under no-load conditions.

3.3.3. Evaluation of the Single Driving Style Identification Model

For the unbalanced data set, the performance of a single driving style recognition model can use indicators such as accuracy, precision, recall, and f-score for evaluation. According to the definition, the f-score considers both precision and recall, so the f-score as a model evaluation metric is better than precision. The Formulas (5)–(7) of each index are as follows:

$$Precision = \frac{TP}{TP + FP} \tag{5}$$

$$Recall = \frac{TP}{TP + FN} \tag{6}$$

$$f\_score = 2 \cdot \frac{Precision \cdot Recall}{Precision + Recall}. \tag{7}$$

The evaluation indicators of the single driving style identification model are shown in Tables 17 and 18. Under heavy-load conditions of the mining truck, the f-scores of different driving styles were greater than 0.9, thus the performance of the identification model was excellent. Similarly, under no-load operation of mining trucks, the accuracies of single driving styles were greater than 88%, thus the performance of the identification model was also excellent. Finally, under conditions of heavy load and no load, according to f-score, the ability of single driving style model identification could conclude that the soft driving style was the best, the normal driving style was the second best, and the aggressive driving style was the weakest.

Table 17. Evaluation index of the single driving style identification model under heavy-load conditions.

Driving Style	Sample Size	Precision	Recall	F-Score	Identification Accuracy
Soft	56	1.00	1.00	1.00	100%
Aggressive	15	0.93	0.87	0.90	80%
Normal	40	0.95	0.97	0.96	95%

Table 18. Evaluation index of the single driving style identification model under no-load conditions.

Driving Style	Sample Size	Precision	Recall	F-Score	Identification Accuracy
Soft	51	0.94	1.00	0.97	88.23%
Aggressive	15	1.00	0.60	0.75	93.33%
Normal	42	0.87	0.93	0.90	92.85%

3.3.4. Importance Analysis of Driving Style Characteristic Parameters

Feature importance was measured by the average impurity (Gini coefficient) attenuation of all decision trees in the random forest method [16], and through the Gini coefficient, the weight values of driving style characteristic parameters of the mining truck under heavy-load and no-load conditions could be calculated. From the analysis of Figure 13, we can see that the weight of throttle pedal angular velocity maximum (wx3_max) and standard deviation (wx3_std) were much larger than other characteristic parameters under heavy-load conditions. Therefore, the driving style characteristic parameters under heavy-load operation conditions were mainly characterized by the parameters of throttle pedal angular velocity, and the relevant characteristic parameters of throttle pedal angular velocity of different driving styles have significant distribution rules (Figure 11a,b). Under no-load conditions of the mining truck, the weight of throttle pedal opening maximum (TPS_max) and standard deviation (TPS_std), and throttle pedal angular velocity maximum (wx3_max) and standard deviation (wx3_std) were far greater than other characteristic parameters. Therefore, the driving styles under no-load conditions were mainly characterized by the characteristic parameters of throttle pedal opening and angular velocity, and the relevant characteristic parameters of throttle pedal opening and angular velocity have significant distribution rules Figure 10a–d).

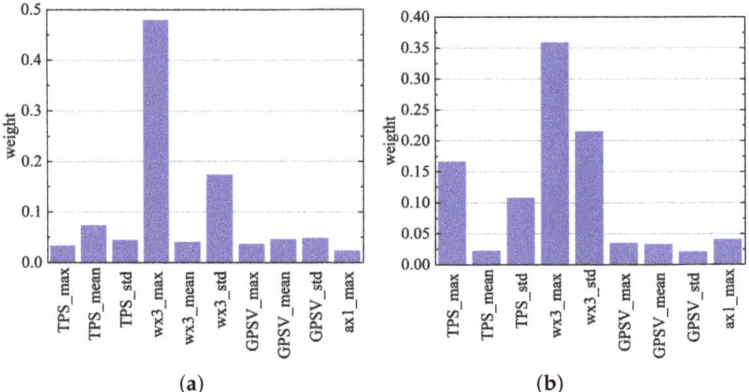

Figure 13. Driving style characteristic parameter weights for mining truck under heavy-load and no-load conditions. (**a**) Weight of driving style characteristic parameters of the mining truck under heavy-load conditions. (**b**) Weight of driving style characteristic parameters of the mining truck under no-load conditions.

3.3.5. The Relationship Between Different Driving Styles and Fuel Consumption

In the Baiyinhua No. 2 mine, diesel oil was generally injected, and the oil tank was filled before day shift and night shift operation, and the fuel consumption is regarded as the daily key assessment index with high reliability. Therefore, the fuel consumption of day shift mining truck operation could be calculated based on the diesel filling before night shift operation, and then the average fuel consumption of a single operation cycle could be calculated based on the number of day shift operation cycles. Based on the identified driving style preference of each driver, the fuel consumption relationship of different driving styles could be roughly calculated.

The fuel consumption of open-pit mine trucks is related to such factors as transportation distance, lifting height, driving style, rolling resistance, and loading. The control variable method was used to compare and analyze the fuel consumption of drivers with the same transportation distance and lifting height in the same haulage road, and minimized the impact of transportation distance, lifting height, rolling resistance, and loading on fuel consumption, and then to calculate the influence of different driving styles on fuel consumption. The drivers with many samples were selected and supplemented with fuel consumption data. Driver No. 3 did not refuel before the night shift operation, and could not estimate the fuel consumption of the day shift and remove it. The fuel consumption statistics of each driver are shown in Table 19. Drivers with numbers of 1, 2, and 4 operated mining trucks to transport rocks to the dumping point with elevation of 1055 through the same haulage road; drivers with numbers of 5 and 6 operated mining trucks to transport rocks to the dumping point with elevation of 1035 through the same haulage road. Among them, the 1055 dumping point and 1035 dumping point were located in adjacent steps, with a height difference of 20 m and a flat plate width of 40 m.

Table 19. Statistics on fuel consumption of different drivers.

Driver Number	Total Fuel Consumption (L)	Cycle Number	Unit Consumption(L)	Dump Site Elevation (m)
1	712	9	79.1	1055
2	903	13	69.4	1055
4	776	12	64.6	1055
5	1031	20	51.5	1035
6	741	13	57	1035

Driver No. 1 has an aggressive driving style with a fuel consumption of up to 79.1 L; Driver No. 2 has a soft driving style, with fuel consumption of 69.4 L, and Driver No. 3 has a normal driving style, with fuel consumption of 64.6 L. Therefore, the fuel consumption of the aggressive driving style was higher than that of the normal driving style and the moderate driving style, and the fuel consumption of the aggressive driving style was about 10% higher than the average fuel consumption. The difference in fuel consumption between the normal and soft driving styles was not large. Therefore, it is necessary to focus on the operation training of the aggressive driving style driver to achieve the purpose of reducing fuel consumption.

4. Conclusions

(1) Based on the same experimental truck and haulage road, the data of driving behavior of different mining truck drivers was collected by sensors. According to Pearson correlation analysis, redundant feature parameters were removed and driving style feature parameters were constructed for the mining truck under conditions of heavy load and no load, then K-means was used to classify driving styles. On this basis, the random forest algorithm was used to identify driving style. Finally, the weight values of driving style features were obtained based on the Gini coefficient. The results show that the accuracy rates of driving style recognition are all more than 90% for the mining truck under conditions of heavy load and no load, which provides a new method for driving style identification of mining truck drivers in open pits.

(2) The characteristic parameters related to throttle pedal angular velocity can characterize the driving style of mining trucks under heavy-load condition; the characteristic parameters related to throttle pedal opening and angular velocity can characterize the driving style of mining trucks under no-load conditions; the driving styles of mining truck drivers in an open-pit mine are not unique, and consist of different driving styles.

(3) The fuel consumption of the aggressive driving style was the largest and 10% higher than the average fuel consumption, while the fuel consumption of the soft driving style was not very different compared to the normal driving style. Therefore, it is necessary to focus on operational training for drivers with an aggressive driving style. These results can be directly applied to the daily fuel consumption management and driving operation training of truck drivers in open-pit mines to achieve the goal of fuel-saving driving.

This experiment required the same experimental truck and haulage road, which conflicted with the on-site mining truck scheduling, therefore, only 11 drivers' driving behavior data were collected in this experiment. The next step is to extract more samples from driving behavior data to further improve the accuracy of the driving style recognition model. The slopes of the outer dump haulage road of the Baiyinhua No. 2 open-pit mine are mostly in the range of 3–8% and road slope has a great influence on drivers' behavior, therefore, the influence of road slope on driving behavior should be considered and incorporated into driving style characteristic parameters.

Author Contributions: Conceptualization, Q.W., R.Z. and Y.W.; methodology, Q.W. and Y.W.; software, Q.W. and Y.W.; validation, Q.W. and Y.W.; formal analysis, Q.W., R.Z., Y.W. and S.L.; investigation, Q.W. and R.Z.; resources, Q.W. and R.Z.; data curation, Q.W. and Y.W.; writing—original draft preparation, Q.W., Y.W. and S.L.; writing—review and editing, Q.W., R.Z., Y.W. and S.L.; visualization, Q.W. and Y.W.; supervision, R.Z., Y.W.; project administration, R.Z., Y.W.; funding acquisition, R.Z. All authors have read and agreed to the published version of the manuscript.

Funding: This research was funded by the National Key Research and Development Program of China grant number 2018YFC0808306.

Acknowledgments: The authors would like to thank the financial support from the National Key Research and Development Program of China (Grant No. 2018YFC0808306).

Conflicts of Interest: The authors declare that there is no conflict of interest related to this paper.

References

1. Akena, R.; Schmid, F.; Burrow, M. Driving style for better fuel economy. *Proc. Inst. Civ. Eng. Transp.* **2017**, *170*, 131–139. [CrossRef]
2. Wang, R.; Lukic, S.M. Review of driving conditions prediction and driving style recognition based control algorithms for hybrid electric vehicles. In Proceedings of the VPCC 2011 Vehicle Power and Propulsion Conference, Chicago, IL, USA, 6–9 September 2011; pp. 1–7.
3. Bingham, C.; Walsh, C.; Carroll, S. Impact of driving characteristics on electric vehicle energy consumption and range. *IET Intell. Transp. Syst.* **2012**, *6*, 29. [CrossRef]
4. Gonder, J.; Earleywine, M.; Sparks, W. Analyzing vehicle fuel saving opportunities through intelligent driver feedback. *SAE Int. J. Passeng. Cars Electron. Electr. Syst.* **2012**, *5*, 450-461. [CrossRef]
5. Kubler, K.A.C. Optimisation of Off-Highway Truck Fuel Consumption through Mine Haul Road Design. Ph.D. Dissertation, University of Southern Queensland, Queensland, Australia, **2015**.
6. Dindarloo, S.R.; Siami-Irdemoosa, E. Determinants of fuel consumption in mining trucks. *Energy* **2016**, *112*, 232–240. [CrossRef]
7. John, M.; Juliana, P. An interactive simulation model of human drivers to study autonomous haulage trucks. *Procedia Comput. Sci.* **2011**, *6*, 118–123.
8. Michael, S.; Brandon, S. Eco-driving: Strategic, tactical, and operational decisions of the driver that influence vehicle fuel economy. *Transp. Policy* **2012**, *22*, 96–99.
9. Af Wahlberg, A.E. Long-term effects of training in economical driving: Fuel consumption, accidents, driver acceleration behavior and technical feedback. *Int. J. Ind. Ergon.* **2007**, *37*, 333–343. [CrossRef]
10. Zarkadoula, M.; Zoidis, G.; Tritopoul, E. Training urban bus drivers to promote smart driving: A note on a Greek eco-driving pilot program. *Transp. Res. Part D Transp. Environ.* **2007**, *12*, 449–451. [CrossRef]
11. Beusen, B.; Broekx, S.; Denys, T.; Beckx, C.; Degraeuwe, B.; Gijsbers, M.; Scheepersa, K.; Govaertsa, L.; Torfsa, R.; IntPanis, L. Using on-board logging devices to study the longer-term impact of an eco-driving course. *Transp. Res. Part D Transp. Environ.* **2009**, *14*, 514–520. [CrossRef]
12. Marina Martinez, C.; Heucke, M.; Wang, F.; Gao, B.; Cao, D. Driving style recognition for intelligent vehicle control and advanced driver assistance: A Survey. *IEEE Trans. Intell. Transp. Syst.* **2018**, *19*, 666–676. [CrossRef]
13. Shinar, D. Aggressive driving: The contribution of the drivers and the situation. *Transp. Res. Part F Traffic Psychol. Behav.* **1998**, *1*, 137–160. [CrossRef]
14. Xie, H.; Tian, G.; Chen, H.; Wang, J.; Huang, Y. A distribution density-based methodology for driving data cluster analysis: A case study for an extended-range electric city bus. *Pattern Recognit.* **2018**, *73*, 131–143. [CrossRef]
15. Si, L.; Hirz, M.; Brunner, H. Big data-based driving pattern clustering and evaluation in combination with driving circumstances. In Proceedings of the 2018 SAE World Congress Experience, Detroit, MI, USA, 10–12 April 2018.
16. Raschka, S. *Python Machine Learning*; Packt Publishing Ltd.: Birmingham, UK, 2015; ISBN 978-1-78355-513-0.
17. Chen, K.; Chen, H.W. Driving style clustering using naturalistic driving data. *Transp. Res. Rec.* **2019**, *2673*, 176–188. [CrossRef]
18. Sun, C.; Wu, C.Z.; Chu, D.F. Driving speed behavior clustering for commercial vehicle based on connected vehicle data mining. *J. Transp. Syst. Eng. Inf. Technol.* **2015**, *15*, 82–87.
19. Han, W.; Wang, W.; Li, X.; Xi, J. Statistical-based approach for driving style recognition using Bayesian probability with kernel density estimation. *IET Intell. Transp. Syst.* **2019**, *13*, 22–30. [CrossRef]
20. Zhu, B.; Jiang, Y.D.; Dend, W.W.; Yang, S.; He, R. Unsupervised clustering of driving styles based on KL divergence. *Automot. Eng.* **2018**, *40*, 1317–1323.
21. Karginova, N.; Byttner, S.; Svensson, M. Data-driven methods for classification of driving styles in buses. In Proceedings of the SAE 2012 World Congress and Exhibition, Detroit, MI, USA, 24–26 April 2012.
22. Bejani, M.M.; Ghatee, M. A context aware system for driving style evaluation by an ensemble learning on smartphone sensors data. *Transp. Res. Part C.* **2018**, *89*, 303–320. [CrossRef]
23. Liu, Y.; Wang, J.; Zhao, P.; Qin, D.; Chen, Z. Research on classification and recognition of driving styles based on feature engineering. *IEEE Access* **2019**, *7*, 89245–89255. [CrossRef]

24. Hao, J.X.; Yu, Z.P.; Zhao, Z.G.; Zhan, X.W.; Shen, P.H. A study on the driving style recognition of hybrid electric vehicle. *Automot. Eng.* **2017**, *39*, 1444–1450.
25. Zhang, B.W.; Liu, X.J.; Fu, J.G.; Wang, J.L.; Su, J.H. Application of multi-function anti-collision warning system in heidaigou open-pit coal mine. *Saf. Coal Mines* **2015**, *46*, 51–54.
26. Sun, E.J.; Zhang, X.K.; Li, Z.X.; Wang, Y. Driver fatigue monitoring at surface mine based on machine vision. *J. Liaoning Tech. Univ. (Natl. Sci.)* **2012**, *31*, 21–25.
27. Wang, H.L.; Lu, S.S. Operation and fuel economy analysis of heavy truck in east open pit mine. *China Coal Ind.* **2017**, *8*,58–60.
28. Shi, B.; Xu, L.; Hu, J.; Tang, Y.; Jiang, H.; Meng, W.; Liu, H. Evaluating driving styles by normalizing driving behavior based on personalized driver modeling. *IEEE Trans. Syst. Man Cybern. Syst.* **2015**, *45*, 1502–1508. [CrossRef]
29. Ma, H.J. Research on Online Trajectory Optimizing Strategy of Bus Engine Operating Condition with the Driving Style Improving. Ph.D. Dissertation, Tianjin University, Tianjin, China, 2018.
30. Chu, D.; Deng, Z.; He, Y.; Wu, C.; Sun, C.; Lu, Z. Curve speed model for driver assistance based on driving style classification. *IET Intell. Transp. Syst.* **2017**, *11*, 501–510. [CrossRef]
31. Augustynowicz, A. Preliminary classification of driving style with objectiverankmethod. *Int. J. Automot. Technol.* **2009**, *10*, 607–610. [CrossRef]
32. Montazeri-gh, M.; Fotouhi, A.; Naderpour, A. Driving patterns clustering based on driving feature analysis. *Proc. Inst. Mech. Eng.* **2011**, *225*, 1301–1317. [CrossRef]
33. Wu, Z.H. *Analysis of Driving Behavior and Traffic Congestion Based on Data Mining*; University of Science and Technology of China: Hefei, China, 2018.
34. Chen, J.R.; Wu, Y.F.; Wu, B. Driver behavior spectrum analyzing method based on vehicle driving data. *J. Comput. Appl.* **2018**, *38*, 1916–1922.
35. Gregorutti, B.; Michel, B.; Saint-pierre, P. Correlation and variable importance in random forests. *Stat. Comput.* **2017**, *27*, 659–678. [CrossRef]
36. Pang, B.; Yue, J.; Zhao, G.; Xu, Z. Statistical downscaling of temperature with the random forest model. *Adv. Meteorol.* **2017**, *2017*, 7265178. [CrossRef]

Article

A Vehicle Type Dependent Car-following Model Based on Naturalistic Driving Study

Ping Wu [1], Feng Gao [1,*] and Keqiang Li [2]

[1] School of Automotive Engineering, Chongqing University, Chongqing 400044, China; wuping@cqu.edu.cn
[2] Department of Automotive Engineering, Tsinghua University, Beijing 100084, China; likq@tsinghua.edu.cn
* Correspondence: gaofengl@cqu.edu.cn; Tel.: +86-189-9618-8196

Received: 8 March 2019; Accepted: 19 April 2019; Published: 23 April 2019

Abstract: In this paper, a car-following model considering the preceding vehicle type is proposed to describe the longitudinal driving behavior closer to reality. Based on the naturalistic driving data sampled in real traffic for more than half a year, the relation between ego vehicle velocity and relative distance was analyzed by a multi-variable Gaussian Mixture model, from which it is found that the driver following behavior is influenced by the type of leading vehicle. Then a Hidden Markov model was designed to identify the vehicle type. This car-following model was trained and tested by using the naturalistic driving data. It can identify the leading vehicle type, i.e., passenger car, bus, and truck, and predict the ego vehicle velocity and relative distance based on a series of limited historical data in real time. The experimental validation results show that the identification accuracy of vehicle type under the static and dynamical conditions are 96.6% and 83.1%, respectively. Furthermore, comparing the results with the well-known collision avoidance model and intelligent driver model show that this new model is more accurate and can be used to design advanced driver assist systems for better adaptability to traffic conditions.

Keywords: driving behavior; car-following; truncated Gaussian Mixture model; Hidden Markov model; vehicle type identification; naturalistic driving study

---

1. Introduction

The topic of car-following behavior has become increasingly important in traffic engineering and safety research [1–3]. Modeling the car-following behavior more effectively and accurately is of great benefit to several application areas, such as simulation of microscopic traffic, development of advanced driving assistance systems, etc. [4–6]. In the existing models, the multi-variables, e.g. relative distance, relative velocity, ego vehicle velocity, leading vehicle velocity, and time headway are required overall or partially to describe the car-following behavior. Moreover, the car-following behavior of a driver is a non-linear system with a high dimension. Thus, a common following model with fixed parameters is no longer suitable for new applications with more intelligence and user friendliness.

In the current researches, classical car-following models are designed to characterize the interactional phenomena between the individual driver and the traffic, e.g., microscopic simulation [7,8]. These types of car-following models include the Gazis-Herman-Rothery (GHR) model, safety distance or collision avoidance (CA) model, linear models, action point models (AP), and Fuzzy logic-based models [1,9]. They can provide a real-time calculation for simulation because of the simplicity. Other kinds of car-following models aim to describe the driver behavior with the same reaction as a real human driver [10,11]. Thus, some empirical models used for simulation were modified to describe the driver behavior [12]. For example, Chen et al. conducted research on headway/spacing between two consecutive vehicles by considering it as a certain stochastic process to achieve better accuracy than CA [13]. Yang et al. modified Gipps' model by formulating the car-following distance as a function of both the distance and the relative velocity between the leading and ego vehicle [14]. After calibration, the field data evaluation results show that this model has a higher accuracy than the original Gipps' model. The Intelligent Driver Model (IDM) is one of the most widely used models, which attempts to eliminate the difference between the real situation and the preferred [15,16].

Recently, with the development of the intelligent learning algorithms based on big data, it is possible to construct a more accurate car-following model based on a great deal of naturalistic driving data [17–24]. Ye et al. found that time headway (THW) is different when a driver follows different vehicles [25]. In other words, the type of leading vehicle has an impact on the driving behavior [4]. However, the aforementioned models ignore this important factor, i.e., leading vehicle type. Thus, Aghabayk et al. established a car-following model that is applicable to a different but fixed vehicle type, i.e., passenger car and heavy vehicle [26]. This model was developed on the basis of the local linear model tree approach with Next Generation Simulation (NGSIM) data obtained from a U.S. freeway under congested traffic conditions. Although various researches on the analysis of car-following behavior have been conducted, little effort has been made to study the influence of leading vehicle type on the car-following process especially in real dynamical driving conditions.

To establish a vehicle type-dependent car-following model, a combined Gaussian mixture model (GMM) and a hidden Markov model (HMM) are used to analyze the naturalistic driving data to find a car-following model considering the leading vehicle type. The GMM is used to fit driver's following behavior by using the expectation maximization algorithm. By this analysis, a joint probability density function describing the relation between ego velocity and relative distance is found. Combining the ego vehicle velocity and relative distance together to form the observation state and considering the leading vehicle type as a hidden one, we designed an identify algorithm, which can estimate the leading vehicle type by using HMM. Furthermore, to predict a driver's following behavior with limited historical data, an algorithm with weighted expectation is proposed to realize its real time application. This model has the following advantages:

(1) It has the ability to identify the leading vehicle type in real-time because the HMM hidden state can be predicted with limited historical data;
(2) The prediction accuracy is ensured by training the model with a large number of naturalistic driving data;
(3) Its responsiveness to dynamical conditions is achieved by estimating the optimum state of a car-following model based on historical data.

The rest of the paper is organized as follows: Section 2 introduces the data set used in this study; the idea and structure of this new car-following model are described in Section 3; how to obtain the model parameters based on the naturalistic data are explained in Section 4; the effectiveness of this model is validated in Section 5; Sections 6 and 7 give out some applications and further discussions about this study; and Section 8 concludes the paper.

2. Car-Following Data Collection and Preprocessing

The naturalistic driving data is collected from a program of China Automotive Engineering Research Institute. In this database, 16 vehicles equipped with data acquisition systems were driven in Beijing, Shanghai, Tianjin, Chongqing, and Chengdu. As shown in Figure 1, these vehicles are equipped with a front-view camera, two side-view cameras and a front millimeter-wave (MMW) radar to measure the environment information including relative velocity, relative distance and leading vehicle type. The vision signal type of camera is LVDS, the resolution of image is 1280 × 720 pixels, and the field angle is H52 and V43. The MMW radar type is ARS 408-21, produced by Continental, and has a detection range of 0–250 m. A Nvidia Jetson TX2 acts as the on-board computer to record and process the signals. The state of ego vehicle is acquired from the vehicle controller area network including the vehicle velocity, steering wheel angle, brake pressure, engine speed, acceleration, and gear position. The data acquisition vehicles have been driven for more than 50,000 km including both highways and city roads.

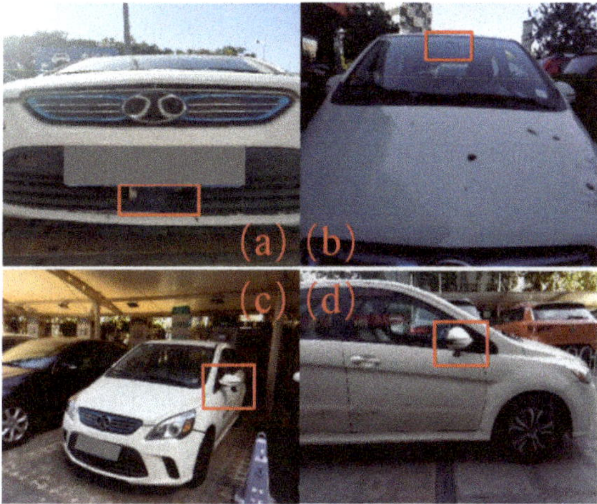

Figure 1. Data acquisition equipment mounted on the vehicle. Figure (**a**) shows the millimeter wave radar. Figures (**b–d**) show the cameras.

In this program, there is no restriction on driving routes. The data acquiring system is activated about 30 s after the engine starts and stops about 30 s before the engine stops. The sampling rate of the naturalistic driving data is 13 Hz. To guarantee that the drivers are not disturbed by the acquisition system, the equipped system is hidden and has no interaction with the driver. The driving behaviors, such as car-following, cut-in, and lane changing, are all recorded in the dataset. To choose the proper data, firstly, we define the car-following scenario, which should satisfy the following five criteria [5,27,28]:

a Velocity range: Ego vehicle velocity, v_{ego}, should be more than 20 km/h, because the condition that the speed is less than 20 km/h contains a lot of stop-and-go scenarios.
b Distance range: Relative distance, y_r, between the rear margin of the leading vehicle and the front margin of the ego vehicle should be less than 120 m. If this distance is greater than 120 m, the preceding vehicle has almost no effect on ego vehicle and this scenario is similar to the free-driving case.
c Restrictions on leading vehicle: The leading vehicle should drive on the same lane with the ego vehicle.
d Road curvature: The radius of the road should be larger than 150 m.
e Time range: The ego vehicle should follow the leading vehicle consistently for more than 10 s. If the time is less than 10 s, there easily exists on-stable car-following scenarios, such as cut-in, cut-out and lane change.

After being filtered by the above five criteria, the extracted dataset of the stable car-following scenario is further divided into three categories according to the leading vehicle type:

Car–car (C-C): a passenger car following a passenger car;
Car–bus (C-B): a passenger car following a bus;
Car–truck(C-T): a passenger car following a truck.

The statistical information of this dataset of car-following scenarios with different leading vehicle types is listed in Table 1.

Table 1. Car-following dataset information.

Statistical Parameter	C-C	C-B	C-T
N	6965	723	305
Tc (s)	242,620	20,003	6655
Tac (s)	34.8	27.7	21.8

N: number of car-following cases. Tc: total car-following time. Tac: average car-following time.

3. Car-Following Model Design

Since a driver uses different following behaviors for different types of leading vehicles [25,29], a new model which can describe this feature is necessary. In this study, GMM is selected for modeling the following behavior of drivers because of the following two advantages:

(a) It has already been demonstrated that GMM is effective in modeling the stochastic features of driver behavior [30,31];

(b) It is a statistical model and the fundamental mechanism or detail of the driver response under internal exciting is not necessary [32].

Furthermore, under real traffic conditions, the leading vehicle may change frequently, so the model should have the ability to identify the leading vehicle type dynamically in real time. Unfortunately, the leading vehicle type can hardly be described by an explicit index with observed signals as its variables, because the interaction mechanism among the driver and environments is still not clear enough now. To overcome this difficulty, HMM is selected because it can describe the high dimension non-linear system and predict hidden states from a series of limited historical data [33]. And so in this study, GMM is used to establish the relation between the relative distance and ego velocity when following different vehicles. Then based on GMM, a HMM is designed to predict the leading vehicle type with historical data to guarantee the real time performance.

3.1. Car-following Behavior Fitted with Gaussian Mixture Model

In the GMM, the relative distance, y_r, and the ego vehicle velocity, v_{ego}, are selected as the exciting of the driver car-following behavior model [26]:

$$x = \left[v_{ego},\ y_r\right] \tag{1}$$

where x is the input. According to the dataset of car-following described by Table 1, three independent GMMs are needed:

$$f_i(x_i;\Theta_i) = \sum_{k=1}^{K} \omega_{i,k} f_{i,k}\left(x_i;\theta_{i,k}\right),\ i \in \{1,2,3\} \tag{2}$$

where i is the type of leading vehicle and its value, 1, 2, 3, denotes passenger car, bus, and truck, respectively. The parameter $\Theta_i = [K, \omega_i, \theta_i]$, where K is the component of GMM, $\omega_{i,k}$ is the weight of the k-th component of type i and satisfying $\sum_{k=1}^{K} \omega_{i,k} = 1$, $\theta_{i,k} = \left[\mu_{i,k}, \Sigma_{i,k}\right]$ is the parameter of component k of type i; $\mu_{i,k}$ and $\Sigma_{i,k}$ are the mean matrix and covariance matrix of the k-th component of type i respectivly.

To identify the parameter values in (2), a revised Expectation Maximum (EM) algorithm is designed to solve the equation because the data used to train the parameters is truncated. It contains two steps called E step and M step respectively [34]:

E step: It is used to compute the posterior probability that x belongs to the k-th component. Since the posterior probability remains unchanged with the truncated data, it is given by

$$p_k^{n+1} = \frac{\pi_k f_k(x_i)}{\sum_k \pi_k f_k(x_i)} \tag{3}$$

where p_k^{n+1} is the probability of the condition that data x_i belongs to the k-th component.

M step: This step is used to compute the maximum Θ. The Lagrange multiplier is used and the original analytic equations are modified as follows to adapt to the truncated dataset:

$$\begin{aligned} \omega_k^{n+1} &= \tfrac{1}{N}\sum_{n=1}^{N} p_k^{n+1}, \mu_k^{n+1} = \frac{\sum_n p_k^{n+1} x^{n+1}}{\sum_n p_k^{n+1}} - m_k, \\ \Sigma_k^{n+1} &= \frac{\sum_n p_k^{n+1}\left(x-\mu_k^{n+1}\right)\left(x-\mu_k^{n+1}\right)^T}{\sum_n p_k^{n+1}} + H_k \end{aligned} \tag{4}$$

The parameters, m_k and H_k, in (4) are calculated by

$$m_k = M^1(0, \Sigma_k; [s-\mu_k, t-\mu_k]), H_k = \Sigma_k - M^2(0, \Sigma_k; [s-\mu_k, t-\mu_k]) \tag{5}$$

where $M^1(\mu, \Sigma; [a,b])$ and $M^2(\mu, \Sigma; [a,b])$ are the moment and the second moment of Gaussian truncated range in $[a,b]$, respectively [34], $s = \min\{\mu_1, \mu_2, \mu_3 \ldots \mu_k\}$ and $t = \max\{\mu_1, \mu_2, \mu_3 \ldots \mu_k\}$.

The termination condition of the EM algorithm is set to be

$$L\left(\Theta_i^{n+1}\right) - L\left(\Theta_i^n\right) < \epsilon \tag{6}$$

where $L\left(\Theta_i^{n+1}\right) = log\left(f\left(x_i; \Theta_i^{n+1}\right)\right)$. The last parameter K of GMM is determined by the Bayesian information criterion (*BIC*) because of its positive correlation with the computation burden [27]. The parameters are trained with the car-following dataset by using (3)–(5) interactively until the termination condition (6) is established.

3.2. Identification of Leading Vehicle Type with Hidden Markov Model

As shown in Figure 2, we define the hidden state, S, as the leading vehicle type, that is, $S = \{S_1, \ldots S_t, \ldots\}, t = 1, 2, \ldots, S_t \in \{1,2,3\}$. The value of hidden state, 1, 2, 3, represents passenger car, bus, and truck, respectively. The observation variable is $O = \{O_1, \ldots, O_t, \ldots\}$, where $O_t = \left\{v_{ego,t}, \ y_{r,t}\right\}$. Then the transfer probability and emission probability of the Markov process are defined as.

$$p(S_t|S_1, \ldots, S_{t-1}, O_1, \ldots, O_{t-1}) = p(S_t|S_{t-1}), p(O_t|S_1, \ldots, S_t, O_1, \ldots, O_{t-1}) = p(O_t|S_t). \tag{7}$$

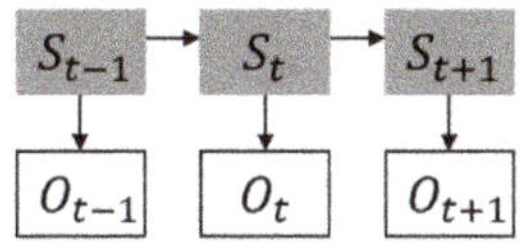

Figure 2. HMM graph model.

Furthermore, the transfer matrix of HMM, $A = \left[A_{i,j}\right] \in \mathbb{R}^{3\times3}$, can be calculated by using the naturalistic driving dataset as

$$A_{i,j} = p(S_t = j, S_{t-1} = i) = \frac{T_{i,j}}{\sum_{j=1}^{3} T_{i,j}}, T_{i,j} = \Sigma\left(M_i \to M_j\right) \tag{8}$$

where $M_i \in \{1,2,3\}$ and the value indicates passenger car, bus, and truck respectively; $M_i \to M_j$ represents a mode transition from M_i to M_j; and $T_{i,j}$ is the total number of transitions. Since O_t is continuous, the emission probability equals to $f(x_i; \Theta_i)$:

$$B(O_t) = \sum_{k=1}^{K} \omega_k B_k(O_t) = \sum_{k=1}^{K} \frac{\omega_k}{(2\pi)^{\frac{r}{2}}|\Sigma_k|^{\frac{1}{2}}} exp\{-\frac{1}{2}(O_t - \mu_k)\Sigma_k^{-1}(O_t - \mu_k)^T\} \tag{9}$$

where $B(O_t)$ is the emission probability and r is the dimension of observation data. The initial probability, π, of the hidden state is defined as

$$\pi = [p(S_t = 1)p(S_t = 2)p(S_t = 3)], p(S_t = i) = \frac{F_i}{\sum_{i=1}^{3} F_i}, \tag{10}$$

where F_i is the number of car-following conditions whose leading vehicle type is i. Then the best state sequence, $S_1^*, \ldots, S_t^*$, to fit the historical data can be calculated by the Viterbi algorithm [35]:

$$S_t^* = \text{argmax}_{S^* \in S}(V_{t,S^*}), \tag{11}$$

$$V_{t,S^*} = \max_{S^* \in S}\left(p\left(O_t \middle| S_{t-1}^*\right)p\left(S_{t-1}^* \middle| S^*\right)V_{t-1,S_{t-1}^*}\right) = \max_{S^* \in S}\left(B_{S_{t-1}^*}(O_t)A_{S^*,S_{t-1}^*}V_{t-1,S_{t-1}^*}\right) \tag{12}$$

The initial state used to solve the above optimal problem iteratively is set to be

$$V_{1,S^*} = p(O_1|S_1)\pi(S_1). \tag{13}$$

Then with a series of historical data, the optimal estimation of leading vehicle type can be obtained by using (7)–(13) [29]. When there is a new sample of car-following data, the observation state, i.e., the ego vehicle speed and relative distance, is calculated by the one-step prediction as

$$\hat{O}_{t+1} = \frac{1}{J}\sum_{j=0}^{2m}\sum_{k=1}^{K} A_k(S_{t+1}, S_t)\omega_k B_k\left(O_{t-j}\right)O_{t-j} = \frac{1}{J}\sum_{k=1}^{K} A_k(S_{t+1}, S_t)E_k(B(O)) \tag{14}$$

where $O = \left\{O_{t-J+1}, \ldots O_t\right\}$ is the historical data, $E_k[B(O)]$ is the expectation of vehicle type k calculated by the data sequence O, and J is the odd integer [36], which is a design parameter discussed in Section 4.2.

In summary, the model parameters, B, A, and π are calculated according to (2), (8), and (10) using the car-following dataset, respectively. When using this model, the current leading vehicle type is identified by (11), using (11) and (14) interactively to predict.

4. Model Training

In this section, the car-following dataset set up in Section 2 is used to train the three GMMs to identify three types of leading vehicles respectively. The dataset is divided into a training and testing set with the ratio of 8:2.

4.1. Training of GMM

The component of GMM, K is given by [37]

$$BIC(K) = K\ln(n_c) - 2\ln(L(\Theta)) \tag{15}$$

where n_c is the number of training data and BIC is an increasing function of the error variance and K. Considering the complexity of the model and the computation burden, a model prefers a lower BIC [38]. Thus, K is optimized by

$$K_{min} = \text{argmin}_K(BIC(K)) \tag{16}$$

Let K increase from 1 to 20 step by step and then fit the components of GMM with the termination condition $\epsilon = 10^{-8}$. The results of BIC are shown in Figure 3.

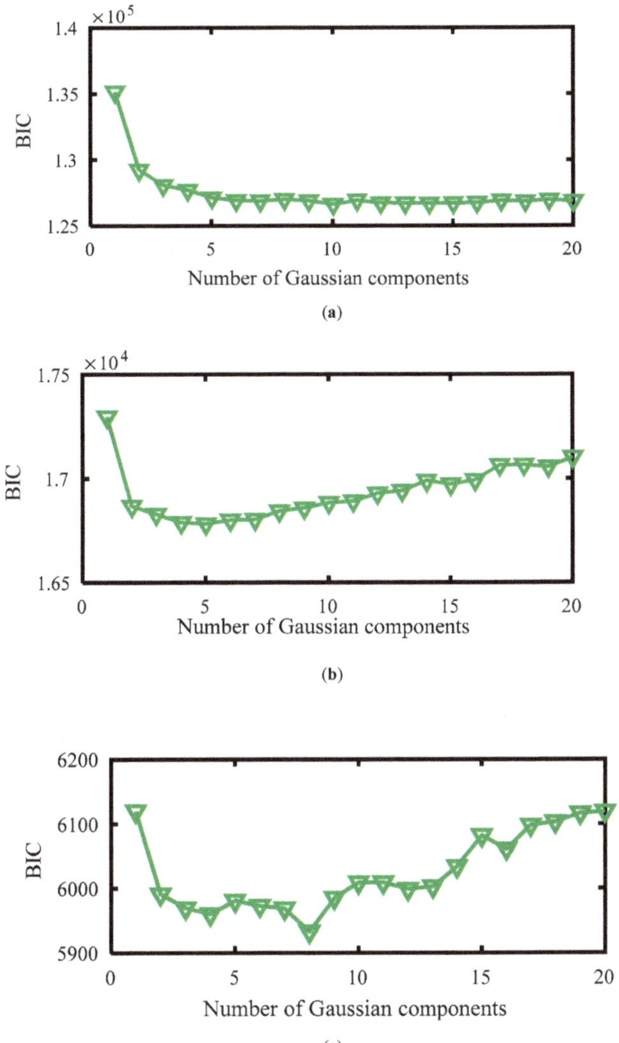

Figure 3. *BIC* value: (**a**) car, (**b**) bus, (**c**) truck.

Normally, *BIC* decreases at the beginning when the value of the Gaussian component is small and will go up until reaching a big constant with the increasing value of the Gaussian component [38]. In Figure 3a, *BIC* decreases at the beginning and then almost holds after $K > 5$. In Figure 3b,c, *BIC* decreases at the beginning and then goes up with the increasing K. The trend of *BIC* in Figure 3a–c is different, which is caused by the different amounts of training data. Normally, a larger amount of training data leads to a larger value of *BIC*. From (16), the value of K should be selected, ensuring the minimum value of *BIC*, and so the value of K for passenger car, bus, and truck is set to 10, 5, and 8, respectively. With the selected K, the raw data and trained GMM are shown in Figure 4.

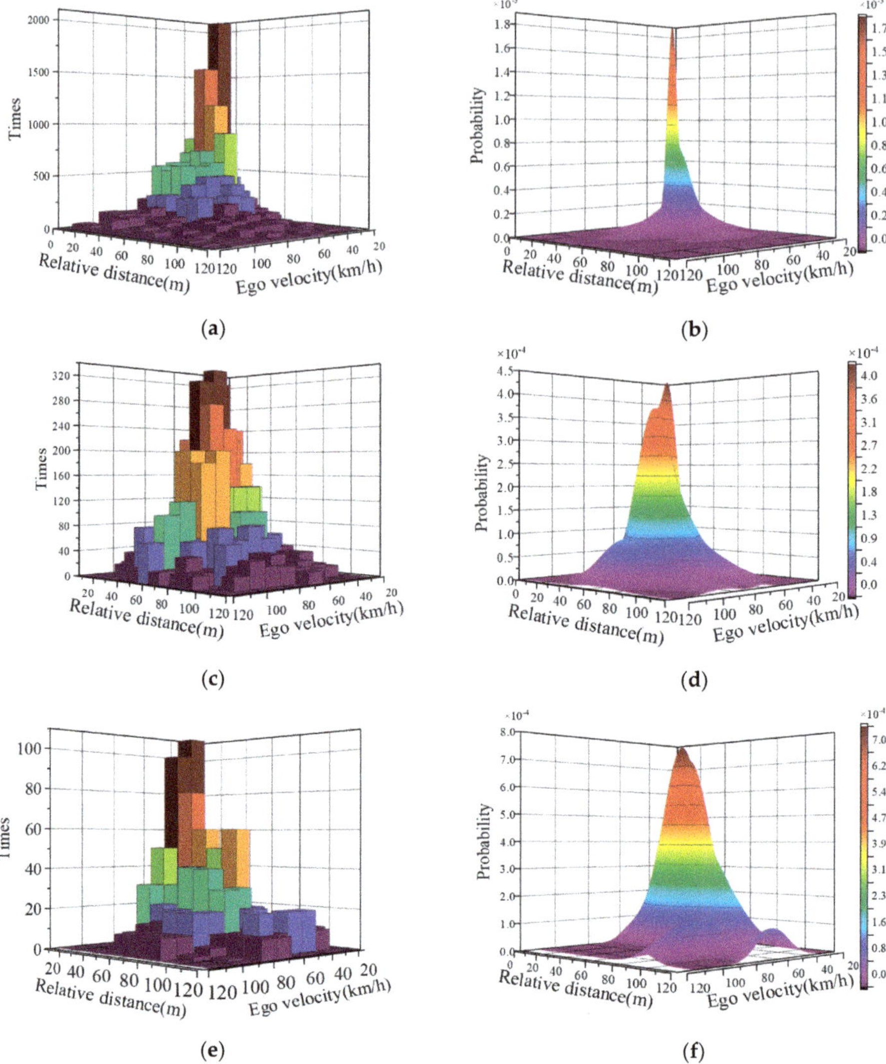

Figure 4. Comparison of raw data and Gaussian mixture model. (**a**) Car: raw data, (**b**) Car: Gaussian mixture model, (**c**) Bus: raw data, (**d**) Bus: Gaussian mixture model, (**e**) Truck: raw data, (**f**) Truck: Gaussian mixture model.

From Figure 4, it is found that three raw data histograms have the same tendency with the fitted GMM surfaces. These distributions are mainly concentrated at the lower relative distance and lower ego velocity. The reason is that the raw data contains more than 75% city road data, about 20% highway data and the rest is country road or others. Furthermore, the distribution of data of different types of leading vehicle is also different. This implies that the driver-following behavior is influenced by the leading vehicle type, which should be considered when setting up the car-following model.

4.2. Prediction of Vehicle Type

With the trained GMMs, in this section, how to obtain the parameters in HMM are discussed. Firstly, the design parameter, J, is to indicate whether global or local features should be focused on [39]. In (14), J is odd when m is an integer. Considering the requirement of dynamic performance, a small m from 1 to 5 is considered and the corresponding $J = [3, 5, 7, 9, 11]$. With the testing data, the root mean squared error (*RMSE*) and root mean squared percentage error (*RMSPE*) are used to evaluate the model accuracy [40]:

$$RMSE = \sqrt{\frac{1}{N}\sum_i (\hat{y}_i - y_i)^2}, RMSPE = \sqrt{\frac{1}{N}\sum_i \left(\frac{\hat{y}_i - y_i}{y_i + \delta}\right)^2} \tag{17}$$

where $\hat{y}_i$ and y_i are the predicted value and the actual driving data respectively, δ is a small positive constant used to prevent zero divisor, and N is the size of data. The model accuracy under different values of J is shown in Table 2. From Table 2, it is found that the minimum *RMSE* and *RMSPE* are both at $J = 3$.

Table 2. Model accuracy with different value of J.

J	Ego Velocity		Relative Distance	
	***RMSE* (km/h)**	***RMSPE***	***RMSE* (m)**	***RMSPE***
3	0.1907	0.0097	0.0750	0.0119
5	0.2627	0.0133	0.1064	0.0169
7	0.3368	0.0170	0.1352	0.0213
9	0.4110	0.0206	0.1616	0.0254
11	0.4813	0.0241	0.1857	0.0291

5. Model Testing

5.1. Identification Accuracy of Leading Vehicle Type

A statistical index is defined to evaluate the accuracy of the presented HMM for identification of a leading vehicle type [5]:

$$\eta = \frac{TT + FF}{TT + TF + FT + FF} \tag{18}$$

where TT is the number of conditions that HMM identifies the correct vehicle type, TF is the number of times that HMM takes the real type as others, FT is the number of times that HMM takes other types as the real one, and FF is the number of times that HMM identifies the non-specified type in the dataset correctly. Firstly, the static condition, under which the leading vehicle remains unchanged but its type is unknown, is selected to test the HMM identification ability. The average accuracy is shown in Table 3.

Table 3. Accuracy of vehicle type identification.

Identification Accuracy	C-C	C-B	C-T	Average
η	98.7%	92.8%	98.2%	96.6%

Under the static condition, HMM shows a high level of performance of identification accuracy, which is 96.6%. When the leading vehicle type holds, the driver receives the same stimulus signal, thus the inner character of observation data, i.e., the relative distance and ego velocity, stay the same. In addition, the three truncated Gaussian mixture distributions have significant differences with each other as shown in Figure 4. That is why the leading vehicle type can be identified with a high level of accuracy.

5.2. Dynamical Condition

When driving on the road, the leading vehicle can hardly always be the same, and so dynamical test conditions are used to further validate the effectiveness of the proposed car-following model. Under this condition, the initial leading vehicle type, the time when the leading vehicle changes, and its duration are all unknown. Such dynamical conditions are more critical than the stable ones. The statistical accuracy is reported in Table 4.

Table 4. Identification accuracy under dynamical conditions.

Identification Accuracy	C-C	C-B	C-T	Average
η	87.6%	81.2%	80.4%	83.1%

From Table 4, it is found that the identification accuracy of the passenger car is higher than others. The reason is that the training data of a passenger car is much larger. Thus, the corresponding component in GMM has a stronger ability and accuracy. The identification accuracy of the bus and truck is almost the same, because the driver has almost the same reaction when the leading vehicle is a truck or bus. Furthermore, compared with the results in Table 3, the accuracy decreases obviously because the change of leading vehicle type causes a time delay to the model. During the identification period, the same reaction may happen to different leading vehicles. Thus, the observation data will hide the missed information, which leads to misidentification. To show the details, a typical identification process is shown in Figure 5.

Figure 5b–e shows a time delay when the leading vehicle changes to another. The driver model needs a period of time to adapt to the new condition. Figure 5c,f shows an error identification, where the truck is misidentified as a passenger car and lasts about 10 s (Figure 5c). This is because the training data of C-T is far less than C-C, which cannot cover all possible driving conditions. In Figure 5f, the bus is misidentified as a passenger car or a truck. Besides the data size, another reason is that the truck and bus have similar shape features, which generates similar stimuli to the driver [4].

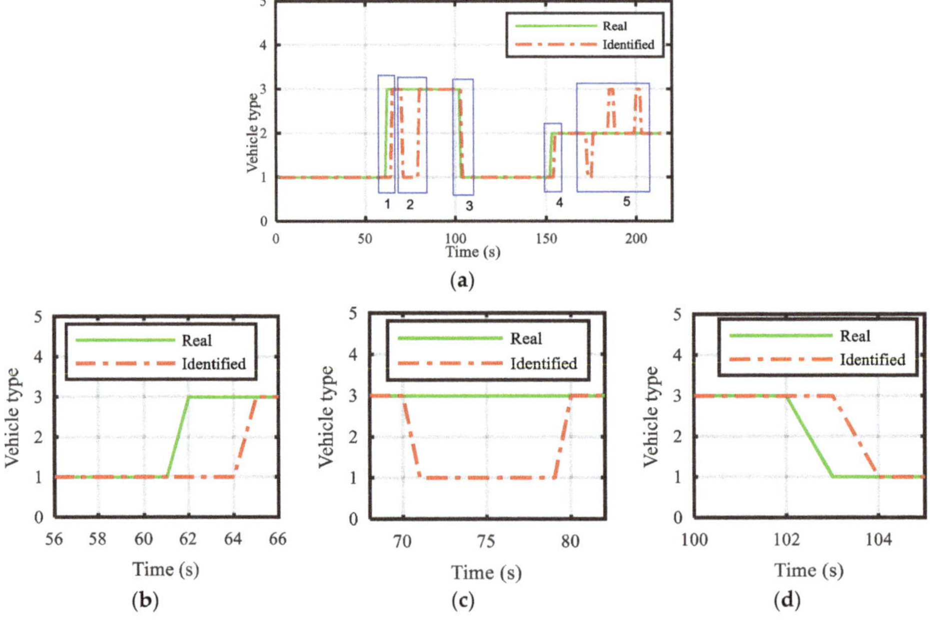

Figure 5. *Cont.*

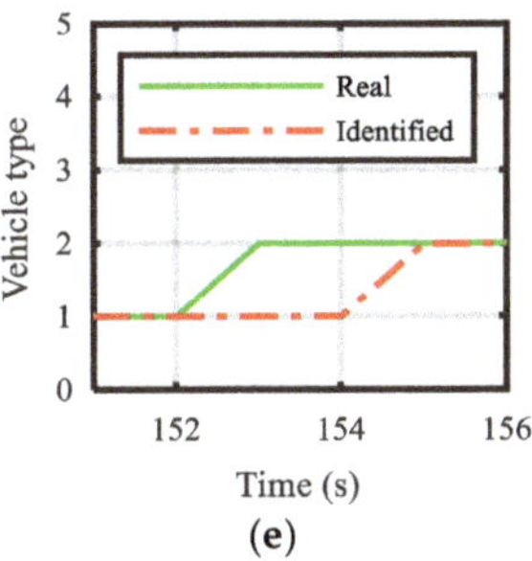

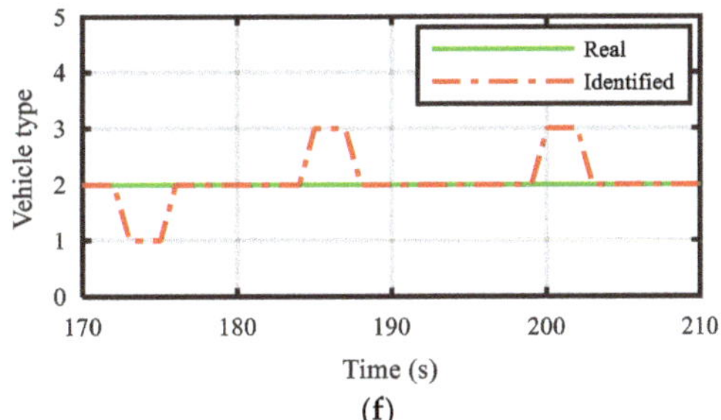

Figure 5. Typical condition test results of leading vehicle type changing. (**a**) is a full figure. (**b**)–(**f**) are zoomed in to the five sections of the full figure in (**a**).

6. Application and Analysis

6.1. Driver Behavior Mimic Application

In order to evaluate the designed model, it has been applied to mimic the driver's behavior. In this application, a 280 s naturalistic driving period is used and the results are shown in Figure 6. The results show that the predicted data is smoother than the real one and the model can predict the driver behavior finely with *RMSE* of velocity, and the relative distance is 0.21 km/h and 0.09 m.

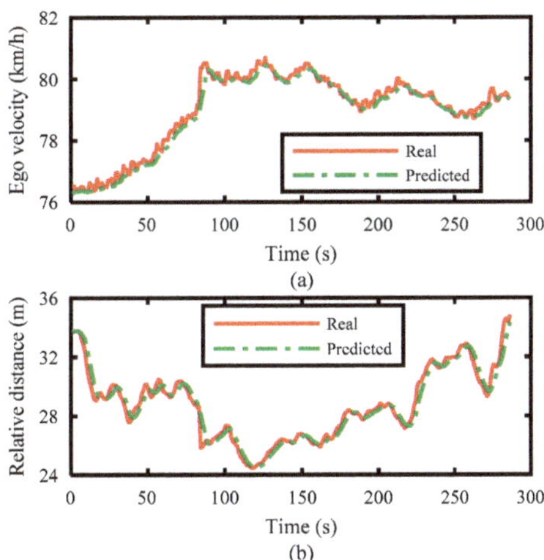

Figure 6. The application results of driver behavior mimic.

6.2. Comparative Analysis

In this section, we compare the proposed model with other existing ones, namely the CA model and IDM. The CA model characterizes the safe distance, which is given by

$$\Delta Y(t-T) = \alpha v_{n-1}^2(t-T) + \beta_l v_n^2(t) + \beta v_n(t) + b_0 \tag{19}$$

where, ΔY is the relative distance, v_{n-1} is the velocity of the ego vehicle and v_n is the leading vehicle velocity. The parameters are set as $T = 0.5$, $\alpha = -0.00028$, $\beta = 0.585$, $\beta_l = 0.00028$, and $b_0 = 4.1$ [1]. IDM can describe the desired following distance, which is given by

$$\Delta Y(v, \Delta v) = Y_0 + vt + \frac{v\Delta v}{2\sqrt{ab}} \tag{20}$$

where, v is the ego vehicle velocity and Δv is the relative velocity. The parameters in (20) are set as $Y_0 = 0.64$, $v_0 = 34$, $T = 1.7$, $a = 2$, and $b = 20.14$ [15,16,24]. The *RMSE* and *RMSPE* of relative distance are given in Table 5.

Table 5. Accuracy comparison between existing models and the proposed model.

Leading Vehicle Type	Model	Relative Distance	
		RMSE (m)	*RMSPE*
Passenger car	CA	1.3481	0.2360
	IDM	0.4192	0.0662
	Proposed	0.0750	0.0119
Bus	CA	2.3510	0.1524
	IDM	0.4986	0.1327
	Proposed	0.0820	0.0054
Truck	CA	14.5195	0.5791
	IDM	4.4358	0.1765
	Proposed	0.2281	0.0091

In Table 5, the accuracy of the proposed model is the best. The other two models have a comparatively larger *RMSE* and *RMSPE*. The reason is that the CA model is designed to keep a safer distance [1] and IDM is designed to eliminate the difference between the real situation and the preferred [15,16]. Therefore, they ignore the influence of the type of leading vehicle on the driver car-following behavior. Thus, they are not an appropriate choice for such application conditions. The comparison results indicate that the proposed model is suitable for mimicking the human driver behavior under such dynamic conditions. Some typical conditions are selected to show the details in Figures 7–9.

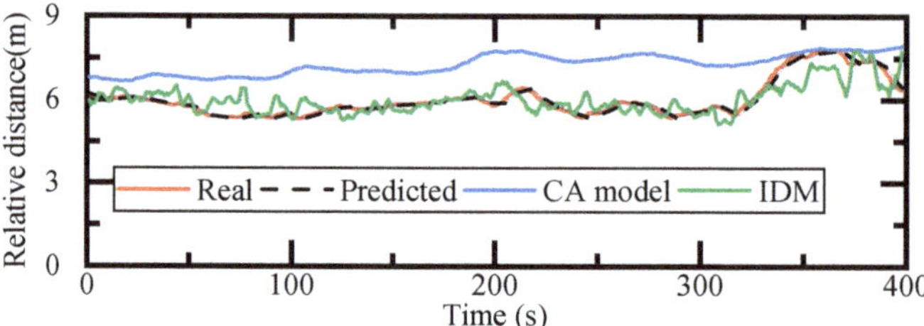

Figure 7. Results when following a passenger car.

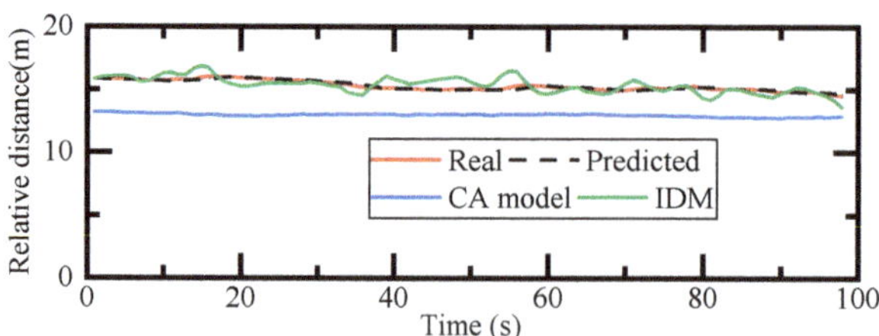

Figure 8. Results when following a bus.

The relative distance error of the proposed model is much smaller than others in Figures 7–9. The output of IDM is closer to the real relative distance in Figures 7 and 8. The CA model has the largest error of relative distance. In Figure 9, both CA and IDM output a smaller relative distance than the driver. The reason is that CA and IDM ignore the influence of leading vehicle type, which is mainly the outline of the leading vehicle [4]. A small-sized vehicle only blocks out a small portion of the traffic environment ahead. Comparatively, a larger vehicle easily causes critical accidents. Therefore, when following a passenger car, the human driver is more aggressive and keeps with the vehicle closer. On the contrary, if the leading vehicle is a bus or a truck, the proposed model keeps a larger relative distance with the preceding vehicle than CA and IDM (Figure 9).

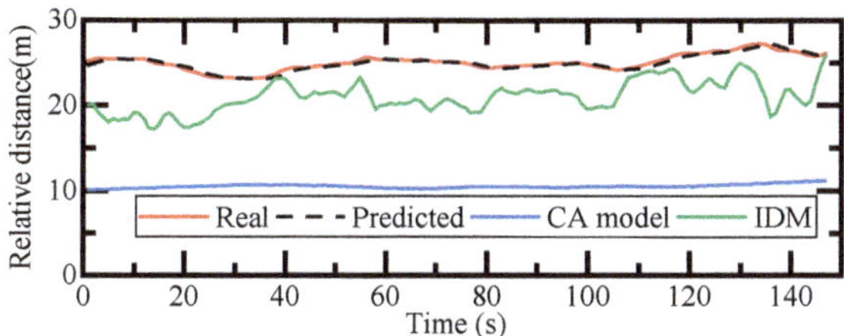

Figure 9. Results when following a truck.

7. Discussion and Future Work

From the above-mentioned analysis, it is found that the proposed model can mimic human driver driving behavior, and the proposed model has a high level of accuracy of leading vehicle type identification. In this paper, we mainly focused on whether the different leading vehicle types have an influence on car-following behavior. However, individuals' driving style were not considered. Potential directions in future work are discussed as follows.

7.1. Influence of Data Components

Driving on the city road and highway road are different driving styles. In this work, the model does not divide naturalistic driving data into city road, highway, and others. In extraction data, raw data contains more than 75% city road data, about 20% highway data, and the rest of the data is country road or others. In this data, the vehicle also keeps a lower velocity and a shorter relative distance. Thus, the proposed model prefers to fit a city road or low velocity driving condition. For the reason given above, the data components need to be divided into suitable parts in future research work.

7.2. Influence of Individual Driving Style

Car-following behavior is influenced by the driver's personal characteristics. First, driving years of a driver can directly affect his or her driving habits. For example, a driver with more experience will demonstrate a more confident driving style. A driving condition which is easy for a seasoned driver, may be perceived to be quite difficult by a novice driver. Such a performance difference will make an influence on the model. Therefore, a historical data-based model method may solve the problem in future work.

7.3. Applications in Future Works

This paper proposed a novel car-following model which considered the leading vehicle type. This model can apply to a human-friendly driver assist system. Each leading vehicle type corresponds to a different car-following strategy. For example, this model can be applied to a human-friendly Autonomous Emergency Braking (AEB) system, thus the driver will be warned at different velocities and different relative distances with the different leading vehicle types. In other words, with the same velocity and relative distance, the AEB system will warn the driver earlier if the leading vehicle is a truck compared to that of a bus or car. The proposed model can also be applied to Adaptive Cruise Control (ACC) system. In the car-following condition with different leading vehicle types, the model can calculate the relative distance and velocity of the leading vehicle to mimic human driver behavior to control the vehicle.

8. Conclusions

This paper has developed a GMM-HMM model for learning, identify leading vehicle type, and predicting ego vehicle velocity and relative distance in a car-following scenario. The proposed model is formulated from sample data. In the proposed model, ego vehicle velocity and relative distance are used. The relation between ego vehicle velocity and relative distance are fitted by a truncated GMM. The leading vehicle type is identified in real-time by a hidden Markov model. The identified vehicle types are passenger car, bus, and truck. Ego vehicle next state is predicted with an expectation of a series of historical driving data. The experiment's results show that the identification accuracy of a single leading vehicle type and changing leading vehicle type are 96.6% and 83.1% on average, respectively. Compared to the proposed model with the existing CA model and IDM, the results show that the proposed model can mimic the driver behavior better than the CA model and IDM. The model is more suitable for a real car-following condition with a changing leading vehicle type.

In summary, the model has two advantages. (1) It can identify the leading vehicle type in real-time with a high level of accuracy; (2) the model has a strong ability to mimic driver car-following behavior in nature driving conditions.

Author Contributions: P.W. and F.G. took the analysis and wrote the paper; K.L. provided the naturalistic driving data.

Funding: This work was supported by National key research and development program under grant 2016YFB0100900 and 2016YFB0101104, Open Fund of State Key Laboratory of Vehicle NVH and Safety (NVHSKL-201705), Industrial Base Enhancement Project (2016ZXFB06002).

Acknowledgments: Thanks to the China Automotive Engineering Research Institute, who provides the naturalistic driving data.

Conflicts of Interest: The authors declare no conflict of interest.

References

1. Brackstone, M.; Mcdonald, M. Car-following: a historical review. *Transp. Res. Part F Traffic Psychol. Behav.* **1999**, *2*, 181–196. [CrossRef]
2. Martinez, C.M.; Heucke, M.; Wang, F.-Y.; Gao, B.; Cao, D. Driving Style Recognition for Intelligent Vehicle Control and Advanced Driver Assistance: A Survey. *IEEE Trans. Intell. Transp. Syst.* **2018**, *19*, 666–676. [CrossRef]
3. Rahman, M.; Chowdhury, M.; Khan, T.; Bhavsar, P. Improving the Efficacy of Car-Following Models with a New Stochastic Parameter Estimation and Calibration Method. *IEEE Trans. Intell. Transp. Syst.* **2015**, *16*, 1–13. [CrossRef]
4. Pariota, L.; Galante, F.; Bifulco, G.N.; Luigi, P. The impact of the leading vehicle type on car-following behaviours. In Proceedings of the International Conference on Models and Technologies for Intelligent Transportation Systems (MT-ITS), Budapest, Hungary, 3–5 June 2015; pp. 30–37.
5. Wang, W.; Xi, J.; Zhao, D. Learning and Inferring a Driver's Braking Action in Car-Following Scenarios. *IEEE Trans. Veh. Technol.* **2018**, *67*, 3887–3899. [CrossRef]
6. Kesting, A.; Treiber, M.; Schönhof, M.; Helbing, D. Adaptive cruise control design for active congestion avoidance. *Transp. Nat. Res. C Emerg. Technol.* **2008**, *16*, 668–683. [CrossRef]
7. Jin, S.; Huang, Z.; Tao, P.; Wang, D. Car-Following Theory of Steady-State Traffic Flow. *Oper. Res.* **1959**, *7*, 499–505.
8. Gipps, P. A behavioural car-following model for computer simulation. *Transp. Nat. Res. B Methodol.* **1981**, *15*, 105–111. [CrossRef]
9. Oh, C.; Kim, T. Estimation of rear-end crash potential using vehicle trajectory data. *Anal. Prev.* **2010**, *42*, 1888–1893. [CrossRef] [PubMed]
10. Bifulco, G.N.; Pariota, L.; Simonelli, F.; Di Pace, R. Development and testing of a fully Adaptive Cruise Control system. *Transp. Nat. Res. C Emerg. Technol.* **2013**, *29*, 156–170. [CrossRef]
11. Kuge, N.; Yamamura, T.; Shimoyama, O.; Liu, A. *A Driver Behavior Recognition Method Based on a Driver Model Framework*; SAE Technical Paper Series; Institute of Transportation Library: Berkeley, CA, USA, 2000.

12. Wu, Z.; Liu, Y.; Pan, G. A Smart Car Control Model for Brake Comfort Based on Car Following. *IEEE Trans. Intell. Transp. Syst.* **2009**, *10*, 42–46.
13. Chen, X.; Li, L.; Zhang, Y. A Markov Model for Headway/Spacing Distribution of Road Traffic. *IEEE Trans. Intell. Transp. Syst.* **2010**, *11*, 773–785. [CrossRef]
14. Yang, D.; Yu, D.; Yang, F.; Pu, Y.; Zhu, L.-L. An enhanced safe distance car-following model. *J. Shanghai Jiaotong Univ.* **2014**, *19*, 115–122. [CrossRef]
15. Kesting, A.; Treiber, M. Calibrating car-following models by using trajectory data methodological study. *Transp. Res. Rec. J. Transp. Res. Board* **2008**, *2088*, 148–156. [CrossRef]
16. Zhu, M.; Wang, X.; Tarko, A.; Fang, S. Modeling car-following behavior on urban expressways in Shanghai: A naturalistic driving study. *Transp. Nat. Res. C Emerg. Technol.* **2018**, *93*, 425–445. [CrossRef]
17. Hao, S.; Yang, L.; Shi, Y. Data-driven car-following model based on rough set theory. *IET Intell. Syst.* **2018**, *12*, 49–57. [CrossRef]
18. Khodayari, A.; Ghaffari, A.; Kazemi, R.; Braunstingl, R. A Modified Car-Following Model Based on a Neural Network Model of the Human Driver Effects. *IEEE Trans. Syst. Man Cybern. Nat. Res. A Syst. Hum.* **2012**, *42*, 1440–1449. [CrossRef]
19. Pourabdollah, M.; Bjarkvik, E.; Furer, F.; Lindenberg, B.; Burgdorf, K. Calibration and evaluation of car following models using real-world driving data. In Proceedings of the IEEE 20th International Conference on Intelligent Transportation Systems (ITSC), Yokohama, Japan, 16–19 October 2017.
20. Wang, X.; Jiang, R.; Li, L.; Lin, Y.; Zheng, X.; Wang, F.-Y. Capturing Car-Following Behaviors by Deep Learning. *IEEE Trans. Intell. Transp. Syst.* **2018**, *19*, 910–920. [CrossRef]
21. Saffarian, M.; De Winter, J.C.F.; Happee, R. Enhancing Driver Car-Following Performance with a Distance and Acceleration Display. *IEEE Trans. Hum. Mach. Syst.* **2013**, *43*, 8–16. [CrossRef]
22. Jin, P.J.; Yang, D.; Ran, B. Reducing the Error Accumulation in Car-Following Models Calibrated with Vehicle Trajectory Data. *IEEE Trans. Intell. Transp. Syst.* **2014**, *15*, 148–157. [CrossRef]
23. Higgs, B.; Abbas, M. Segmentation and Clustering of Car-Following Behavior: Recognition of Driving Patterns. *IEEE Trans. Intell. Transp. Syst.* **2015**, *16*, 81–90. [CrossRef]
24. Zhu, M.; Wang, X.; Wang, Y. Human-like autonomous car-following model with deep reinforcement learning. *Transp. Nat. Res. C Emerg. Technol.* **2018**, *97*, 348–368. [CrossRef]
25. Ye, F.; Zhang, Y. Vehicle Type–Specific Headway Analysis Using Freeway Traffic Data. *Transp. Rec. J. Transp. Nat. Res.* **2009**, *2124*, 222–230. [CrossRef]
26. Aghabayk, K.; Sarvi, M.; Forouzideh, N.; Young, W. New Car-Following Model considering Impacts of Multiple Lead Vehicle Types. *Transp. Rec. J. Transp. Nat. Res.* **2013**, *2390*, 131–137. [CrossRef]
27. Chen, B.; Zhao, D.; Peng, H. Evaluation of automated vehicles encountering pedestrians at unsignalized crossings. In Proceedings of the IEEE Intelligent Vehicles Symposium (IV), Los Angeles, CA, USA, 11–14 June 2017; pp. 1679–1685.
28. Dozza, M.; Bärgman, J.; Lee, J.D. Chunking: A procedure to improve naturalistic data analysis. *Anal. Prev.* **2013**, *58*, 309–317. [CrossRef]
29. Aghabayk, K.; Sarvi, M.; Young, W. Understanding the Dynamics of Heavy Vehicle Interactions in Car-Following. *J. Transp. Eng.* **2012**, *138*, 1468–1475. [CrossRef]
30. Pariota, L.; Bifulco, G.N.; Brackstone, M. A Linear Dynamic Model for Driving Behavior in Car Following. *Transp. Sci.* **2016**, *50*. [CrossRef]
31. Jiang, R.; Hu, M.-B.; Zhang, H.; Gao, Z.-Y.; Jia, B.; Wu, Q.-S. On some experimental features of car-following behavior and how to model them. *Transp. Nat. Res. B Methodol.* **2015**, *80*, 338–354. [CrossRef]
32. Lefèvre, S.; Carvalho, A.; Borrelli, F. A Learning-Based Framework for Velocity Control in Autonomous Driving. *IEEE Trans. Autom. Sci. Eng.* **2016**, *13*, 1–11. [CrossRef]
33. Angkititrakul, P.; Terashima, R.; Wakita, T. On the Use of Stochastic Driver Behavior Model in Lane Departure Warning. *IEEE Trans. Intell. Transp. Syst.* **2011**, *12*, 174–183. [CrossRef]
34. Lee, G.; Scott, C. EM algorithms for multivariate Gaussian mixture models with truncated and censored data. *Comput. Stat. Data Anal.* **2012**, *56*, 2816–2829. [CrossRef]
35. Viterbi, A. Error bounds for convolutional codes and an asymptotically optimum decoding algorithm. *IEEE Trans. Inf. Nat. Theory* **1967**, *13*, 260–269. [CrossRef]
36. Yuan, Y.; Liu, X. Municipal waste generation forecasting based on improved hidden markov model. *Ind. Eng. Manag.* **2014**, *19*, 52–56.

37. Liu, S.; Sa, R.; Maguire, O.; Minderman, H.; Chaudhary, V. Spot counting on fluorescence in situ hybridization in suspension images using Gaussian mixture model. *Image Process.* **2015**, *9413*. [CrossRef]
38. Zhang, X.-B.; Qian, R.-H. Research on technical state evaluation of Vehicle Equipment based on BIC cluster analysis. In Proceedings of the 2nd International Conference on Big Data Analysis (ICBDA), Beijing, China, 10–12 March 2017; pp. 303–306.
39. Nhita, F.; Saepudin, D.; Adiwijaya; Wisesty, U.N. Comparative Study of Moving Average on Rainfall Time Series Data for Rainfall Forecasting Based on Evolving Neural Network Classifier. In Proceedings of the 3rd International Symposium on Computational and Business Intelligence (ISCBI), Bali, Indonesia, 7–9 December 2015.
40. Meng, Q.; Weng, J. An improved cellular automata model for heterogeneous work zone traffic. *Transp. Nat. Res. C Emerg. Technol.* **2011**, *19*, 1263–1275. [CrossRef]

Article

The Multi-Station Based Variable Speed Limit Model for Realization on Urban Highway

Soobin Jeon [1], Chongmyung Park [2] and Dongmahn Seo [1,*]

[1] Department of Computer Engineering, School of Software Convergence, Catholic University of Daegu, Gyeongsan-si 712010, Korea; marsberry@cu.ac.kr

[2] Department of Computer and Communications Engineering, School of Information Technology, Kangwon National University, Chuncheon-si 200701, Korea; chongmyung.park@gmail.com

* Correspondence: sarum@cu.ac.kr; Tel.: +82-010-4053-5447

Received: 24 March 2020; Accepted: 6 May 2020; Published: 13 May 2020

Abstract: Intelligent transport systems (ITS) are a convergence of information technology and transportation systems as seen in the variable speed limit (VSL) system. Since the VSL system controls the speed limit according to the traffic conditions, it can improve the safety and efficiency of a transport network. Many researchers have studied the real-time VSL (RVSL) algorithm based on real-time traffic information from multiple stations recording traffic data. However, this method can suffer from inaccurate selection of the VSL start station (VSS), incorrect VSL calculations, and is unable to quickly react to the changing traffic conditions. Unstable VSL systems result in more congestion on freeways. In this study, an enhanced VSL algorithm (EVSL) is proposed to address the limitations of the existing RVSL algorithm. This selects preliminary VSL start stations (pVSS), which is expected to end congestion using acceleration and allocates final VSSs for each congestion interval using selected pVSS. This controls the vehicles that entered the congestion area based on the selected VSS. We used four metrics to evaluate the performance of the proposed VSL (VSS stability assessment, speed control stability assessment, travel time, and shockwave), which were all enhanced when compared to the standard RVSL algorithm. In addition, the EVSL algorithm showed stable VSL performance, which is critical for road safety.

Keywords: variable speed limit; traffic control; travel time; intelligent transportation system

---

1. Introduction

Intelligent transportation systems (ITS) are integrated traffic systems based on the convergence of civil engineering and information technology fields. Such systems aim to provide a low cost/high efficiency traffic management system that enhances traffic safety. Several studies have been performed with the aim of reducing traffic congestion and alleviating the risk of car accidents on urban expressways. Both variable speed limit (VSL) and ramp metering technologies are representative methods for improving traffic conditions on expressways [1]. When vehicle accidents and traffic shockwaves are present, sudden variations in traffic flow are inevitable, which increase the risk of further accidents. In the VSL system, speed indicators are installed on the roads at regular intervals and the system collects traffic information in real time. Based on this information, the VSL system controls traffic flow by applying suitable speed limits in the regions suffering from traffic congestion. Hence, the VSL system can alleviate traffic congestion. In addition, a major advantage is its fast response to sudden events, such as road repair works and vehicle accidents. Since the VSL system enhances road safety, it is an important solution for upcoming ITS [2,3].

Existing VSL systems [4] used traffic information collected from single stations, which are devices to detect vehicles as a like magnetic loop detector, deployed in the different sections of road. This approach has already contributed to improving traffic congestion. However, in the case of urban expressway environments where the vehicles are travelling at high speed, an appropriate VSL system should consider the traffic conditions detected by multiple stations since traffic in one area can affect other parts of the expressway. Hence, recent VSL studies proposed the real-time variable speed limit (RVSL) method [5–8], which focuses on traffic control between multiple stations.

Especially, Jo et al. (JVSL) [7] is one of the most recently studied RVSL models and shows a very good performance from the freeway in Twin City, Minnesota. Although this algorithm shows good performance, it could not be applied to the highway due to some problems as below. It is operated next. First, it was found that the congestion area of the freeway contained multiple VSL sections. In addition, it defines the VSL start station (VSS) and controls the speed limit of vehicles in real time. However, often, the JVSL cannot determine the correct VSS. Errors occur in the calculation of the VSL for each station, which can shut down the entire system. This often happens with legacy RVSL systems. Since the purpose of VSL is to improve the safety on roads, if the VSL system cannot stably control traffic flow, traffic congestion will not be alleviated and could be aggravated over time.

Based on the VSS, the VSL algorithm controls the speed of the vehicles entering the congested section of the road. To control the speed, the VSL displays the modified speed value on a dynamic message sign (DMS), which is visible to drivers and located at a different location than the station. Thus, the VSL system must control the speed of vehicles based on the location of the DMS. However, the JVSL model calculates the VSL value of each station based on the position of the VSS instead of the location of the DMS. Since the stations and DMS are in different areas, an accurate VSL value cannot be displayed on each DMS device.

In this study, an enhanced VSL algorithm (EVSL) is proposed to address limitations of the existing JVSL algorithm. The proposed algorithm enhances the safety on the freeway because it provides a stable VSL for multiple stations. The proposed algorithm is composed of three parts. First, the moving average speed is calculated to achieve the stable speed data of each station. Second, a new method is proposed for making VSS assignments to accurately determine the VSS. Third, a new method calculates the VSL value of each station based on the VSS and DMS locations, instead of using only the station location. Lastly, the new VSL is applied to the DMS of each station. To evaluate the proposed algorithm, we constructed a simulation environment presenting real road conditions (outer ring road in the south of Seoul, South Korea). The PTV VISSIM software was used to simulate our proposed algorithm. Calibration of the system using the raw data was performed to make it suitable for use in a similar real traffic environment. The proposed algorithm was compared to the existing JVSL algorithm considering the efficiency and stability. The proposed VSL method showed better performance for all four metrics used to evaluate the performance (VSS stability assessment, speed control stability assessment, travel time, and shockwave).

The remainder of this paper is organized as follows. Section 2 reviews Existing JVSL algorithms and discusses the problem of them. Section 3 proposes an enhanced VSL algorithm to solve JVSL problems. Section 4 describes the experimental environment to simulate a real traffic environment. Section 5 offers measurement and evaluation of the performance of the proposed enhanced VSL algorithm. Section 6 concludes the paper.

2. Control Design and Problems of JVSL

The VSL system is being employed to reduce traffic accidents on urban freeways. Traffic accidents and shockwaves on the road suddenly affect the speed of vehicles, which can cause additional accidents [9]. To solve this problem, the VSL system collects real-time traffic information, which is used to determine an appropriate speed limit for each section of the freeway suffering traffic congestion. With an active VSL system in place, the vehicles reduce their speed before entering the congested section, which reduces the risk of accidents and improves road safety. The VSL system is composed of three elements including sensors monitoring the traffic conditions, the control system processing the sensing data, and the DMS devices displaying the VSL. Since the VSL enhances the stability of traffic flow and reduces driver stress, it is widely applicable. Specifically, to improve the stability of freeway conditions, several studies measured traffic data and analyzed them in real time. These studies focused on data such as the speed, density, and volume of traffic. In addition, some studies proposed an integrated system including both VSL and ramp metering on interconnections [10,11].

Previous studies developed VSL algorithms based on information received about the road conditions, which contributed somewhat to road safety. However, since these studies were based on data from single stations, inadequate performance was observed for multiple stations and they required 5–10 min to update the VSL. To provide a quick reaction to the rapidly changing traffic conditions, the VSL system should monitor variations in real time. Hence, a VSL update time of more than 5 min is expected to cause other traffic problems. Hence, the RVSL system was proposed [7,8].

The JVSL is one of real-time VSL algorithms based on the interworking model between stations on urban freeways. Stations are installed on the freeway with an interval of 1–2 km and collect the traffic information. The JVSL algorithm determines the VSL value based on the information such as speed, density, and traffic volumes. In addition, to quickly react to variations in traffic conditions, the system collects traffic information at 30 s intervals and calculates the VSL value. The JVSL is composed of four stages. As shown in Figure 1, first, the bottleneck station (BS) where the traffic congestion occurred and its location are identified. Second, based on this information of BS, the VSS location is examined, which represents the tail station of the traffic congestion area, as shown in (Figure 1 ②). Third, based on the VSS, this system determines the number of stations controlled by the VSL (Figure 1 ③). Lastly, this displays the speed value calculated at each station in the DMS. For several years, the JVSL algorithm has been studied on urban freeways based on the real-time traffic data from multiple stations. However, it showed several problems, which caused malfunctions in the algorithm at various stages of the calculations. In this study, to examine the limitations of the JVSL method, we analyze each stage of the process.

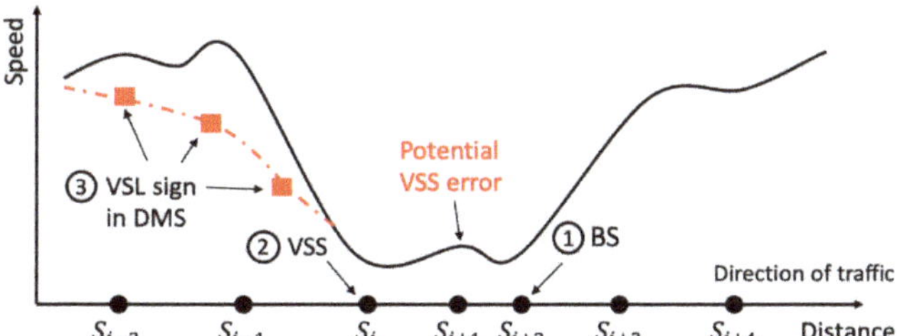

Figure 1. Schematic diagram showing the variable speed limit (VSL) algorithm of the real-time variable speed limit (RVSL) system.

2.1. Bottleneck Station (BS) Detection

As shown in Figure 1, to detect the VSS, the JVSL searches the whole area suffering traffic congestion. First, this detects and identifies the start point (start location) of traffic congestion as the BS (Figure 1). This calculates the speed difference between two stations for a period of 1.5 min in order to identify the BS. However, since the JVSL algorithm selects the BS using only one simple parameter (speed), it cannot quickly react to ever-changing traffic conditions. These critical points cause many traffic control problems for the ITS.

Figure 2 shows three types of speed graphs being operated by the JVSL system in a particular zone, where the traffic speed (red line) are shown at locations of stations. Calculated VSL speeds are shown in blue circles, and displayed VSL speed is adjusted to be displayed in units of 5 km/h, which are shown in black circles. In addition, the VSS location and traffic congested section are identified. The horizontal axis denotes the stations (2.5 km spacing), while the vertical line shows the speed of each station. The VSS is the selected station connected to the red solid line. For example, in Figure 2a, S7 is the VSS. These figures highlight three problem areas from the JVSL.

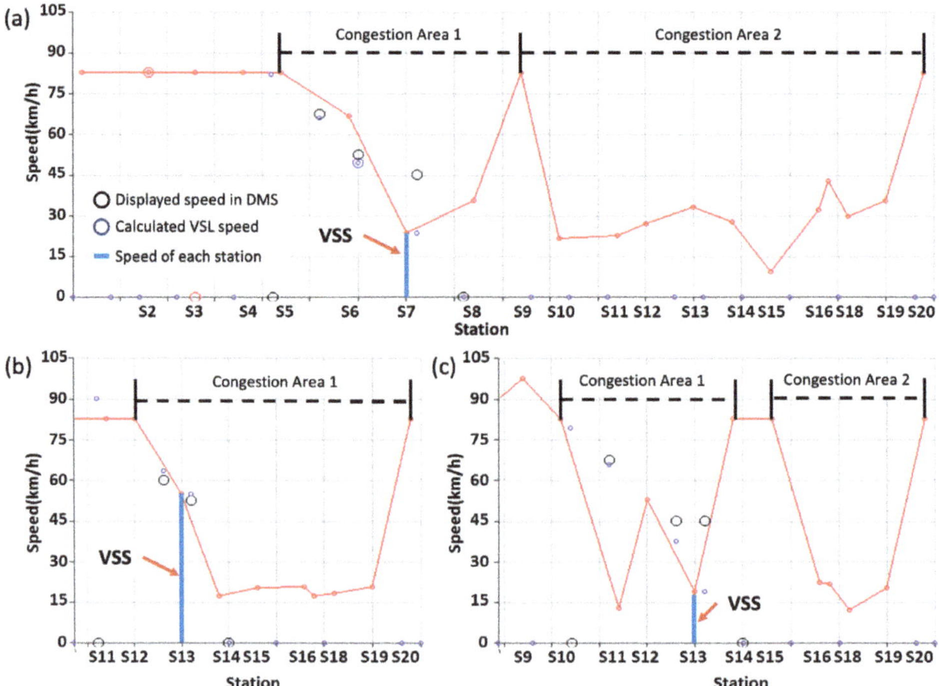

Figure 2. Example graphs of incorrect Variable Speed Limit Start Station (VSS) decision in Jo et al. (JVSL) algorithm [7]. (**a**) and (**c**) are scenarios where only 1 VSS is selected for 2 congestion zones. (**b**) is a scenario where the wrong VSS location is determined. Each graph includes speed (red line) of each station in the congestion section of the freeway.

The example cases in Figure 2a,c include two types of congestion areas, which has low speed flow under 30 km/h. In fact, when the congestion began, these contained only one congestion area, but, over time, it was changed to two congestion areas. Since the JVSL cannot anticipate this variation in traffic flow in real time, only one VSS (congestion area 1) was selected in each case. However, after 1.5 min, analysis of the speed difference is performed again, and the JVSL can select a different VSS in the congestion area 2. However, since urban freeways have rapidly varying traffic conditions, this slow reaction capacity of the current JVSL causes safety problems on the road.

In this study, we propose a new method for rapidly detecting the VSS position. The proposed method searches for locations that we expect to be the end of congestion to detect the correct VSS location quickly instead of identifying the full location suffering traffic congestion (Section 3.2.2). In addition, the proposed method can quickly detect the VSS from the predicted locations and can rapidly react to real-time variations in the traffic conditions (Section 3.2.3).

2.2. VSL Start Station (VSS) Decision

To detect the VSS, as shown in Figure 1, the JVSL searches the speed (U) of congestion sections from the BS. While searching the process, if the selected station satisfies Equation (1), that station is designated as VSS. U denotes the speed. S denotes the station. i denotes the station number. t denotes the time. Minimum VSL is the lowest speed value to control the vehicle speed in the VSL system. Most of the algorithms varying from studies are set to 30 km/h [4].

$$If\ U_{i-1} - U_i > threshold\ then\ VSS = S_i$$

where

$$If\ U_{i-1} > Minimum\ VSL_t\ then\ threshold = \frac{16km}{h}$$
$$If\ U_{i-1} \leq Minimum\ VSL_t\ then\ threshold = Minimum\ VSL_{t-1} \quad (1)$$

JVSL decides the VSS using the speed difference between the current station (U_i) and the next station (U_{i-1}), where the threshold value is 16 km/h, as shown in Equation (1). However, since the traffic flow conditions on the road are unpredictable, if an unproven fixed value is used as the threshold value, inaccurate VSS values could be chosen.

First, even in the congested section, the speed difference could be over the threshold value. For example, the speed of S12 (Station 12) is in the congestion area 1 of Figure 2. The VSS in which the VSL should operate is S11 in the congestion area, but S13 is selected as the VSS. Figure 2b also selects the wrong S13 as the VSS rather than S14. Therefore, this system displays an inappropriately high VSL value in the wrong place. Inaccurate VSS decisions lead to problems in the VSL calculation process.

Second, as shown in Figure 2a,c, the VSS cannot be decided from congestion area 2 because the VSL system is suspended for unknown issues (Figure 2a) or spends plenty of time determining and creating the new VSL. The VSL should be constant over the congested section and vehicles entering this region should be continuously controlled. If the VSL system is suspended in the middle of congestion control or is re-started, it can have a severe negative impact on traffic flow.

Our proposed method (Section 3.2.2) addresses the inaccurate VSS decisions made by the JVSL by using both the speed and acceleration from real traffic data to predict possible VSS values. Lastly, an accurate VSS is determined and used to control the congestion areas.

2.3. Number of Control Stations and VSL Calculation

The JVSL calculates the number of stations controlled by the VSL. First, by comparing the fixed speed limit with the real traffic speed at the VSS, the scope of controlled stations is designated. If the speed of the VSS is higher than the speed limit, as the difference between the speed decreases, the scope of VSS control also decreases. Second, when a shockwave is calculated, the total congestion situation on the road should be examined. Lastly, the number of control stations is decided using the VSL system.

The JVSL establishes the number of control stations with a minimum of two and a maximum of three. If many stations are controlled by the VSL, the travel time will increase. Hence, an upper limit is set. Since the distance between stations is generally 0.3–2 km, it is acceptable to determine the scope of VSL control, according to the number of stations. However, in this case, the maximum distance over which a reduced speed is applied can reach up to 6 km and greatly affect the total travel time. Therefore, it has been proposed that the maximum distance over which the VSL is applied is 3.2 km [7].

As mentioned above, the VSL calculation in the JVSL depends on the number of control stations, where the location and speed of the VSL are calculated using strategy 1, 2, and 3. In order to provide accurate VSL control, the DMS location should be included in the process of calculating the VSL. However, JVSL calculates VSL speed based on location of station and display it in the DMS, which is in a completely different location from the station. These are caused to display an incorrect speed limit to the vehicles entering to the congestion area and reach more traffic accidents than when the VSL is not in operation.

In Section 3.3, based on the VSS, the moving time, speed, and acceleration to downward DMS within 3.2 km are analyzed. The optimal VSL distance and value are selected, along with a VSL value for each DMS. Using this approach, the new VSL system operates based on the DMS instead of the stations, while the vehicle speed can be controlled within the optimal VSL distance.

3. Enhanced Variable Speed Limits (EVSL)

VSL systems can decrease the risk of accidents by detecting areas suffering traffic congestion on urban freeways and applying speed limits on those sections. When high speed vehicles approach those with lower speed, the accident rate increases significantly. Hence, the VSL system detects the sections with the greatest speed differences between vehicles on the freeway and initiates speed control. It is a goal of VSL systems that total travel times are not significantly decreased when the speed reduction is applied on a freeway. To accomplish this, it is necessary to apply an appropriate speed over an accurate VSL operation area.

In Section 2, the existing JVSL method was introduced and its limitations were discussed.

- The method for selecting the congestion area is too simple. It does not exploit enough traffic data to allow it to rapidly react to the real traffic conditions, which results in inaccurate congested sections and VSS detection.
- Method for determining the VSS using only speed data is too inaccurate and simple. Inaccurate VSS decisions lead to problems in the VSL calculation process that generates the wrong speed limit or system suspend.
- Method for adjusting the maximum VSL distance risks delaying the total travel time. The maximum distance over which a reduced speed is applied can reach up to 6 km and greatly affect the total travel time.
- Method for calculating the VSL value of each station is based on the location of the VSS instead of DMS. Since the stations and DMS are in different areas, an accurate VSL value cannot be displayed on each DMS device.

Since this method has some critical problems mentioned above, it is impossible to operate the JVSL algorithm to the real urban road for vehicle safety. Therefore, in this section, we aim to apply a new algorithm to the freeway after solving them.

The enhanced VSL algorithm proposed in this case is based on selection of both the DMS and station and is composed of three steps, as shown in Figure 3. First, it calculates the moving average speed to calculate the speed for each station. Second, the VSS is decided based on a new method. Third, the VSL is calculated based on both the location of the VSS and DMS. Lastly, the calculated VSL values are applied to the relevant DMSs. These steps are discussed in more detail in the following sections.

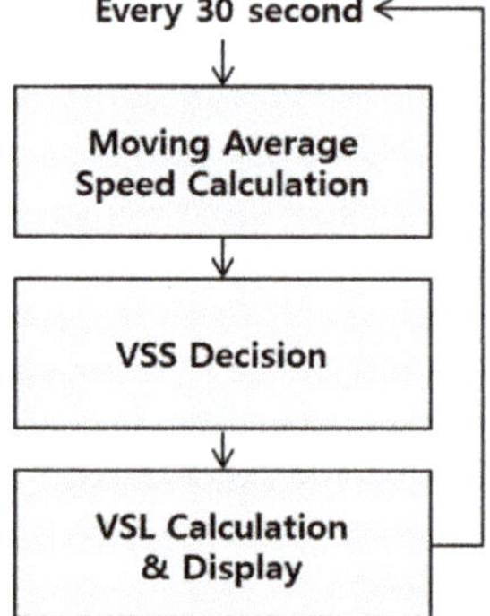

Figure 3. Flow chart of the enhanced variable speed limit (EVSL) algorithm.

3.1. Moving Average Speed Calculation

The proposed algorithm quickly reacts to the actual traffic conditions on the freeway. To achieve this, the algorithm takes traffic data every 30 s from each station and updates the VSL based on recent traffic information. The process of collecting traffic information by the traffic control system to use data in VSL is as follows.

- The traffic detection management devices installed at each station on the freeway collects vehicle data from sensors.

- The collected data is transmitted to the central server every 30 s including traffic volume, percent of Occupancy, average speed, and density calculated from each management device.
- The proposed algorithm is performed every 30 s using vehicle information recorded in a central database.

We obtain the average speed (u), density (k), and volume (q) for 30 s from the central database. Figure 4 shows the parabolic relationship between the traffic flow (volume) and density [12]. This figure shows that, as the traffic flow increases, the values of flow and density also increase. In addition, the accuracy of speed increases, since the number of vehicles that can be referred to rises as the flow grows.

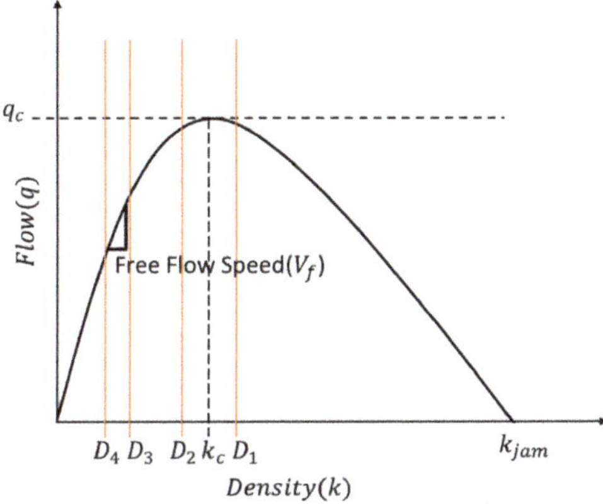

Figure 4. Relation between density and flow.

However, after the maximum flow value is reached, the density value decreases. We realize that traffic congestion begins at this position. We denote this position as the critical density (k_c) at which the road is occupied by the maximum density of vehicles possible. After k_c, although the density value increases, the traffic flow begins to decrease because traffic congestion results in low vehicle speed.

If the density exceeds k_c and reaches jam density (k_{jam}), the flow value is 0. This means that traffic has stopped. From this observation, we define three traffic states [12]: very low flow, free flow ($k < k_c$), capacity flow ($k = k_c$), and congestion flow ($k > k_c$). Therefore, as the density approaches zero, the number of vehicles passing through the sensor decrease, and the accuracy of the speed also decreases. In this case, we can assume the cases where the speed data could be zero when collecting traffic data from sensors every 30 s. There are two cases where the speed is zero. First, there is no vehicle passing the station in the period of 30 s, and the speed and density value are zero. In this case, the proposed algorithm determines the station as an area where the vehicle does not pass, sets the speed of the station to the speed limit (100 km/h), and does not select the VSL operating zone. Second, the density reaches jam density (k_{jam}) as mentioned in the above paragraph. In this case, we set the speed value considering Equations (2) and (8).

VSL aim to ensure traffic safety and reduce traffic congestion in high traffic environments. Therefore, VSL will tend to be operated at free flow ($k < k_c$) or capacity flow ($k = k_c$), and, in this area, the probability of an error calculating the wrong speed is less than low flow and traffic jam, since the speed is calculated using sufficient vehicle data. Basically, the proposed algorithm basically uses a moving average speed. If the road state has a constant speed (uniformly accelerated motion) for 1 min 30 s, the sensors constantly measure a constant speed value and we can get relatively high reliability speed data from them. Therefore, the average data for 1 min (two periods: 30 s × 2) is used as speed data in the general case, as shown in Equation (2).

However, it may be difficult to obtain accurate speed data if the speed trend of the station has not been constant such as the lowest volume capacity (lowest vehicle density and highest speed) and the congestion area (highest vehicle density and lowest speed: k_{jam}) due to low reliability of the sensors. Especially, the sensor cannot measure the accurate data in the congestion area because performance of the vision detector is poor at low speeds and the loop detector also has low speed measurement accuracy at high density traffic flow due to sensor characteristics [13,14]. Therefore, the algorithm finds a stable moving average interval (MAI) based on road conditions after deciding whether the measurement area is congested or not based on the density of vehicles, as shown in window size 2 of Equation (2).

$$MAI = \begin{cases} 2, & u_{t-2,i} < u_{t-1,i} < u_{t,i} \text{ or } u_{t-2,i} > u_{t-1,i} > u_{t,i} \text{ or } u_{t-2,i} = u_{t-1,i} = u_{t,i} \\ 6, & 35 \leq k \\ 4, & 35 > k \geq 25 \\ 3, & 25 > k \geq 15 \\ 4, & 15 > k \geq 10 \\ 6, & 10 > k \end{cases} \quad (2)$$

To classify the window size of the average speed based on the density at the congested area, we classify each density area based on average k_c of the freeway where the VSL is operated in. First, k_c of each station is obtained from Equation (3). It is calculated by the Green Shield Model using the maximum traffic flow (Q_c) and free flow speed (V_f) [15].

$$k_c = \frac{Q_c}{v_f} \quad (3)$$

Q_c is obtained from Equation (4). k_{jam} refers to the density of traffic congestion. To calculate Equations (3) and (4), V_f and k_{jam} should be calculated first. They can be obtained from real road data. However, it is difficult to calculate these two values accurately. Therefore, we calculate their approximate values from Equations (5) and (6) based on the speed (v) and density (k) values obtained for the corresponding road. However, the v and k values vary depending on whether the traffic state is smooth or congested. In this study, traffic data are collected for a real road, and the k and k_{jam} values are obtained from Equations (5)–(7).

$$Q_c = \frac{v_f \times k_{jam}}{4} \quad (4)$$

$$k_{jam} = \frac{v_f}{|b|} \quad (5)$$

$$v_f = \overline{v} - b\overline{k} \quad (6)$$

$$b = \frac{\sum\left(k_i - \overline{k}\right)(v_i - \overline{v})}{\sum(k_i - k)^2} \quad (7)$$

We set the window size for calculating the moving average of the speed according to the reliability of the sensor if the trend of the speed for 1 min and 30 s is not constant. The window size is set based on the density since the reliability of the sensor is determined by the number of vehicles passing through the sensor. First, the density section, k_c, where congestion begins is determined. Each station on the road has a different k_c value. Figure 5 shows the average speed and q-k graph for the congestion area of the Seoul Outer Ring freeway from January 2016 to January 2017. Referring to q-k scatter graph, k_c distribution of each station can be seen to have a value of about ±5, based on the average density value 30. Therefore, we determined the interval of k_c in Figure 4 as 25 (D_1) – 35 (D_2) based on empirical data in order to generalize the k_c values of all stations. Second, in the case of a highly reliable free flow speed area that includes a large number of vehicles as in Figure 5, the density value for selecting the window size is determined according to the flow (q) value. In this paper, we set the free flow speed area from q_c to a minimum value of flow by referring to the q-k scatter values in Figure 5. Therefore, D_4 in Figure 4 is determined as the density value of 10 at the minimum value position of the free flow area. Third, D_3 for setting the confidence area in the free flow uses the value of 75% from the reference value of D_4 [15]. In this paper, we use the value of D_3 as 15. Lastly, the window size is determined as in Equation (2). The lowest density and highest density areas with very low vehicle information are 6 (3 min), the low density and high-density areas are 4 (2 min), and the free flow speed area is 3 (1 min 30 s).

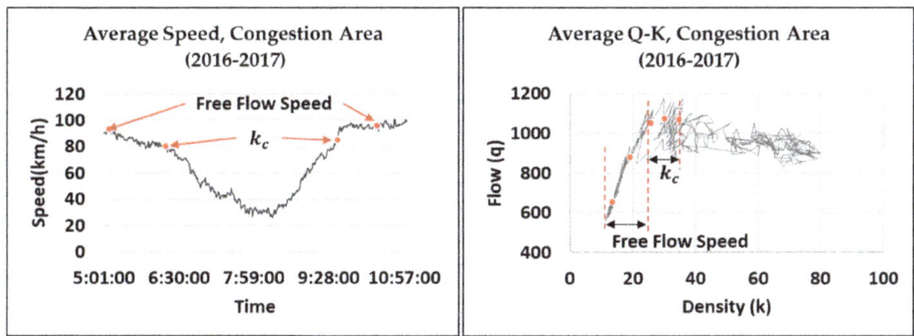

Figure 5. The average speed and q-k graph for the congestion area of the Seoul Outer Ring freeway from January 2016 to January 2017.

The moving average speed is calculated using Equation (8). u denotes the speed, t denotes the time, i denotes the station number, and k denotes the density.

$$u_{avg}(km/h) = \frac{\sum_{n=0}^{MAI-1} u_{t-n,i}}{MAI} \quad (8)$$

3.2. VSL Start Station (VSS) Decision

The station at the start of the congestion area is designated as the VSS, and the VSL system controls the speed of vehicles from this point (where the high-speed traffic meets the slower vehicles), as shown in Figure 1. Based on the selected VSS, the VSL value is established for each station, as shown in Figure 1. Therefore, an accurate decision of the VSS is necessary to ensure stable performance.

Our EVSL algorithm provides a more accurate and efficient manner of determining the VSS location compared to the standard RVSL system. This is achieved by using the vehicle acceleration between stations to assist in the detection of pVSS locations and selection of a final VSS. High acceleration values are usually observed at pVSS locations.

The algorithm analyzes past traffic data for the pVSS locations and compares them to the real acceleration values in order to identify locations with significant variations in traffic flow. Generally, at the start of the congested region, the traffic flow changes abruptly and large differences in the speed between stations are observed. Therefore, the predicted VSS locations can be detected accurately. Lastly, this system selects the final VSS after a merging operation is performed on overlapping sections, which are assumed to be a single congested section among the pVSS locations.

3.2.1. Acceleration Calculation

The pVSS locations are selected as described in the previous section. When acceleration values between stations are calculated, accurate speed differences can be predicted for each station. Figure 6 shows the method for calculating the acceleration at each station. Basically, EVSL calculates an acceleration using the distance and speed between station S_i and the preceding station S_{i-1}, as shown in Figure 6, where Equation (9) shows the corresponding acceleration equation. u denotes the speed, i denotes the station number, and d denotes the distance between stations.

$$Acceleration\ of\ Station_i(km/h^2) = \frac{u_i^2 - u_{i-1}^2}{2 \times d} \tag{9}$$

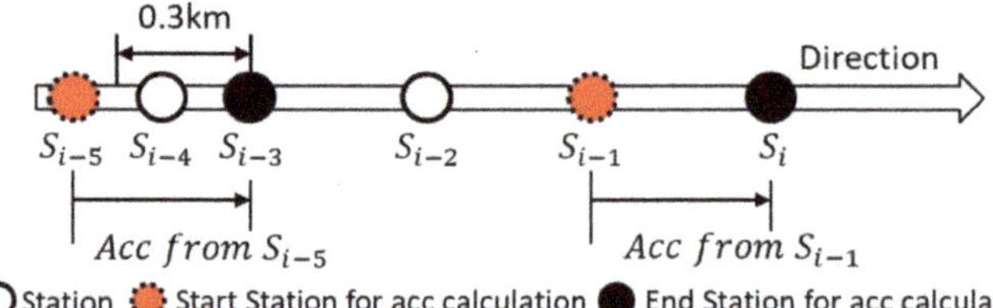

Figure 6. Schematic diagram showing the method for calculating acceleration at each station.

If the distance between two stations is less than 300 m, when vehicles are travelling over 80 km/h, the speed difference between two stations is too small to detect. Hence, we calculate the acceleration over a minimum distance of 300 m. As shown in Figure 6, the downstream station (S_{i-4}) installed within 300 m from the upstream (S_{i-3}) station is merged with the upstream station (S_{i-3}) and excluded from the acceleration calculation and pVSS determination. Therefore, the acceleration for S_{i-3} is calculated with respect to S_{i-5} instead of S_{i-4}.

3.2.2. Preliminary VSS (pVSS) Decision

First, our EVSL algorithm selects candidate VSSs that are considered to be tail portions of congestion areas to determine an accurate VSS location. In generally, VSS is the tail portion of the congestion area. If a vehicle approaches the VSS, its speed should rapidly decrease. Therefore, as shown in Figure 7, we can estimate that two areas (between S_{i-2} and S_{i-3} and between S_{i-4} and S_{i-5}) are the pVSS, which is a high deceleration congestion estimated area. Hence, S_{i-2} and S_{i-4} are selected as pVSS locations following the criteria given by Equations (10) and (11), which use the acceleration and speed values between the current and previous stations.

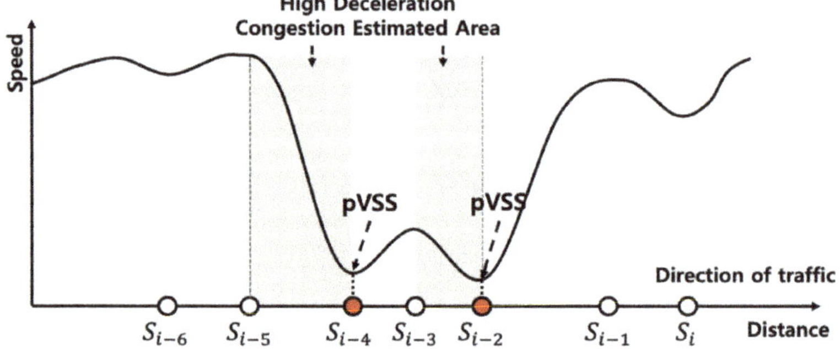

Figure 7. Schematic diagram showing the method for determining the preliminary VSL start station (pVSS) locations from the station location and speed.

To select the pVSS, the threshold value of acceleration and speed was investigated [16]. This approach is based on reports of accidents occurring on the I-35W freeway in Minnesota, USA. We judged that there would be no problem even if the data investigated in Reference [16] was used because the road environments of the Seoul outer ring freeway composed of the experimental environment and the I-35W were similar. This includes the number of lanes (I-35W: 3–5, Seoul outer ring: 3–5), speed limit (I-35W: 104 km/h, Seoul outer ring: 100 km/h), and more. Figure 8 shows the distribution of acceleration and speed for the accidents occurring for 2009.9–2009.12 on the I-35W freeway, which has a normal speed limit of 104 km/h (65 mile/h). It can be seen that 60% of the accidents occurred below 90 km/h (55 mile/h) and below acceleration values of -2400 km/h^2 (-1500 mile/h^2^2). In this case, when the current station ($S_{t,i}$) is not the VSS at t − 1, if the speed of $S_{t,i}$ is < 90 km/h and the acceleration is below -2400 km/h^2, $S_{t,i}$ could be selected as a pVSS in Equation (10). S denotes the Station, t denotes the time, i denotes the station number, u denotes the speed, and a denotes the acceleration.

$$S_{t,i} = pVSS, \ \ if \ u_{(t-2,t-1,t),i} \leq 90 \ {}^{km}/_{h} \ and \ a_{(t-2,t-1,t),i-1} \leq -2400 \ {}^{km}/_{h^2} \tag{10}$$

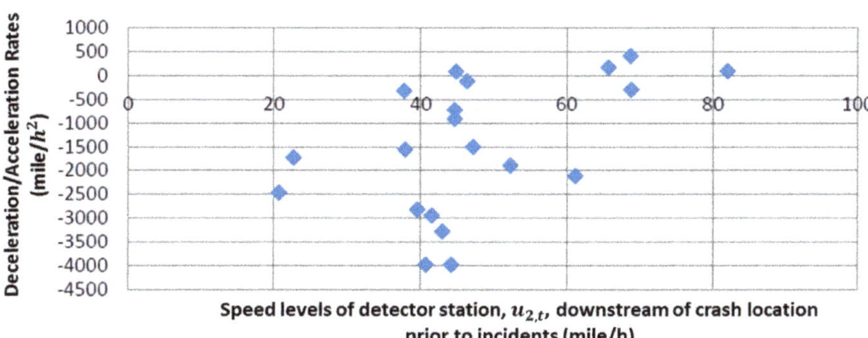

Figure 8. Measured accelerations of traffic flow prior to accidents on the I-35W, Minnesota, USA (2009).

In addition, when $S_{t,i}$ was assigned as the VSS at t − 1, if it satisfies Equation (11), it could continuously be assigned as a pVSS.

$$S_{t,i} = pVSS, \ \ if \ a_{(t-2,t-1,t),i-1} \leq -1200 \ {}^{km}/_{h^2} \tag{11}$$

In our EVSL algorithm, to reduce the error of VSS calculation arising from the short update period and allow stable selection of the pVSS, if a station satisfies the conditions of Equation (10) or (11) for 1.5 min, it qualifies as a pVSS.

3.2.3. VSS Decision

The VSS is finally selected from the group of pVSS. Figure 9 shows the congestion area, locations of pVSS, and the method for determining the VSS. In Figure 9a, there are two pVSSs that can be final VSS in the congestion area. However, if the VSS is chosen in the congestion area as shown by the pVSS, S_{i-2} in Figure 9a, this may worsen the congestion than before. In addition, since the speed of the congestion area is already low, application of a VSL in this area is not necessary. Therefore, such unnecessary pVSS locations should be eliminated and an accurate VSS is required to control the speed of the vehicles approaching the congested area.

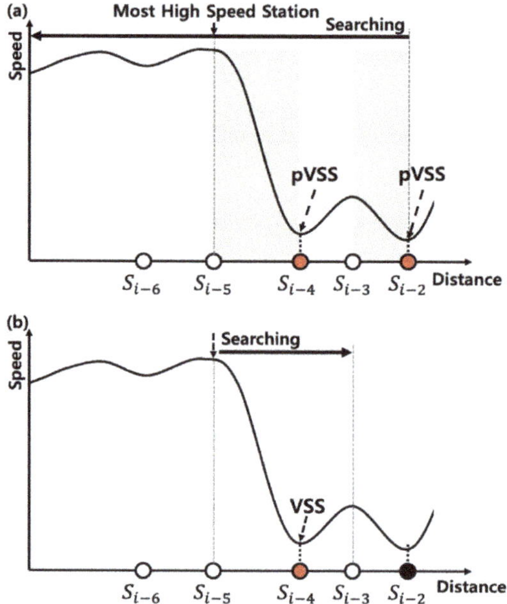

Figure 9. Schematic diagram showing the method for determining the variable speed limit start station (VSS). (**a**) The location of pVSS and upward search in the opposite direction of traffic flow to find the scope of VSL. (**b**) downward searching to select VSS.

First, to achieve this, upward searching opposite of the direction of traffic flow is performed from the first downstream pVSS location. Searching proceeds until the condition of Equation (12) is satisfied and part of the congestion area is defined by locating the first station that is not affected by traffic congestion. We used a threshold value of congestion of 40 km/h following the definition of the Korea Expressway Corporation [17]. S denotes the Station, t denotes the time, i denotes the station number, u denotes the speed, and a denotes the acceleration.

$$S_{t,i} = End\ of\ Congestion,\ \ if\ a_{t,i} > 0\ {}^{km}/_{h^2}\ \text{and}\ u_{t,i} \geq\ 40\ {}^{km}/_{h} \tag{12}$$

Next, as shown in Figure 9b, downward searching is performed from the station selected by Equation (12) until the condition of Equation (13) is satisfied. As shown in the figure, the search operation is performed up to the station near the congestion area.

$$Downstream\ Searching\ Stop\ Condition,\ \ if\ a_{t,i} > 0\ {}^{km}/_{h^2} \tag{13}$$

Lastly, while the search process of Equation (13) is performed, the station satisfying the condition shown in Equation (14) is updated as the VSS. The VSS is updated until the search process is finished, which results in the pVSS near the end of the congestion area being designated as the VSS.

$$S_{t,i} = VSS,\ \ if\ a_{t,i} \leq -1200\ {}^{km}/_{h^2} \tag{14}$$

3.3. VSL Calculation and Display

The VSL system displays the corrected speed for each section of road via DMS installed on the freeways. To improve on the existing JVSL algorithm that assumed that both the DMS and station were at the same location, our new algorithm first defines the location of the DMS and VSL values corresponding to specific DMS calculated based on the VSS. This results in more suitable speed being displayed on each DMS, where the vehicles entering the congestion area show a reduced speed on the DMS located upstream from the VSS. Figure 10 shows the relationship between the VSS and DMS, where the empty circles denote the station and filled circle denote the VSS. As shown in Figure 10a, the VSS is connected to all DMS systems located upstream, which defines the VSL distance. As shown in Figure 10b, if the current VSS and the upstream VSS connect with the same DMS. The lower VSS has priority.

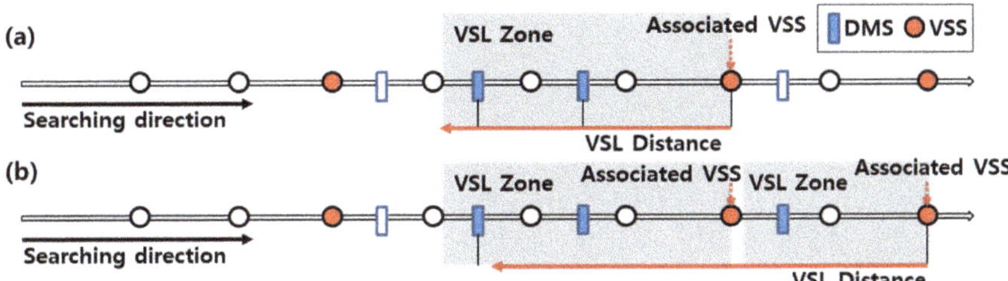

Figure 10. Schematic diagram showing the relationship between the dynamic message sign (DMS) and VSL start station (VSS). (**a**) A single VSS scenario for associating with DMS, (**b**) Multiple VSS scenario for associating with DMS.

As shown in Figure 11, the VSL distance is the total distance from the VSS over which the VSL is applied, which was limited to 3.2 km in our algorithm in an attempt to not affect the travel time of vehicles within a travel distance [4].

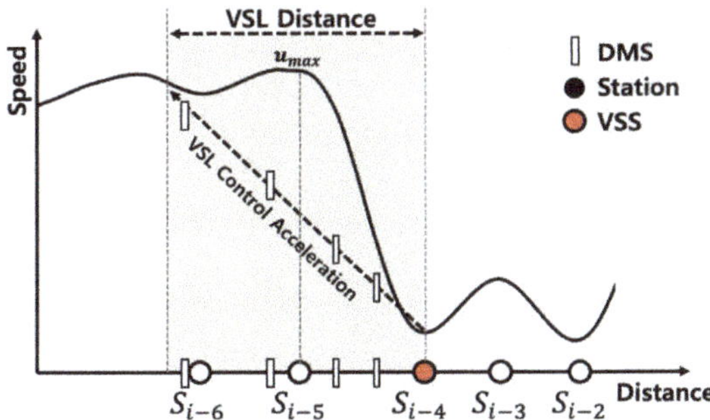

Figure 11. Schematic diagram showing the method for controlling the variable speed limit (VSL).

A new equation for the VSL distance is proposed, which is based on the speed, moving time, and acceleration, as shown in Equation (15). Based on the current speed at the VSS, if VSL control acceleration ($a_{t,i}$: VCA) is known, the total distance to reach the speed limit ($u_{limit,i}$) can be calculated. u denotes the speed, *limit* denotes the speed limit, i denotes the station number, t denotes the time, and a denotes the acceleration.

$$VSL\ Distance\ (km) = \frac{{u_{limit,i}}^2 - {u_{t,i}}^2}{2 \times \alpha_{t,i}} \tag{15}$$

As shown in Figure 11, the VCA is used to calculate and decide the VSL distance and speed in each DMS. In previous studies, the acceleration value was fixed as $-2400\ \mathrm{km/h^2}$ [4]. However, the speed and distance vary according to the speed limit and traffic conditions. If the VSL is determined using a fixed acceleration, wrong traffic information could be provided to vehicles on the freeway. Therefore, a new method for calculating the VCA is required, which updates the value in real time. The new equation for the VCA($a_{t,i}$) is shown by Equation (16), which exploits the speed difference between the speed of the current VSS and the station that has the maximum speed within the 3.2 km VSL distance from current VSS. In addition, this equation uses the travel time ($tt_{t,i}$) between 3.2 km from the VSS to consider the current traffic condition. Lastly, VCA is calculated using travel time from the VSS and the speed difference, which is shown in Equation (16). u denotes the speed, i denotes the station number, t denotes the time, a denotes the acceleration, d denotes the distance, and γ denotes the limited station. Equation (17) decides the limit of stations. Since the length of the VSL is limited to 3.2 km, if distance between current VSS and the station is more than 3.2 km, subsequent stations are not applied to the VCA calculation.

$$\alpha_{t,i}(^{km}/_{h^2},\ \text{VCA}) = \frac{MAX\left(u_{t,i},\ u_{t,i-1}, \cdots u_{t,i-\gamma}\right) - u_{t,i}}{tt_{t,\ i\ to\ i-\gamma}} \quad (16)$$

$$\gamma = i,\ \left|d_{t,i} - d_{t,vss}\right| > 3.2\ km \quad (17)$$

Based on VCA, the speed of each DMS connected to the VSS is calculated using Equation (18). To calculate the VSS speed, VCA($a_{t,i}$) and the distance between VSS and DMS are used, and the final speed of each DMS can be achieved. u denotes the speed, i denotes the station number, t denotes the time, $a_{t,i}$ denotes the VCA, q denotes the DMS, and L denotes the distance between VSS and DMS.

$$u_{t,q} = \sqrt{u_{t,VSS}{}^2 + 2 \times \alpha_{t,i} \times L} \quad (18)$$

The VSL displayed on the DMS has the following restrictions: the minimum speed close to the congested section is 40 km and the maximum speed is 80 km (at this value, the vehicle begins to reduce its speed). Since the distance between the DMS locations is included in the calculation, the speed at the DMS location is displayed at 5 km/h intervals following the definition of the Korea Expressway Corporation [17].

4. Experimental Environment for Microscopic Simulation

The proposed algorithm was tested on sections of an outer ring expressway located in the south of Seoul, Republic of Korea. This is a 25-km section located between Byeollae-IC and West-Hanan-IC (Figure 12) with seven interchanges and two intersections connected to two freeways. This includes a total of 23 stations, which are installed every 400 m to 2200 m and have the vision detections to obtain the vehicle information in each lane, as shown in Figure 12. In the experiment, we used the Verkehr In Städten—SIMulationsmodell 6 (VISSIM 6, Karlsruhe, Germany, 2013) simulator supported by the Planung Transport Verkehr (PTV) Corporation [18], which is a micro-simulator that can be implemented and tested for various road environments. To control the traffic flow and VISSIM simulator, we implemented the Korea highway traffic analysis (KHTA) system, which controlled VSSIM and provided similar real road conditions in our experiments [19]. After the simulation was complete, the performance of the VSL algorithm was evaluated using performance metrics. The experiment was executed using the time zone suffering the most severe traffic congestion (5 AM to 11 AM). In addition, to simulate the pattern of various vehicles, five types of random seeds (Seed 10, 15, 17, 20, and 25) were used to create vehicle patterns. In the simulation, open traffic data provided by the Korea Expressway Corporation were used [20]. In addition, the system was calibrated to tailor it for similar real road environments using the method proposed by Berkeley using the changing vehicle-following behavior value [21].

VISSIM provides the various calibration parameters to control the vehicle behavior and regulation on the road for the simulation. The calibration of parameters is the most important during the simulation process because the vehicles in the simulator behave differently depending on the calibration parameters. Therefore, it is necessary to adjust the vehicle behavior in the simulator similarly to the real traffic flow through the calibration of the simulator parameters. In Figure 12, the bottlenecks of the measurement section are Guri IC, Topyung IC, Gang-il IC, Sang-il IC, and Seo-hanam IC, which are interchanges that many vehicles enter from another urban freeway or highway.

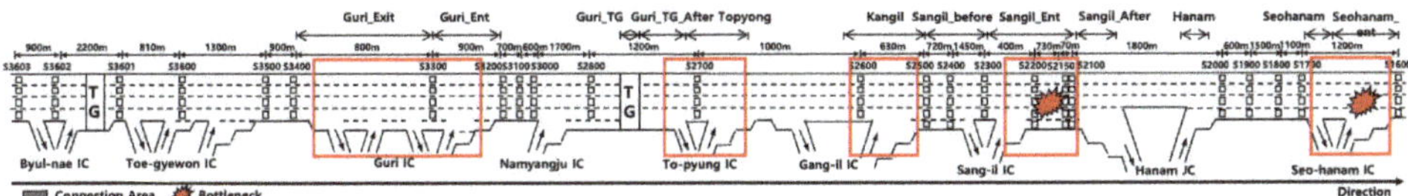

Figure 12. Schematic diagram showing the section of the road (Seoul Outer Ring Road) tested in the simulations.

Tables 1 and 2 are calibration parameters to make the bottlenecks similar to a real traffic road in the simulation environment. Table 1 has calibration parameters for traffic flow control based on the Wiedemann model, which has a model version released in 1974 and 1999 [22,23]. These models have 10 types of calibration parameters, which are named Calibration and Comparison (CC) and classified by distance, speed, acceleration value, and more. We adjusted only three parameters to adjust the traffic flow, which are CC1 (Headway Time), CC8 (Standstill Acceleration), and CC9 (Acceleration at 80 km/h). CC1 is desired headway time between the lead and following vehicles. CC8 is the desired acceleration from standstill. CC9 is the desired acceleration at 80.45 km/h. Since these are key parameters to affect the traffic flow and make a bottleneck, we use only them to adjust traffic flow of the simulator environment.

Table 1. Calibration parameters (CC1, CC8, CC9) for Traffic Flow Control of Verkehr In Städten - SIMulationsmodell (VISSIM).

Parameter for Each Station			
Link Type	**CC1 Headway Time**	**CC8 Standstill Acceleration**	**CC9 Acceleration At 80 km/h**
Freeway	1.0 s	3.5 m/s^2	1.5 m/s^2
Guri Exit	1.2 s	1.28 m/s^2	1.5 m/s^2
Guri Entrance	1.85 s	2.28 m/s^2	1.5 m/s^2
Guri Toll Gate (TG)	0.5 s	3.28 m/s^2	1.8 m/s^2
Guri TG After	1.0 s	4.19 m/s^2	1.5 m/s^2
Topyong	2.3 s	1.85 m/s^2	1.0 m/s^2
Kangil	1.75 s	2.0 m/s^2	1.5 m/s^2
Sangil Before	1.0 s	3.19 m/s^2	1.5 m/s^2
Sangil Entrance	2.4 s	2.28 m/s^2	1.19 m/s^2
Sangil After	1.8 s	2.58 m/s^2	1.19 m/s^2
Hanam	1.8 s	3.19 m/s^2	1.5 m/s^2
Seohanam	2.5 s	1.0 m/s^2	0.95 m/s^2
Seohanam Entrance	2.5 s	1.5 m/s^2	1.0 m/s^2

Table 2. Calibration parameters for lane change control of VISSIM.

Parameters for Lane Change			
Necessary Lane Change	**Maximum Deceleration**	**−1 m/s^2 per Distance**	**Accepted Deceleration**
Own	−3.0 m/s^2	77.33 m	−0.91 m/s^2
Trailing Vehicle	−3.25 m/s^2	70 m	−1.19 m/s^2
Wait Time Before Diffusion		Minimum Headway	
23.13 s		0.09 m	

Table 2 includes calibration parameters to adjust a behavior method for the lane change of the vehicle in a simulator. The ego vehicle and trailing vehicle is affected from the Own vehicle. Maximum deceleration is max deceleration of more than two vehicles and −1 m/s^2 per distance is preparing distance to reduce the speed of each vehicle. An accepted deceleration is the minimum deceleration of each vehicle. Above parameter values are already measured from Shin at al. [24], which simulates the same simulation section. Therefore, we fully follow the measurement values of the calibration parameters obtained above.

5. Performance Evaluation

5.1. VSS Stability Assessment

A major advantage of VSL is its fast response to sudden events, such as road repair works and vehicle accidents. Since the VSL system enhances road safety, it is an important solution for a traffic environment. In addition, it aims to reduce the total travel time through traffic flow control. Existing RVSL algorithms provide the VSL system, which focuses on traffic control between multiple stations in the freeway. However, they select and use the wrong VSS, which is in the wrong place due to the VSS selection error mentioned in Section 2.2. Hence, this wrong VSS occurs with various errors of the whole VSL system, which cannot control and worsen the traffic flow.

Since the new EVSL was developed to improve accuracy of the VSS selection and reduce propagation of errors in the total VSL system, we compared the VSS detection results using both the JVSL and the EVSL algorithms, as shown in Table 3. Total simulation time is about 6 h at each random seed, and the error rates in the selection of the VSS location were measured at 30 s intervals. The error rates were calculated as the VSS locations ±2 stations from the position where the correct VSS should be located. The location of correct VSS was set manually by the human before re-simulation.

Table 3. VSL start station (VSS) decision error result.

Random Seed	JVSL			EVSL		
	Total VSL	Error	Error Rate	Total VSL	Error	Error Rate
	(30 s Interval)		(Percent)	(30 s Interval)		(Percent)
SEED 10	505	112	22.2%	555	1	0.2%
SEED 15	512	121	23.6%	571	0	0.0%
SEED 17	488	97	19.9%	532	0	0.4%
SEED 20	511	109	21.3%	540	1	0.2%
SEED 25	503	85	16.9%	545	0	0.0%
Average	504	105	21%	549	1	0.1%

In Table 3, the numeric unit of the Total VSL and Error is 'tick,' which has an interval of 30 s. Total VSL refers to the VSL working time among total time (720 ticks) and the error is the error issued time. The experimental results for the five different random seeds used are shown. It can be seen that the JVSL algorithm showed a VSS decision error around every 1 h (105 ticks/30 s), corresponding to a high error rate of 21%, which has a negative effect on the calculation of the final VSL (in addition to the traffic volume and accident risk). In contrast to the JVSL algorithm, the EVSL algorithm had a very low error rate below 0.5% due to the selection of the correct VSS and suitable VSL. Hence, it is expected that the new algorithm would enhance traffic safety.

The continuity of the VSL system is important, as discontinuities in the system can aggravate traffic problems rather than improve flow. Figure 13 shows the variation in the selected VSS location among all stations on the road at 5-min intervals for both the JVSL and our EVSL algorithm, where the continuity of VSS is represented by a line graph. A station number of 0 means that the VSL is not operative. Figure 13a shows the results using JVSL, where many instances of zero values occurred during operation since the VSS was not selected within the required time and the algorithm was stopped due to the wrong algorithm of the VSS calculation. In contrast, Figure 13b shows the results of the EVSL algorithm, where good continuity was observed. Hence, this system is expected to provide stable VSL control in practical situations.

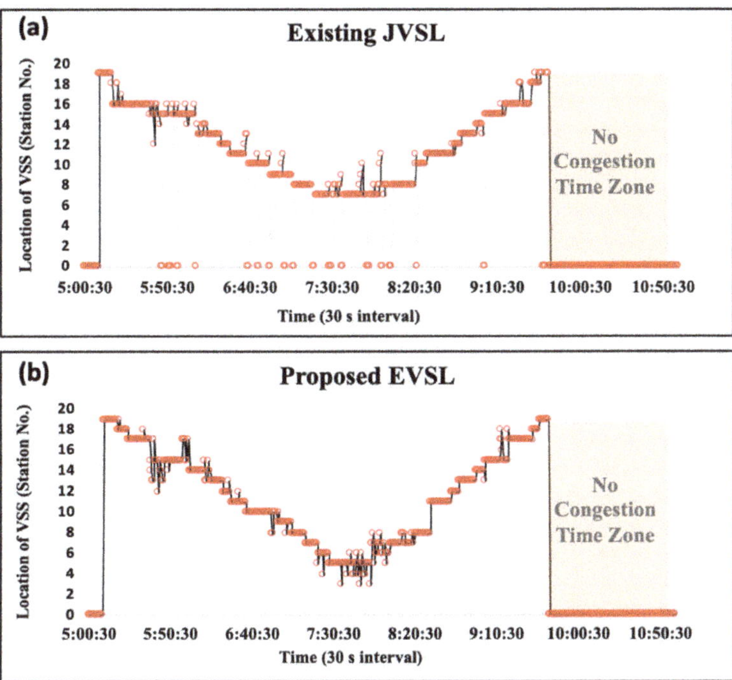

Figure 13. Comparison of the VSL start station (VSS) continuity for the (**a**) existing JVSL and (**b**) EVSL algorithm. The x axis is the time (30 s interval). The y axis is the location of VSS. The number is the station number on the road. Zero (0) with dot line means there is no VSS station on the road.

5.2. Speed Control Stability Assessment

The wide distribution of speed can result in large variations in speed, which increases the risk of accidents on the freeway. The VSL system can keep the road safety by controlling the speed in the sections with the highest speed variations. To evaluate the safety of the proposed EVSL, we measured the speed variance on the tested freeway, as shown in Figures 14 and 15. To compare accurate situations applying the no VSL, JVSL, and EVSL, we experienced speed variance on the corresponding stations for the period from the beginning time to the ending time of the simulation.

As mentioned in Section 2.2, the JVSL has a problem to select the wrong congestion area. In addition, since the JVSL determines the distance to apply a VSL algorithm to freeway by the number of stations, instead of the station distance, it applies the VSL to longer sections of freeway than necessary. Hence, since the JVSL provides the wrong speed value at the wrong place as mentioned above, the speed variance of the JVSL is higher speed variance than that in the case of no VSL and EVSL being applied over the time period of 5:45 AM to 6:45 AM when the traffic congestion starts. The new EVSL algorithm showed slightly lower speed variance results than others over the same time period with proper congestion area selection and precise speed control. In the time period of 7:00 AM to 7:35 AM, the speed variance of the EVSL is higher than the JVSL because the EVSL has a traffic congestion later than the JVSL.

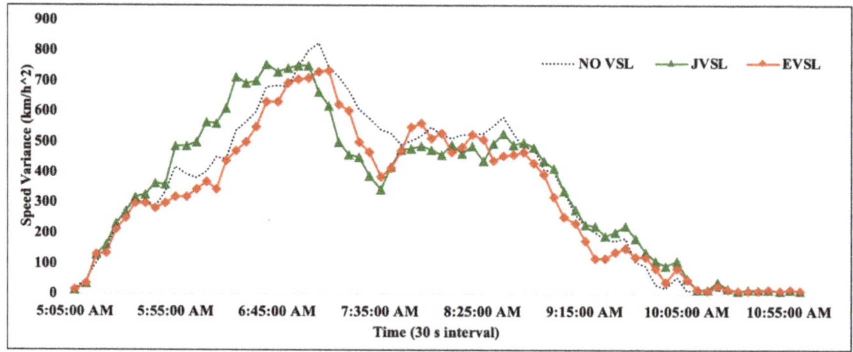

Figure 14. Speed variance over time. The x axis is the time (30 s interval). The y axis is the speed variance (km/h^2).

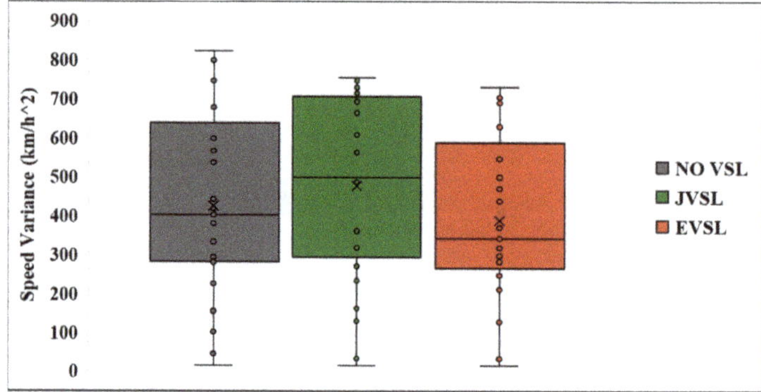

Figure 15. Variation analysis of total speed variance.

Figure 15 shows the comprehensive analysis of total speed variance over the time period of 5 AM to 11 AM. We can see that the JVSL shows a relatively high rate of speed variance in a wide distribution. On the other hand, since the EVSL shows a low rate of speed variance and narrower distribution than others, it proves that the EVSL performs more stable VSL calculations than others and controls traffic conditions more stably.

Figures 16 and 17 shows the average and max deceleration chart over the time period of 5 AM to 12 AM, which compares the existing JVSL, new EVSL, and no VSL. The deceleration, which is measured to evaluate the stability of traffic flow, is calculated by a difference between stations. We can expect that a high deceleration between stations means the vehicles are entering to the congestion area quickly and the drivers suddenly break a lot due to unstable traffic conditions. We prove the EVSL algorithm decrease the average deceleration of congestion area over the time period between 5:45 AM and 6:45 AM (the time when the congestion begins), as shown in Figures 16 and 18A. Especially, the EVSL algorithm shows the lower max deceleration than others from 5:45 AM to 8:00 AM, as shown in Figures 17 and 18B, because it sets the VSS to the correct location, which is the tail of congestion area and controls the vehicle speed through a correct calculation of the VSL speed.

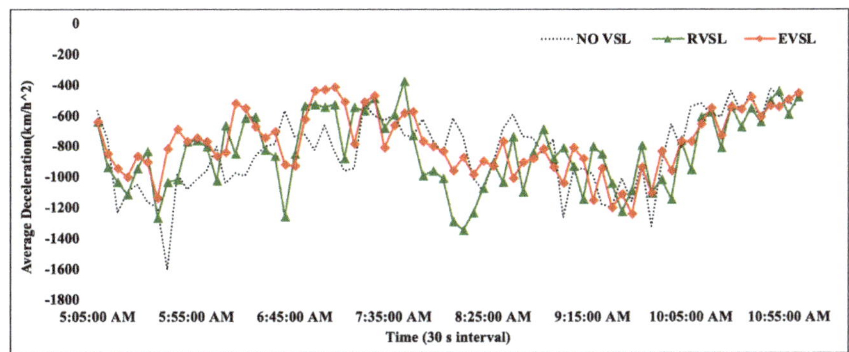

Figure 16. Average deceleration comparison between No variable speed limit (VSL), real-time VSL (RVSL), and enhanced VSL (EVSL). The x-axis is the time (30 s intervals). The y-axis is average deceleration (km/h^2).

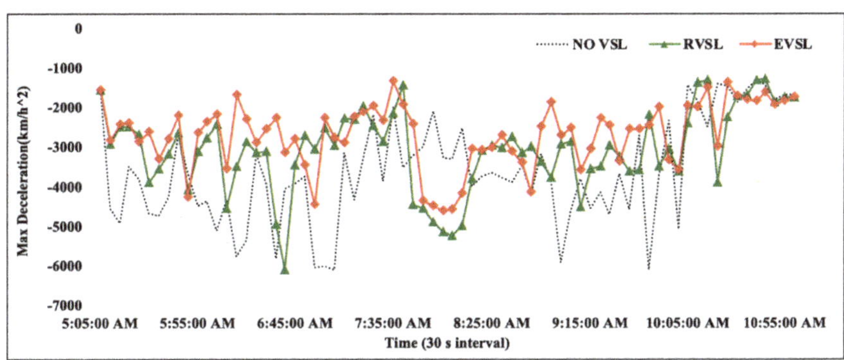

Figure 17. Max deceleration comparison between No VSL, RVSL, and EVSL. The x-axis is the time (30 s intervals). The y-axis is max deceleration (km/h^2).

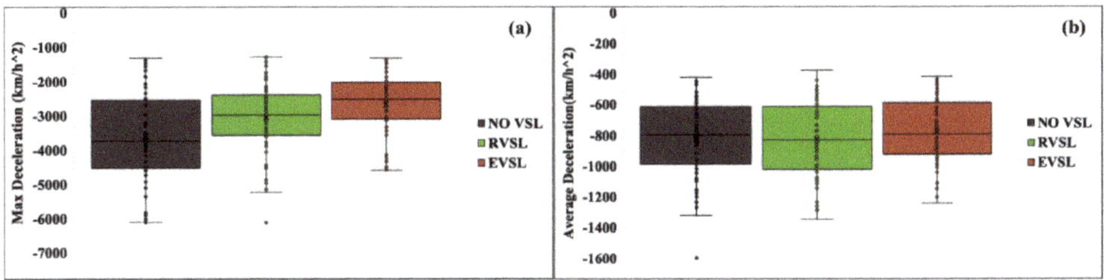

Figure 18. Max (**a**) and average (**b**) deceleration comparison between RVSL and EVSL.

5.3. Total Travel Time and Shockwave

Traffic control algorithms focus on the total travel time from the start point of the vehicle to the destination. The VSL system reduces the speed of vehicles before a congestion area in order to enhance road safety. However, if the VSL is applied over an excessively wide area, the real travel time could be increased unnecessarily without improving safety. The JVSL algorithm can significantly affect the total travel time because of the risk of operating irregular VSL due to a VSS decision error. Figures 19 and 20 compare the total travel times using the JVSL and the EVSL algorithm to the case of not using a VSL system. The JVSL resulted in a longer travel time (an increase of 1–3 min) compared to no VSL system. However, the EVSL algorithm showed a total travel time like no VSL.

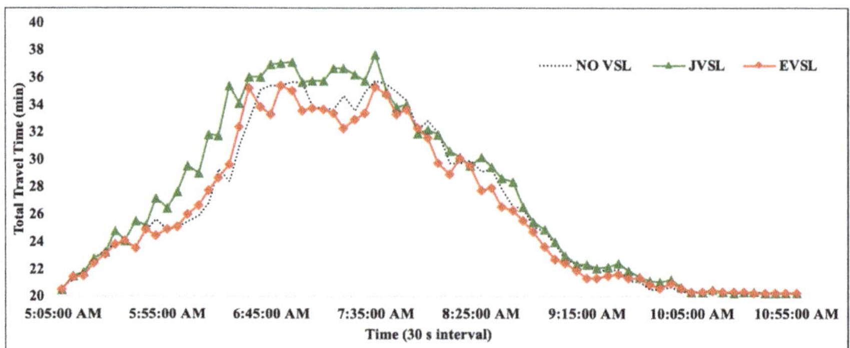

Figure 19. Total travel times for different traffic control algorithms. The x-axis is the time (30 s interval). The y-axis is the total travel time (min).

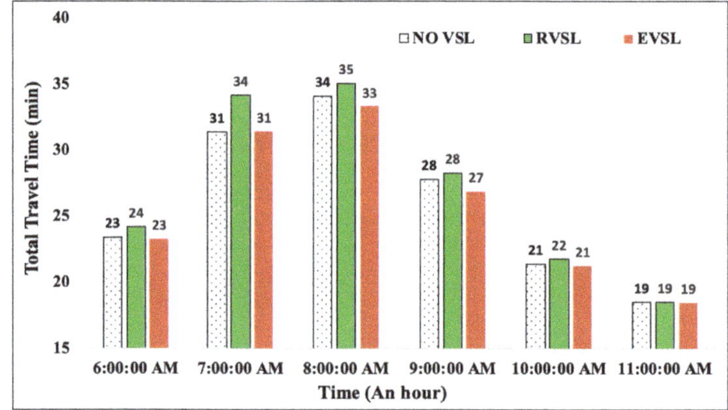

Figure 20. Total travel times for different traffic control algorithms by an hourly interval.

It has been proposed that the use of JVSL systems reduces the shockwave phenomenon in congestion areas. However, this conclusion was based on studies of short sections (1–2 km) of road. In the case of longer sections (20–30 km), there exists insufficient research since there are too many related variables and controlled sections to analyze. As shown in Figure 20, when compared to the non-VSL case, the new algorithm shows a reduced travel time (1–2 min for the peak time between 6:00 AM and 8:00 AM). This suggests that the shockwave phenomenon is alleviated. This phenomenon is continued up to the section of the traffic congestion. Since the new VSL algorithm operated stably with an accurate VSS, it is expected to reduce the risk of traffic accidents and the shockwave phenomenon.

6. Conclusions and Future Work

A new EVSL algorithm was proposed to overcome the limitations of the existing JVSL system and improve safety on freeways. The new algorithm accurately locates the VSS position, which enables it to correctly identify the sections of road to apply the required VSL. In addition, as based on the DMS distance, it effectively established the VSL range for traffic congested sections. Simulation results compared the JVSL system with the new algorithm. The existing algorithm showed many limitations due to its inaccurate selection of the VSS location, which resulted in aggravation of traffic congestion. In particular, analysis of speed control stability assessment and total travel time showed that the JVSL applied incorrect VSL operations in specific sections. Hence, the speed variance of the JVSL is higher speed variance for about 12% than that in the case of no VSL and EVSL being applied over the time period of 5:45 AM to 6:45 AM where the traffic congestion starts. A delay in the total travel time of an average of 1–2 min was observed comparing with no VSL. In contrast, the proposed EVSL algorithm produced an accurate VSS location and VSL value, decreasing the speed variance by 9% and the total travel time by 1–2 min when compared to the case of no VSL system. Hence, the proposed EVSL is expected to reduce the risk of accidents and alleviate shockwave phenomenon in the traffic congestion area. Lastly, reduced shockwaves also decreased the total travel time. Our algorithm can enhance freeway safety, along with traffic efficiency.

We also took advantage of the data provided by Mn/DoT to obtain the threshold of the speed and acceleration for pVSS finding. They provide detailed distribution maps of speed and acceleration in the event of an incident. However, a threshold based on domestic data is needed to find the correct pVSS because the EVSL system is designed based on Korean freeways. Therefore, in a future work, we have a plan to collect incident data based on domestic data and the tail portion data of the congestion area based on the EVSL algorithm in operation in Korea. Based on the collected data, future studies will determine the appropriate threshold of speed and acceleration for domestic roads.

In addition, we will study a traffic control system where VSL is integrated with ramp metering where vehicles entering the freeway are controlled, as this is expected to provide a more effective traffic control solution.

Author Contributions: S.J. performed the research and wrote the paper. C.P. & D.S. provided guidance for the research and revised the paper. All authors have read and agreed to the published version of the manuscript.

Funding: The study was funded by the Korea government (MSIT: Ministry of Science and ICT) (NRF-2020R1G1A1010166).

Acknowledgments: The National Research Foundation of Korea(NRF) grant supported this work.

Conflicts of Interest: The authors declare no conflict of interest.

References

1. Qureshi, K.N.; Abdullah, A.H. A survey on intelligent transportation systems. *Middle-East J. Sci. Res.* **2013**, *15*, 629–642.
2. Figueiredo, L.; Jesus, I.; Machado, J.T.; Ferreira, J.R.; De Carvalho, J.M. Towards the development of intelligent transportation systems. In Proceedings of the 2001 IEEE Intelligent Transportation Systems, Oakland, CA, USA, 25–29 August 2001; pp. 1206–1211.
3. Man, I.T.K.; Engineer, C.T. Intelligent Transport Systems. In Proceedings of the Better air Quality Motor Vehicle Control & Technology Workshop, Hongkong, 18–20 September 2000.
4. Lee, C.; Hellinga, B.; Saccomanno, F. Evaluation of variable speed limits to improve traffic safety. *Transp. Res. Part C Emerg. Technol.* **2006**, *14*, 213–228. [CrossRef]
5. Allaby, P.; Hellinga, B.; Bullock, M. Variable speed limits: Safety and operational impacts of a candidate control strategy for freeway applications. *EEE Trans. Intell. Transp. Syst.* **2007**, *8*, 671–680. [CrossRef]
6. Rämä, P. Effects of weather-controlled variable speed limits and warning signs on driver behavior. *Transp. Res. Rec.* **1999**, *1689*, 53–59. [CrossRef]
7. Jo, Y.; Choi, H.; Jeon, S.; Jung, I. Variable speed limit to improve safety near traffic congestion on urban freeways. In Proceedings of the 2012 IEEE International Conference on Information Science and Technology, Wuhan, Hubei, China, 23–25 March 2012.
8. Wang, M.; Daamen, W.; Hoogendoorn, S.P.; van Arem, B. Connected variable speed limits control and car-following control with vehicle-infrastructure communication to resolve stop-and-go waves. *J. Intell. Transp. Syst.* **2016**, *20*, 559–572. [CrossRef]
9. Hegyi, A.; De Schutter, B.; Hellendoorn, H. Model predictive control for optimal coordination of ramp metering and variable speed limits. *Transp. Res. Part C Emerg. Technol.* **2005**, *13*, 185–209. [CrossRef]
10. Frejo, J.R.D.; Papamichail, I.; Papageorgiou, M.; De Schutter, B. Macroscopic modeling of variable speed limits on freeways. *Transp. Res. Part C Emerg. Technol.* **2019**, *100*, 15–33. [CrossRef]
11. Jeon, S.; Jung, I. Coordinated Ramp Metering for Minimum Waiting Time and Limited Ramp Storage. *IEICE Trans. Fundam. Electron. Commun. Comput. Sci.* **2016**, *99*, 1843–1855. [CrossRef]

12. Immers, L.H.; Logghe, S. *Traffic Flow Theory*; Faculty of Engineering, Department of Civil Engineering, Section Traffic and Infrastructure: Kasteelpark, Arenberg, Germany, 2002.
13. Chen, C.; Petty, K.; Skabardonis, A.; Varaiya, P.; Jia, Z. Freeway performance measurement system: mining loop detector data. *Transp. Res. Rec.* **2001**, *1748*, 96–102. [CrossRef]
14. van Erp, P.B.; Knoop, V.L.; Hoogendoorn, S.P. Estimating the vehicle accumulation: Data-fusion of loop-detector flow and floating car speed data. In Proceedings of the 97th Transportation Research Board Annual Meeting, Washington, DC, USA, 7–11 January 2018.
15. Mathew, T.V.; Rao, K.V.K. *Introduction to Transportation Engineering*; Indian Institute of Technology: Bombay, India, 2006.
16. Kwon, E.; Park, C. *Development of Freeway Operational Strategies with IRIS-in-Loop Simulation*; Minnesota Department of Transportation: Saint Paul, MN, USA, 2012.
17. Korea Highway Corridor Information, Korea Expressway Corporation, Gimcheon. 2016. Available online: http://data.ex.co.kr (accessed on 24 March 2020).
18. PTV. *VISSIM Version 5.2 Manual, Innovative Transportation Concepts*; PTV Planung Transport Verkehr AG: Karlsruhe, Germany, 2009.
19. Han, Y.T.; Jeon, S.B.; Shin, S.J.; Seo, D.M.; Jung, I.B. Traffic Analysis and Simulation System for Korea Highway. *KIISE Trans. Comput. Pract.* **2016**, *22*, 426–440. [CrossRef]
20. Korea Highway Open Data, Korea Expressway Corporation. Available online: http://data.ex.co.kr (accessed on 24 March 2020).
21. Gomes, G.; May, A.; Horowitz, R. Congested freeway microsimulation model using VISSIM. *Transp. Res. Rec.* **2004**, *1876*, 71–81. [CrossRef]
22. Higgs, B.; Abbas, M.; Medina, A. Analysis of the Wiedemann car following model over different speeds using naturalistic data. In Proceedings of the Procedia of RSS Conference, Indianapolis, IN, USA, 1–22 September 2011.
23. Fellendorf, M.; Vortisch, P. Validation of the microscopic traffic flow model VISSIM in different real-world situations. In Proceedings of the 80th Annual Meeting of the Transportation Research Board, Washington, DC, USA, 7–11 January 2001.
24. Shin, S.J.; Lee, C.S.; Han, Y.T.; Jeon, S.B.; Seo, D.M.; Jung, I.B. Calibration for Simulating a ITS Algorithm in Korea Highway. In Proceedings of the Korea Information Processing Society Conference, Busan, Korea, 4–5 November 2016; pp. 443–446.

Communication

Floating Car Data Adaptive Traffic Signals: A Description of the First Real-Time Experiment with "Connected" Vehicles

Vittorio Astarita *, Vincenzo Pasquale Giofré, Demetrio Carmine Festa, Giuseppe Guido and Alessandro Vitale

Department of Civil Engineering, University of Calabria, Via Bucci Cubo 46/B, 87036 Rende, Italy; vincenzo.giofre@unical.it (V.P.G.); dc.festa@unical.it (D.C.F.); giuseppe.guido@unical.it (G.G.); alessandro.vitale@unical.it (A.V.)

* Correspondence: vittorio.astarita@unical.it

Received: 11 December 2019; Accepted: 1 January 2020; Published: 7 January 2020

Abstract: The future of traffic management will be based on "connected" and "autonomous" vehicles. With connected vehicles it is possible to gather real-time information. The main potential application of this information is in real-time adaptive traffic signal control. Despite the feasibility of using Floating Car Data (FCD), for signal control, there have been practically no real experiments with all "connected" vehicles to regulate traffic signals in real-time. Most of the research in this field has been carried out with simulations. The purpose of this study is to present a dedicated system that was implemented in the first experiment of an FCD-based adaptive traffic signal. For the first time in the history of traffic management, a traffic signal has been regulated in real time with real "connected" vehicles. This paper describes the entire path of software and system development that has allowed us to make the steps from just simulation test to a real on-field implementation. Results of the experiments carried out with the presented system prove the feasibility of FCD adaptive traffic signals with commonly-used technologies and also establishes a test-bed that may help others to develop better regulation algorithms for these kinds of new "connected" intersections.

Keywords: adaptive traffic signals; Intelligent Transportation Systems; Floating Car Data; traffic management; connected and autonomous vehicles

---

1. Introduction

Connected vehicles will become an important element of the forthcoming Internet of Things (IoT) and of Intelligent Transportation Systems (ITS) [1]. The various sensors that "connected" vehicles will carry will bring an enormous amount of data that could be useful for real-time traffic regulation.

Connected vehicles may help to solve traffic congestion problems which were hardly solved with traditional methodologies. Traditional traffic engineering practice was based on attempts to shift demand on transit systems [2] and on better road traffic control by adopting tools such as: traffic simulation [3–8] dynamic network loading equilibrium and dynamic models [9–12] and the study and attempt to affect user route choice [13–17].

All these methodologies, that often do not require the use of advanced technologies, were applied by road administrations with varying results. Some city administrations (especially in Italy) do not always deal properly with the task of providing well-adjusted traffic signal settings. Traffic signal regulation very often is not real-time adjusted and is very badly performed with outdated fixed time traffic signal settings. This can cause traffic congestion that is a serious problem in cities and also a great cause of air pollution [18–21].

The introduction of "connected" vehicles and the use of spread out sensors data could help city administrations to manage traffic in cities better and launch a new generation of adaptive traffic signals which can be controlled on the base of real-time traffic data obtained directly from "connected" probe vehicles.

Many scientific works have been presented where "connected" vehicles act as floating probe vehicles. These systems are usually defined as based on: "Floating Car Data" (FCD). First FCD-based papers have proposed the use of specific Radio-frequency identification (RFID) systems [22,23]. Other papers have proposed the use of dedicated local wireless networks using vehicle to infrastructure (V2I) communication data [24]. Smartphones also have been used in different applications to connect vehicles [25,26] to infrastructures.

The use of the existing wireless mobile phone networks could prove to be cost efficient since ad hoc developed wireless systems could have to bear sensibly higher costs. For this reason, and since smart phones are so pervasive, some scientific works have explored the use of a smart phone as a means to obtain FCD.

The concept of FCD obtained by counting the flow of smartphones on vehicles was anticipated in the patent [27] and was implemented starting in 2008 by the use of Bluetooth protocol as a means to detect the smartphones. Some scientific works were published on Bluetooth as a means of gathering FCD [28–32] and the methodology is currently still applied in many cities.

The first works on smartphone-based FCD did not take advantage of satellite positioning systems: [33–37]. Subsequently, other scientific works were based on using Global Navigation Satellite Systems (GNSS) which nowadays are available on every mobile phone.

Today, mobile phones can act as sensors and transmitters; and with them it is possible to estimate travel times [38,39], speeds [40,41], assess safety [42,43], give assistance in safer driving [44,45], estimate fuel consumption [46–49] measure road pavement quality and recognize specific problems [50–52] and also model route choice behavior [53].

In [54] buses are used as FCD and in [55] low frequency FCD are applied to estimate traffic conditions. Some papers have also explored the use of FCD data in traffic safety to evaluate the characteristics of different driving styles and for real-time incident detection: [43,56,57].

The satellite localization system of smartphones was also used to obtain vehicle trajectories and estimate the traffic signal timings in real-time. The idea is to assist the drivers to achieve a better driving approach at traffic signals with increased safety and reduced fuel consumption [58,59].

Many works have been presented on the specific issue of this paper: the use of FCD data coming from smartphone to regulate adaptive traffic signals in real-time [60–69]. In some cases, the adaptive traffic signal system is used in combination with driver assistance, so that both driver behavior and traffic signals are real-time adjusted.

In all the above-cited studies on adaptive traffic signals there is no real experimental implementation with real vehicles. Practically all preceding FCD studies for adaptive signals were simulation-based. In this paper we intend to illustrate the first case of FCD-based adaptive traffic signal implemented in real-time in the field with real vehicles.

This paper presents both the software and the hardware development of a dedicated system for traffic signal real time regulation based on data coming from "connected" vehicles.

The steps that were taken to obtain the experimental results that are described below were:

- Development of a simulation platform for the evaluation and development of control algorithms.
- Development of a connected traffic signal.
- Development of a specific smart- phone application.
- Development of a specific traffic signal control server.
- Planning and mapping of the experimental site.
- Conclusive implementation of all the different parts of the system in the final experiment.

As a result of these steps a three-legs intersection was regulated in real-time on the basis of FCD data. This paper describes this first experiment and the controlling system.

Many research activities can be stimulated by this paper, yet we believe that the more important contribution is in the detailed description of all the steps and concepts that allowed carrying out the first successful experiment, with a traffic signal regulated by FCD of 100% of connected vehicles, since this information can be very valuable to reproduce our results in other situations.

New experiments can, in fact, involve more experts to accept the feasibility of this concept, to grasp the advantages of FCD adaptive traffic signals and develop better systems and algorithms for these kinds of new "connected" intersections.

2. Materials and Methods

2.1. The Proposed Cooperative FCD-Based Adaptive Traffic Signal System

In many countries the traffic signal systems that are adopted are low-technology fixed-time control units, where the signal timing is preset according to historic traffic counts. More advanced countries deploy high-priced real-time traffic signals that apply a real-time control technology based on data coming from sensors embedded in the road. Sensors installation is time-consuming and expensive, moreover, sensors have a very high maintenance cost.

The proposed system relies on a mobile phone application that can transform any vehicle into a "connected" vehicle. Traditional sensors use magnetic loop detector measurements to obtain traffic flow counts. In our system information on traffic flows is not directly obtained, instead it is extrapolated from "connected" vehicle trajectories. The data obtained is different and to some extent more detailed since from vehicle trajectories it is possible to evaluate directly the queue length for every maneuver at the intersection [70].

Moreover, the presented system can be considered a cooperative intelligent transportation system (one of the first being actually implemented in the field) since it requires the cooperation of connected vehicles in communicating their position in real time. The following issues regarding cooperation and competition have been thoroughly discussed in [63]:

- In traffic signal regulation there is competition between drivers coming from different approaches of the same intersection, they compete for the available green time. In FCDATS there is also competition between "connected" vehicles and traditional vehicles, "connected" vehicles in fact could receive more green time at the expense of not "connected" vehicles that cannot be considered by FCDATS systems.
- Cooperation is naturally present in FCDATS since a "connected" vehicle sends useful information on the current state of the traffic network. This information could advantage also other vehicles which are not connected. The findings of previous works [63,64,66] based on simulation showed that when a percentage of 30% is reached all vehicles take advantage from the shared information coming from "connected" vehicles.

For these two issues FCDATS could be defined as "coopetitive" systems since both competitive and cooperative aspects have a great importance. Drivers have the choice of participating in the system by "connecting" the driven vehicle or to stay out of the system and take advantage of other drivers that are connected. This kind of issue has a great importance when the system rewards "connected" vehicles or when there are costs to go through to participate in the system (green light fees in a pay-per-green scheme or, in any case, the costs of acquiring new equipment or a vehicle that can connect to the system).

All issues regarding competition-cooperation in FCDATS can be studied in simulation. In our experiment all the vehicles were connected since the scope was that of proving the feasibility and ease of implementation of these systems and the consistency of simulated results with real on-field results.

The general structure of the experimented system is presented in Figure 1, where the following physical parts are depicted:

- the traffic lights of the intersection that are actuated by a local controller;
- the local electronic controller (Raspberry microcomputer) that is connected on the internet and that acts as a web server regulating the traffic signals at request;
- a central server that collects data from "connected" vehicles" and establishes the best traffic signal cycle according to data processing and the algorithm that are described in the following;
- the "connected" vehicles with a smartphone application that receive data relative to positions and speeds from the GNSS system and transmit them to the central server (on the common local wireless phone data network).

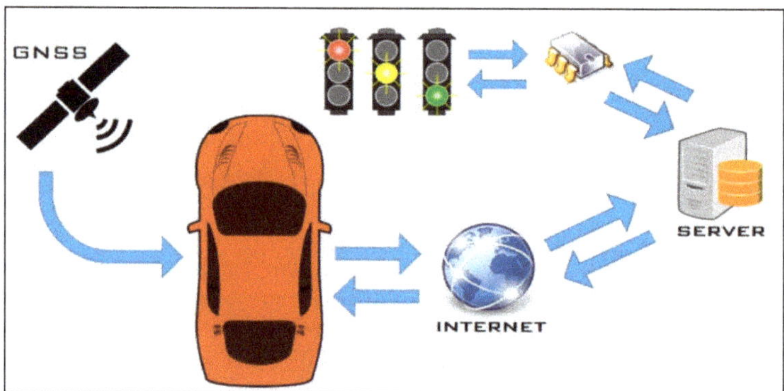

Figure 1. General structure of the proposed system.

2.2. *Development of a Simulation Platform and a Dedicated ITS Laboratory*

A complete simulation platform had to be developed to test the system offline. The simulation platform allowed testing algorithms and understanding the behavior of the proposed system in a simulated reality.

All previous studies (cited in the introduction) on adaptive traffic signals based on FCD have been developed only in simulation without having in mind a real deployment in the field. Moving from simulation to a real implementation in the field requires including, in the simulation platform, all the details that are often not considered in academic works.

The main novelty introduced in our simulation platform is the module that simulates GNSS errors. Simulation of GNSS errors has been introduced in the paper [71] and at the moment there are no other academic works on this issue. This issue and other details in simulation are very important when the objective is that of a final real implementation in the field. A control algorithm that could work in a simulation environment bringing perfect results could possibly fail in a real implementation as shown in the following pages.

To test our system we had previously to implement and develop specifically our dedicated microsimulation model TRITONE [72–75], since other traffic microsimulation packages do not have the possibility of easily implementing FCDATS control algorithms and there is no simulation package where the GNSS error is simulated (a possibility for the researcher who wants to replicate our results without using TRITONE is that of using SUMO [76], which is open source, adding necessary modules as described in the following).

The microsimulation TRITONE was the core of our simulation platform and one of the two modules of our ITS laboratory. The ITS laboratory was designed with two physically separated computers (having in mind the final implementation in the field):

- a first computer that was host for the TRITONE package combined with the GNSS error simulator. These two softwares together created a virtual reality environment based on microsimulation;

- a second computer that was designed to be the central signal control server which was designed to regulate traffic signals from vehicles trajectories. The central server was structured so that it would work, in real time, both in simulation and connected with real traffic signals.

The whole laboratory structure is depicted in Figure 2, more details on the microsimulation software TRITONE can be found in the above-cited works.

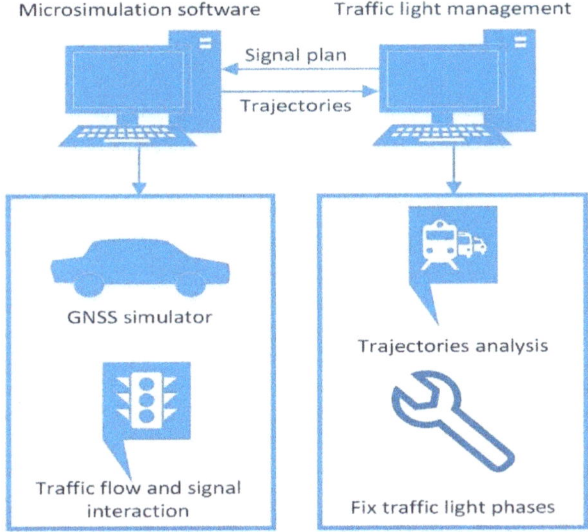

Figure 2. Intelligent Transportation Systems (ITS) laboratory for traffic signal algorithms testing and development.

The GNSS error simulator module is based on the results of experimental surveys on the field with GPS receivers on common smartphones [71]. GNSS device error (GPS) was found to be correlated to different causes but the most important was the occlusion of the sky caused by buildings. For this reason, we differentiated three different scenarios:

- Scenario type A, where the satellite localization is almost perfect as the sky is almost completely clear.
- Scenario type B, where moderate urbanization can create sight disturbances that can partially reduce satellite signals (buildings on only one side of the street).
- Scenario type C, where urbanization creates disturbances with tall buildings at least 18 m on both sides of the street (urban canyon).

The experiment illustrated in this paper was conducted in a scenario which could be classified as being in all types of scenario: A, B and C since buildings were present (some higher than 18 m), but buildings were localized only on one side of the intersection. Taller buildings were localized at a distance of around 40 m from the central line of lanes; smaller buildings (5 m high) were localized at 10 m from the central line of lanes.

Following the procedure presented in [71] the error in the position of a vehicle is considered as the sum of two errors a distance error and an angle error. Distance errors are generated according to the Rayleigh distribution while the uniform distribution is used to generate angle errors. The first position on the network of every vehicle that is generated in the microsimulation is corrected adding these two errors:

The first distance error of every vehicle is generated according to the following distribution:

$$f(x)_{Rayleigh} = \frac{x}{\sigma^2} e^{-\frac{x^2}{2\sigma^2}} \quad (1)$$

where the variance is chosen among three different values corresponding to the three above indicated case scenarios.

The first angle error is generated according to a uniform distribution:

$$f(x)_{Uniform} = \frac{1}{n}, (n = 2\pi) \quad (2)$$

After the first error is generated when the vehicle appears in the network the procedure continues updating the distance error following a normal distribution with mean and standard deviation that are relative to one of the three scenarios and also relative to the error absolute value in the preceding time step (this is an empirical procedure that considers autocorrelation with little approximation). The experimental data showed, in fact, that autocorrelation translates into a different standard deviation of the error at time t+1 according to the absolute value of the error at time t.

A similar procedure is carried out for the angle simulation. In this case the autocorrelation is simply reproduced by using a normal distribution for each of the three scenarios. The angle error simulation completely follows theory with no approximation (this happens since the angle distribution is a uniform distribution).

With the indicated procedure distance and angle errors are added to the exact simulated position obtained from the simulation model. In the laboratory the trajectories that are generated in this way are transferred with internet protocol from the simulation computer to the traffic light (signal) management computer server. The server elaborates again trajectories with a map-matching algorithm. Elaborated position data are then fed into the adaptive traffic signal algorithm (FCDATS).

The traffic signal management server was used both in the laboratory testing phase and in the final experiment. The flow of information for the final experiment is depicted in Figure 3, where, in this case, the information is exchanged between the traffic signal management server and the external reality.

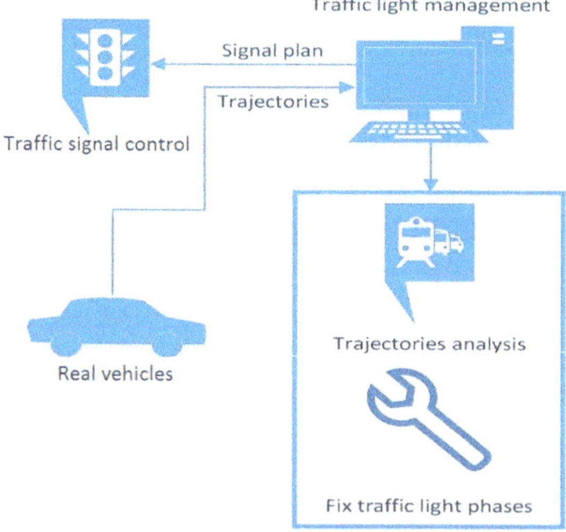

Figure 3. Traffic signal management in the on-field experiment.

2.3. Evaluation and Development of Control Algorithms

With the above described ITS laboratory some algorithms for FCDATS were tested on test networks. It was important to test algorithms in simulation since the control system can work on trajectories of "connected" vehicles and most of the literature on adaptive traffic signals is based on traditional traffic flow measures. Algorithms were tested on test networks to establish the advantages of the proposed system and the effects of different percentages of "connected" vehicles.

Mainly two control algorithms have been simulated in the ITS laboratory before testing them in the on-field experiment against a static optimized cycle:

- a greedy algorithm [66];
- an algorithm based on Nash bargaining equilibrium [63].

2.3.1. Greedy Algorithm

The first algorithm that was tested was a greedy algorithm that counts the number of "connected" vehicles that are present on every approach of the intersection and allocates green giving priority to the approach loaded with the greater number of vehicles. The position of the last "connected" vehicle is used to calculate the queue length and enough green time is assigned to let the queue dissipate considering also the other vehicles that may have arrived since the moment this last "connected" vehicle stopped. The estimation procedure is similar to that applied in [77].

In other words, in the proposed greedy algorithm connected vehicles are treated like priority vehicles. More details on the implementation of this greedy algorithm are in [66].

2.3.2. An Algorithm Based on Nash Bargaining Equilibrium

The Nash bargaining algorithm applied to traffic signal regulation was first developed in [78,79].

The bargaining problem has been investigated in game theory where there is some resource that is generated by reaching an agreement between different parties. The solution to the bargaining problem is how this resource must be divided among the agreeing parts. The bargaining game theory was introduced in regulating traffic signals on the assumption that the signal phases are like different players that need to find an agreement on how to split the total green time of the intersection. Phases will end up finding dynamical agreements which tends to an equilibrium point that is consistent with the Nash bargaining equilibrium solution.

The Nash bargaining equilibrium solution in traffic signal regulation represents a tendency for the system towards the agreement that rational players should reach. In general, it could be said that the Nash equilibrium approach is an axiomatic approach which simplifies the problem since there is no need to evaluate in detail the process of the game. The best agreement is the equilibrium point and all actions should be devoted to move the system toward that point. More details on the implementation of this algorithm can be found in [63].

2.3.3. Fallback on Fixed Cycle, Green Requirements

With both the greedy algorithm and the Nash bargaining algorithm the control strategy goes back to the fixed cycle when there is no "connected" vehicle" at the intersection. Fallback on a fixed cycle is also programmed in the local signal controller in case of communication malfunctions. In a real commercial application, it is desirable that the local signal controller can perform a more sophisticated check on the signal commands coming from the server: in case an error is detected the local signal controller will take charge over the central server. In our experiment, we imposed only the minimum and maximum requirements for the green time of each phase. The minimum green time was 4 s and the maximum green time was established at 180 s.

2.3.4. Obtained Results

From simulations the following results were obtained:

- The overall performance in terms of total average travel time is good since average travel time is always decreasing with the increase in the percentage of "connected" vehicles.
- The overall performance in terms of total average travel time can go from a mere 5% reduction (when few vehicles are connected) to over 70% reduction (with 100% of connected vehicles and with a badly regulated intersection). This depends on different conditions of the intersection considered, it must be noted that the reduction in travel times was calculated in our simulations on real urban intersections with real measured flows and signal cycles. Moreover, we did not confront our system with a traditional real time adaptive traffic signal system.
- With a low percentage of "connected" vehicles (under 30%) these "connected" vehicles would take advantage of the system and would obtain priority over other drivers who would be slightly delayed at the traffic signals. It is important to note that in all our simulations the total travel time reduction for "connected" vehicles was always greater than the small delay that traditional vehicles have to suffer;
- With a percentage of over 30% of "connected" vehicles the system works very efficiently and offers convenient cycle lengths and phase times for all users;
- The system can potentially reduce overall traffic congestion and pollution emissions;
- The system works also for multiple intersections and can automatically coordinate green times at different intersections [64].

These results demonstrated that according to simulations, the reduction in overall travel time justifies the implementation of the proposed system by city administrations that would be able to regulate traffic signals using inexpensive Floating Car Data (FCD).

2.4. Development of a Connected Traffic Signal

The traffic signal used for our experiment was directly connected to the internet with an Ethernet cable and was designed with two parts: the traffic lights and the local electronic controller.

The traffic lights in our prototype system have been mounted on a dedicated pole structure and completed using a single strip multicolor LED (for the three colors display in the same lamp). Each one of the three lamps (one for each of the three approaches of the intersection) was controlled by three relays actuated by the local electronic controller. The local electronic controller was composed of a Raspberry Pi3B+ card equipped with its own ethernet connection port and powered by a 5 V-3 A power supply. This local controller was connected to 9 controlled 5 V relays which activate the individual traffic light lamps (Figure 4).

The local controller device is not equipped with a monitor, but in the start-up phase, after the system is operational, a tenth relay is used to activate an operational light. From that moment, the device acts as a web server that is ready to receive the commands of the central system through a binary string. The web server is based on an open source architecture running apache and a PHP script. This web server commands the lights accordingly to this binary string sent by the traffic signal control management server.

The relays to be activated are represented in this string by a simple bit which consequently serves to actuate the colors to be assigned to the lanterns. In detail, the Raspberry with its Rasbian operating system and Apache web server runs a series of PHP scripts that are able to recognize the instructions of the central server and decide which General Purpose Input/Output (GPIO) to activate or deactivate.

The exchange of information between the local controller and the server takes place only every time a phase is changed, or an extension of the green is established, therefore with variable frequency.

The local controller receives control requests from the central server through internet connection and directly activates the relays which switch on the right colors of three traffic lights.

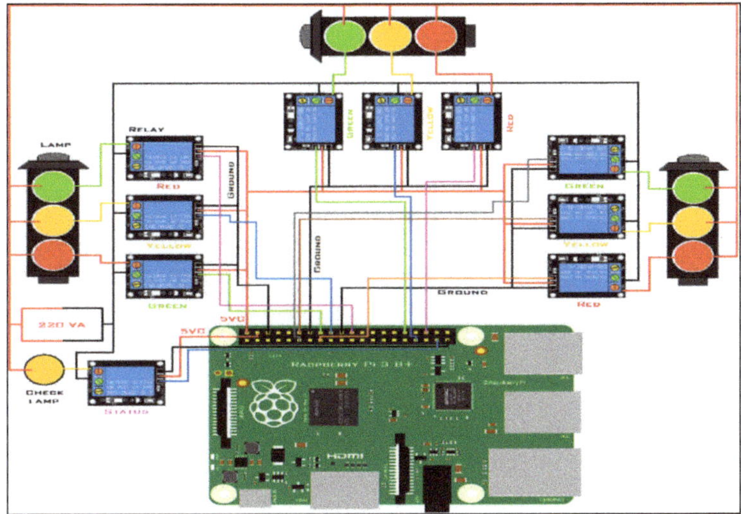

Figure 4. Local controller and connections with traffic lights.

2.5. Development of a Specific Smartphone Application

A specific smartphone application was developed to connect the vehicles with the traffic signal control server. The "connected" vehicles in this way can provide the system with their own position through this dedicated smartphone application, capable of retrieving information from the GNSS chip and transferring it, via a standard wireless phone internet connection and a GET/POST protocol, to the traffic central control system.

The application developed for the smartphones was written for the Google Android system. Constantly, at 1Hz frequency, it requires satellite location information, which is returned in WGS84 format to be sent to the central system as a string. The exchange of information between smartphones and the central system also occurs at a frequency of 1 Hz. The information that is transferred is the latitude, longitude and speed.

2.6. Development of a Specific Traffic Signal Control Server

The central control system was implemented on a Dell PowerEdge 2650 server connected on a 1000 GHz optical fiber. The hardware of this device was equipped with a 2.8 GHz Intel® Xeon ™ processor with 32 bit technology, 512 MB ECC DDR SDRAM and 5 HDD 146 GB. On the software side, the system is based on the Microsoft Windows Server 2008 system, MySQL database and web server scripts compiled using Delphi language. The server is programmed to receive the positions of the various smartphones and send control requests to the local traffic signal control unit.

The traffic central control system stores the received positions of the vehicles on the network, recreating the trajectories of approach to the intersection with a map-matching algorithm. These trajectories are obtained with the positions of the vehicle in which the speed was positive (discarding the points of vehicles that are stopped in a queue). This is done in order to establish optimally the direction of origin of the vehicles. The system then uses this data to estimate the number of vehicles queuing in the various maneuvers and generates the new traffic signal plan using one of the two dedicated algorithms described above. All this procedure is repeated every second, deciding whether to prolong the green of the active phase or to grant the green to another phase. Once a change of phase is decided by the traffic control system the information is sent via internet to the local management system of the traffic signal system using the GET/POST protocol in the form of a 9-bit binary string. Each new phase is introduced following the sequence green -> yellow -> all red -> red for the previous active phase, the sequence red-> all red > green for the new active phase and the sequence red-> all red -> red phase for the phases which were not active before and after the phase change. The light changes are established so as to guarantee the correct and safe passage from one phase to another.

2.7. Planning and Mapping of the Experimental Site

A dedicated experimental site, which is depicted in Figure 5, was set up in a parking lot area of the University of Calabria.

Figure 5. The experimental site: a three-legs intersection.

The intersection is a three-legs intersection with single lanes approaches so that the three traffic lights control all maneuvers of an approach at the same time for a single shared lane. The traffic signal phases are the three depicted in Figure 6a. Drivers were instructed to repeat the same path to avoid the formation of cluster of vehicles moving along the same path. It was possible to identify 9 different paths among which 5 were assigned to a total number of 5 drivers. Four of these different paths are presented in Figure 6b.

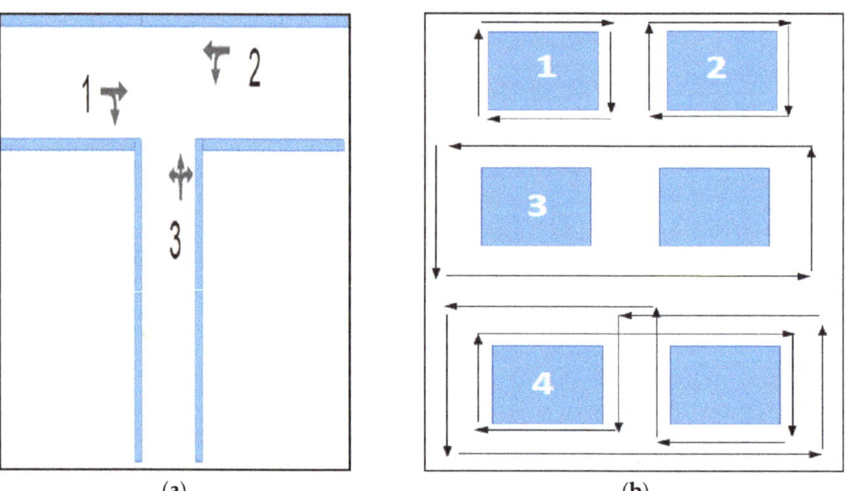

Figure 6. (**a**) Phase diagram; (**b**) Four of the five different paths implemented in the proposed first experimental setting.

The same experimental setting could be used for more complicated intersections with a different lane structure. Since the main objective of this first implementation is mostly the demonstration of feasibility of such a prototype system it was decided to keep the intersection lay out as simple as possible.

Before starting with the experiment involving drivers in the intersection, the informatics platform was tested to avoid the DeadLock problem. Moreover, some test runs were completed with "connected" vehicles to test the GNSS signal strength and the ability of the traffic signal control server to load the trajectories sent by the mobile phone application.

2.8. The Whole Assembled System

The whole prototypal system is depicted in Figure 7. The connections between the traffic signal management server and the local controller are cable-based while the connections directed to the central server from the mobile applications are through wireless internet data 3G phone lines (blue lines).

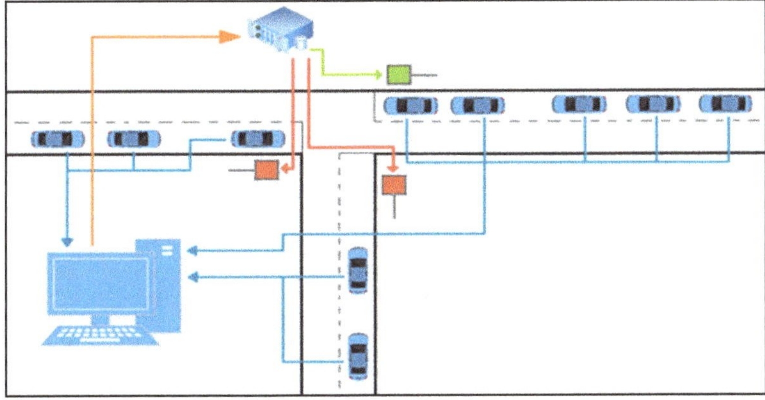

Figure 7. The whole prototypal system for the experiment.

Different coded software in different languages has been applied on the different instances of hardware as described in the following Figure 8 where the data flow is clearly depicted.

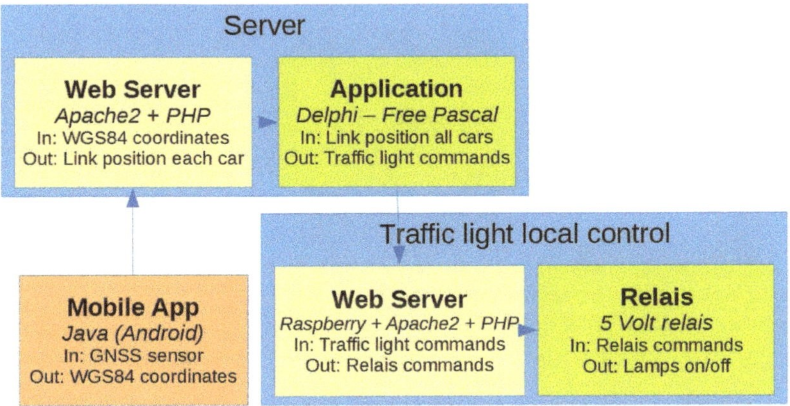

Figure 8. Data flow and general structure of the different software modules in the proposed system.

3. Results

In this section the experiment that was thoroughly prepared and tested in the simulation laboratory is presented. Before the experiment was carried out there were some doubts and the possible causes of failure in the experiment were identified:

- the algorithms would be unable to work with real trajectories since the localization errors could be different from the laboratory simulated errors;
- the server would fail to receive all the trajectories in real time;
- the traffic signal controller would fail to activate the phases according to the signal control server.
- the signal control server would fail.
- the mobile phone applications (which were uploaded on drivers' smartphones just before the experiment) would suffer from compatibility issues or not work properly.
- the driver would not be able to follow the given paths.

The experiment had, in fact, the following failures: the mobile phone applications showed compatibility issue when coupled with driver's smartphones and in the end only 5 vehicles could participate in the experiment (at the experiment site there were more vehicles than working smartphones); one of the three test runs we conducted (when the Nash bargaining equilibrium algorithm was used) failed, possibly since the signal control server did not work properly.

Apart from these two failures, the experiment was a complete success. Results were beyond expectations in terms of performance of the system as shown in the following.

The experiment was conducted and three test runs of the system were completed: with a fixed traffic light plan, with a greedy algorithm and with the Nash bargaining based algorithm. Unfortunately, as indicated above, the Nash bargaining-based algorithm did not work at the experiment site, we have not surely identified the specific cause of failure which could be related to the implementation of the specific algorithm in the control server or with the failure of one of the components of the system (software, hardware or communication link). For this reason, the following results are relative only to the static cycle and the adaptive greedy algorithm case.

In this work, we have included a brief description of the Nash bargaining equilibrium-based algorithm since the fact that it worked perfectly in the simulation laboratory and it did not work in the real experiment might be of interest for the reader and reveals the concept that some algorithms that have to be implemented in a real FCDATS road setting might need some special adjustment before they work as expected from simulation.

All the three experimented scenarios were also simulated in the described laboratory applying the Gazis-Herman (GH) car-following model in the simulation software Tritone. The main parameters of the simulations and of the car-following models were established by applying the GEH function.

The performance measures that were used to make a confrontation between different scenarios were: total outflow, average travel time and average speed.

It must be noted that the vehicles circulated in the experimental intersection in such a way that every driver had an assigned circular path to repeat again and again. In this way the formation of stable platoons of vehicles was avoided. As a result of this methodology the traffic flow that arrived at the intersection was a function of the chosen traffic signal regulation. A badly regulated traffic signal would, in fact, increase the waiting times and consequently reduce the total traffic flow through the intersection. For this reason, the total outflow can be seen as a good measure of performance.

Two drones were used during all the experiment to document all the scene; the drones were suspended high above the intersection and in a fixed position, in this way two videos were taken and allowed the measure of performance for the traffic system to be extracted (see video at Supplementary Materials).

According to simulations, the results in terms of system performances were expected to be as depicted in the following Table 1.

Table 1. Simulation results for the whole intersection.

	Total Outflow [veh/h]	Average Travel Time [s]	Average Speed [km/h]
Static cycle	381	97	2.26
FCDATL	549	44	4.97
Percentage gain	44.09%	54.60%	120.28%

The simulation showed a percentage increase of 44% for the total outflow of the intersection in veh/h and a percentage reduction of around 55% for the average travel time.

As anticipated, the experiment went beyond expectations as depicted in the following Table 2. The increase in total outflow was 116% instead of 44% obtained in simulation, the reduction in terms of average travel time was around 73% instead of around 55% from the simulated reality. The increase in average speed was up to 271% instead of the simulated 120%.

Table 2. Experimental results for the whole intersection.

	Total Outflow [veh/h]	Average Travel Time [s]	Average Speed [km/h]
Static cycle	209	156	1.43
FCDATL	450	42	5.33
Percentage gain	115.97%	73.11%	271.83%

In terms of performance the experiment was a success and the use of 100% "connected" vehicles has guaranteed that for every approach at the intersection there was an increase of performances.

Total measures of performances are also shown in the following Figures 9 and 10.

Figure 9. Simulation and experimental results of static cycle.

Figure 10. Simulation and experimental results of floating car data adaptive traffic signal (FCDATS).

The results for the single links of the networks corresponding to the three different approaches (indicated in Figure 6a) are presented in the following Tables 3 and 4.

Table 3. Simulation results for maneuver.

Approach n°	Static Cycle			FCDATL		
	Outflow [veh/h]	Average Travel Time [s]	Average Speed [km/h]	Total Outflow [veh/h]	Average Travel Time [s]	Average Speed [km/h]
1	102.0	34.24	2.44	148.0	15.96	5.24
2	91.0	39.91	1.94	149.0	15.23	5.08
3	188.0	22.90	2.53	252.0	12.88	4.50

Table 4. Experimental results for maneuver.

Approach n°	Static Cycle			FCDATL		
	Outflow [veh/h]	Average Travel Time [s]	Average Speed [km/h]	Total Outflow [veh/h]	Average Travel Time [s]	Average Speed [km/h]
1	41.7	68.20	1.32	119.9	14.73	6.11
2	75.1	40.94	2.02	138.0	13.97	5.93
3	91.8	46.55	1.08	192.5	13.17	3.83

Data of the single maneuvers shows how from the simulation a slightly better performance of the intersection with higher outflows and speeds and lower travel times was expected. Results from simulations are quite similar, but different, from real experimental values.

More experimental runs with a different level of congestion would be needed to assess the capacity of microsimulation statistically to foresee the performance of control algorithms.

4. Discussion

This paper demonstrates the feasibility of floating car data adaptive traffic signal (FCDATS) with current technologies such as common smartphones connected on 3G wireless mobile phone networks. All the research in this area is based on simulation and this work is the first attempt to experiment existing technologies for real-time adaptive regulation of traffic signals with "connected" vehicles. In this paper, the benefit of using FCD in a cooperative system for adaptive traffic signal is proved with a real implementation with real "connected" vehicles.

As an example of previous works: the Colombo project [62] where "developed algorithms for traffic surveillance and traffic lights control" are evaluated, only in simulation, within the COLOMBO system with the aid of complex modules such as the micro-simulator SUMO and the vehicular emissions model PHEM.

This work was also based on simulation, though simulation was used not as a means to evaluate the proposed system but as a means to support the development of the system which reached its final stage of a real experimentation with real "connected" vehicles.

Sustainability of the traffic system is the final goal of the proposed system. This can be obtained by a reduction of fuel consumption and pollutants emissions connected with the enhancement of traffic signal performances and drivers' comfort. The following results were obtained for the proposed system:

- Especially with low traffic volumes, drivers who are driving through intersections regulated by FCD data will practically always receive the green light.
- Combined results of previous works based on simulation and of the experiment of this work prove that FCD for adaptive traffic signal control not only can be very effective in reducing average travel time and pollution emissions but it is also absolutely feasible with current widespread technologies.
- Another result of this work is that simulation as a tool to evaluate FCDATS control algorithms brings results which are very similar to the results obtained in the real on-field experiment. The result is only partial since a definitive proof would need experimenting different conditions in terms of traffic flows, percentage of connected vehicles and different intersection layouts. Once

this result is completely achieved it will be possible to test new algorithms and control logics just in simulation (the use of simulation is obviously less resource consuming). Once a good simulation test-bed is established a deployment of the proposed system in real city intersections can be easily performed with a good confidence that the FCDATS tested system can help to regulate traffic signals better.

- The experiment we performed involved the collaboration of more than 20 people at the experimental site involving: personnel occupied in the outline of the vehicle paths and intersection that was outlined in a big parking area, drivers, drone pilots and system operation management. This work was completed without specific funding and future development of this research is definitely possible in the direction of more experiments with different conditions.
- Adaptive real-time control of traffic signals is currently performed with data coming from induction loop detectors. The information that is elaborated is based on traffic flow counts at some given locations. The use of FCD data allows using trajectories of vehicles in adaptive traffic signal control. This information is much richer since detailed data of speeds and positions of single vehicles can be used. In other words, the proposed system is not only cheaper and simpler in the implementation it is also potentially more advanced since optimization of the traffic signal can be performed on a potentially more extended data set.
- In this work in a realistic setting with a real case study, the authors also investigated the "Deadlock" problem that is very frequent in simulation contests. Deadlock is a situation in which the informatics platform is busy because two or more competing actions are each waiting for the other to finish. In our experiment there was no case of deadlock and we met no difficulty managing the data coming from smartphones. This problem might still reveal itself with higher traffic flows and an interesting development of the research would be to exclude this possibility with other specific experiments.

The introduction of this system in a real city could happen with some drivers subscribing to the system by downloading the application on the smartphone and other drivers not adopting the system. Since the system can only consider "connected" vehicles this would require an extensive analysis of how such a system would work when traditional not "connected" vehicles are present. The contemporary presence of both connected and not connected vehicles involves concepts of cooperation and competition. Concepts of cooperation and competition have been studied using simulation in: [63,64]. The performed simulations demonstrated that there are benefits for all drivers when the percentage of connected vehicles reaches around 30% and that even under this percentage the drivers that are not using the system would not receive too much delay. Simulation showed that independently from the penetration rate of the system when FCD adaptive traffic signals (FCDATS) are implemented there is an overall reduction of travel times and reduced pollution emissions and fuel consumption. Producing these results also on the road with the proposed experimental prototype system and the proposed experimental setting is possible but goes beyond the scope of this first experimental test that mainly aimed to demonstrate the feasibility of these kinds of system and to propose a methodology that also others can reproduce and extend.

New developments of this research will possibly carry out more experimental runs with different intersection layouts, a different percentage of connected vehicles and different control algorithms.

Supplementary Materials: The following are available online at https://drive.google.com/file/d/16EYfgP__JjXflRqcowbSuqKOyRbaO4-l/view?usp=sharing. There are two videos available taken with two different drones of all the experimental runs.

Author Contributions: V.A., V.P.G. and G.G. were responsible for conceptualization and methodology, V.P.G. carried on all the software development, D.C.F. and A.V. performed supervision, review and editing. A.V. performed supervision at the experimental site. All authors have read and agreed to the published version of the manuscript.

Funding: This research received no external funding.

Acknowledgments: The authors wish to thanks Gianfranco Salfi and Claudio Capalbo for managing drone video footage, all the students that participated in the experiment, the rector of the University of Calabria and all the other UNICAL staff that made this experiment possible without external funding.

Conflicts of Interest: The authors declare no conflict of interest.

References

1. Mokaddem, Y.E.; Jawab, F. Researches and Applications of Intelligent Transportations Systems in Urban Area: Systematic Literature Review. *ARPN J. Eng. Appl. Sci.* **2019**, *14*, 639–652.
2. Marzano, V.; Tocchi, D.; Papola, A.; Aponte, D.; Simonelli, F.; Cascetta, E. Incentives to freight railway undertakings compensating for infrastructural gaps: Methodology and practical application to Italy. *Transp. Res. Part A Policy Pract.* **2018**, *110*, 77–188. [CrossRef]
3. Astarita, V.; Florian, M.; Musolino, G. A microscopic traffic simulation model for the evaluation of toll station systems. In Proceedings of the IEEE Conference on Intelligent Transportation Systems, Oakland, CA, USA, 25–29 August 2001.
4. Astarita, V.; Giofré, V.; Guido, G.; Vitale, A. Investigating road safety issues through a microsimulation model. *Procedia Soc. Behav. Sci.* **2011**, *20*, 226–235. [CrossRef]
5. Young, W.; Sobhani, A.; Lenné, M.G.; Sarvi, M. Simulation of safety: A review of the state of the art in road safety simulation modelling. *Accid. Anal. Prev.* **2014**, *66*, 89–103. [CrossRef]
6. Astarita, V.; Giofré, V.P. From traffic conflict simulation to traffic crash simulation: Introducing traffic safety indicators based on the explicit simulation of potential driver errors. *Simul. Model. Pract. Theory* **2019**, *94*, 215–236. [CrossRef]
7. Osorio, C.; Punzo, V. Efficient calibration of microscopic car-following models for large-scale stochastic network simulators. *Transp. Res. Part B Methodol.* **2019**, *119*, 156–173. [CrossRef]
8. Martinez, F.J.; Toh, C.K.; Cano, J.C.; Calafate, C.T.; Manzoni, P. A survey and comparative study of simulators for vehicular ad hoc networks (VANETs). *Wirel. Commun. Mob. Comput.* **2011**, *11*, 813–828. [CrossRef]
9. Gentilea, G. New formulations of the stochastic user equilibrium with logit route choice as an extension of the deterministic model. *Transp. Sci.* **2018**, *52*, 1531–1547. [CrossRef]
10. Gentile, G. Solving a Dynamic User Equilibrium model based on splitting rates with Gradient Projection algorithms. *Transp. Res. Part B Methodol.* **2016**, *92*, 120–147. [CrossRef]
11. Cantarella, G.E.; Di Febbraro, A.; Di Gangi, M.; Giannattasio, O. Stochastic Multi-Vehicle Assignment to Urban Transportation Networks. In Proceedings of the MT-ITS 2019—6th International Conference on Models and Technologies for Intelligent Transportation Systems, Cracow, Poland, 5–7 June 2019.
12. Cantarella, G.E.; Watling, D.P. A general stochastic process for day-to-day dynamic traffic assignment: Formulation, asymptotic behaviour, and stability analysis. *Transp. Res. Part B Methodol.* **2016**, *92*, 3–21. [CrossRef]
13. Trozzi, V.; Gentile, G.; Kaparias, I.; Bell, M.G.H. Effects of Countdown Displays in Public Transport Route Choice Under Severe Overcrowding. *Netw. Spat. Econ.* **2015**, *15*, 823–842. [CrossRef]
14. Kucharski, R.; Gentile, G. Simulation of rerouting phenomena in Dynamic Traffic Assignment with the Information Comply Model. *Transp. Res. Part B Methodol.* **2018**, *126*, 414–441. [CrossRef]
15. Marzano, V.; Papola, A.; Simonelli, F.; Papageorgiou, M. A Kalman Filter for Quasi-Dynamic o-d Flow Estimation/Updating. *IEEE Trans. Intell. Transp. Syst.* **2018**, *19*, 3604–3612. [CrossRef]
16. Papola, A.; Tinessa, F.; Marzano, V. Application of the Combination of Random Utility Models (CoRUM) to route choice. *Transp. Res. Part B Methodol.* **2018**, *111*, 304–326. [CrossRef]
17. Papola, A. A new random utility model with flexible correlation pattern and closed-form covariance expression: The CoRUM. *Transp. Res. Part B Methodol.* **2016**, *94*, 80–96. [CrossRef]
18. Martin, C. NO IDLE MATTER: Signal-Controlled or Signal-Free. *TRAFFIC Technol. Int.* **2007**, 56–59.
19. Steve Huntingford Traffic Lights Cause More Harm than Good. Available online: https://www.telegraph.co.uk/cars/comment/traffic-lights-cause-more-harm-than-good/ (accessed on 12 October 2018).
20. Cassini, M. In your car no one can hear you scream! Are traffic controls in cities a necessary evil? *Econ. Aff.* **2006**, *26*, 75–78. [CrossRef]
21. Cassini, M. Traffic lights: Weapons of mass distraction, danger and delay. *Econ. Aff.* **2010**, *30*, 79–80. [CrossRef]

22. Ban, X.J.; Li, Y.; Skabardonis, A.; Margulici, J.D. Performance evaluation of travel-time estimation methods for real-time traffic applications. *J. Intell. Transp. Syst. Technol. Plan. Oper.* **2010**, *14*, 54–67. [CrossRef]
23. Wright, J.; Dahlgren, J. *Using Probe Vehicles Equipped with Toll Tag as Probes for Providing Travel Times*; California PATH Program, Institute of Transportation Studies, University of California at Berkeley: Berkeley, CA, USA, 2001; ISSN 1055-1417.
24. Priemer, C.; Friedrich, B. A decentralized adaptive traffic signal control using V2I communication data. In Proceedings of the IEEE Conference on Intelligent Transportation Systems, St. Louis, MO, USA, 4–7 October 2009.
25. Astarita, V.; Festa, D.C.; Giofrè, V.P. Mobile Systems applied to Traffic Management and Safety: A state of the art. *Procedia Comput. Sci.* **2018**, *134*, 407–414. [CrossRef]
26. Wahlstrom, J.; Skog, I.; Handel, P. Smartphone-Based Vehicle Telematics: A Ten-Year Anniversary. *IEEE Trans. Intell. Transp. Syst.* **2017**, *18*, 2802–2825. [CrossRef]
27. Astarita, V.; Danieli, G.; D'Elia, S. Sistema di Rilievo del Deflusso Veicolare per il Monitoraggio e Controllo del Traffico Basato Sulla Registrazione della Localizzazione dei Terminali di Telefonia Mobile. Patent 0001322707, 26 January 2000.
28. Wasson, J.S.; Sturdevant, J.R.; Bullock, D.M. Real-time travel time estimates using media access control address matching. *ITE J. (Institute Transp. Eng.)* **2008**, *78*, 20–23.
29. Young, S. *Bluetooth Traffic Monitoring Technology: Concept of Operation & Deployment Guidelines*; University of Maryland: College Park, MD, USA, 2008.
30. Barceló, J.; Montero, L.; Marqués, L.; Carmona, C. Travel Time Forecasting and Dynamic Origin-Destination Estimation for Freeways Based on Bluetooth Traffic Monitoring. *Transp. Res. Rec. J. Transp. Res. Board* **2010**, *2175*, 19–27. [CrossRef]
31. Barceló, J.; Montero, L.; Bullejos, M.; Serch, O.; Carmona, C. A kalman filter approach for exploiting bluetooth traffic data when estimating time-dependent od matrices. *J. Intell. Transp. Syst. Technol. Plan. Oper.* **2013**, *17*, 123–141. [CrossRef]
32. Bachmann, C.; Roorda, M.J.; Abdulhai, B.; Moshiri, B. Fusing a bluetooth traffic monitoring system with loop detector data for improved freeway traffic speed estimation. *J. Intell. Transp. Syst. Technol. Plan. Oper.* **2013**, *17*, 152–164. [CrossRef]
33. Ygnace, J.L.; Drane, C.; Yim, Y.B.; De Lacvivier, R. *Travel Time Estimation on the San Francisco Bay Area Network Using Cellular Phones as Probes*; California PATH Program, Institute of Transportation Studies, University of California at Berkeley: Berkeley, CA, USA, 2000; ISSN 1055-1417.
34. Bolla, R.; Davoli, F.; Giordano, A. Estimating Road Traffic Parameters from Mobile Communications. In Proceedings of the 7th World Congress on Intelligent Systems, Turin, Italy, 6–9 November 2000; pp. 1107–1112.
35. Rose, G. Mobile Phones as Traffic Probes: Practices, Prospects and Issues. *Transp. Rev.* **2006**, *26*, 275–291. [CrossRef]
36. Astarita, V.; Bertini, R.L.; d'Elia, S.; Guido, G. Motorway traffic parameter estimation from mobile phone counts. *Eur. J. Oper. Res.* **2006**, *175*, 1435–1446. [CrossRef]
37. Sohn, K.; Hwang, K. Space-based passing time estimation on a freeway using cell phones as traffic probes. *IEEE Trans. Intell. Transp. Syst.* **2008**, *9*, 559. [CrossRef]
38. Herrera, J.C.; Work, D.B.; Herring, R.; Ban, X.; Jacobson, Q.; Bayen, A.M. Evaluation of traffic data obtained via GPS-enabled mobile phones: The Mobile Century field experiment. *Transp. Res. Part C Emerg. Technol.* **2010**, *18*, 568–583. [CrossRef]
39. Bar-Gera, H. Evaluation of a cellular phone-based system for measurements of traffic speeds and travel times: A case study from Israel. *Transp. Res. Part C Emerg. Technol.* **2007**, *15*, 380–391. [CrossRef]
40. Guido, G.; Vitale, A.; Saccomanno, F.F.; Carmine, F.D.; Astarita, V.; Rogano, D.; Gallelli, V. Using Smartphones As a Tool to Capture Road Traffic Attributes. *Appl. Mech. Mater.* **2013**, *432*, 513–519. [CrossRef]
41. Guido, G.; Saccomanno, F.F.; Vitale, A.; Gallelli, V.; Festa, D.C.; Astarita, V.; Rogano, D.; Carvelli, D. Extracting vehicle tracking data from smartphone sensors. In Proceedings of the 92th Annual Meeting of Transportation Research Board, Washington, DC, USA, 13–17 January 2013.
42. Jiang, Y.; Zhang, J. Influence of Smartphone Apps with Driving Safety Related Diagnosis Functions on Expressway Driving Speed Changes. *J. Transp. Eng. Part A Syst.* **2017**, *144*, 04017069. [CrossRef]
43. Guido, G.; Vitale, A.; Astarita, V.; Saccomanno, F.; Giofré, V.P.; Gallelli, V. Estimation of Safety Performance Measures from Smartphone Sensors. *Procedia Soc. Behav. Sci.* **2012**, *54*, 1095–1103. [CrossRef]

44. Jiang, Y.; Zhang, J.; Wang, Y.; Wang, W. Drivers' behavioral responses to driving risk diagnosis and real-time warning information provision on expressways: A smartphone app–based driving experiment. *J. Transp. Saf. Secur.* **2018**, 1–29. [CrossRef]
45. Astarita, V.; Guido, G.; Giofrè, V.P. Co-operative ITS: Smartphone based measurement systems for road safety assessment. *Procedia Comput. Sci.* **2014**, *37*, 404–409. [CrossRef]
46. Astarita, V.; Guido, G.; Mongelli, D.; Giofrè, V.P. A co-operative methodology to estimate car fuel consumption by using smartphone sensors. *Transport* **2015**, *30*, 307–311. [CrossRef]
47. Astarita, V.; Guido, G.; Mongelli, D.W.E.; Giofrè, V.P. Ecosmart and TutorDrive: Tools for fuel consumption reduction. In Proceedings of the 2014 IEEE International Conference on Service Operations and Logistics, and Informatics, SOLI, Qingdao, China, 8–10 October 2014.
48. Astarita, V.; Festa, D.C.; Mongelli, D.W.E. *EcoSmart:* An Application for Smartphones for Monitoring Driving Economy. *Adv. Mat. Res.* **2014**, *827*, 360–367. [CrossRef]
49. Astarita, V.; Festa, D.C.; Mongelli, D.W.E.; Mongelli, N.; Ruffolo, O.; Servino, A. *EcoSmart: A Survey and Sector Analysis of Mobile Phone Application Market for Fuel Consumption Reduction*; Trans Tech Publications Ltd.: Stafa-Zurich, Switzerland, 2014; Volume 519–520, ISBN 9783038350194.
50. Astarita, V.; Caruso, M.V.; Danieli, G.; Festa, D.C.; Giofrè, V.P.; Iuele, T.; Vaiana, R. A Mobile Application for Road Surface Quality Control: UNIquALroad. *Procedia Soc. Behav. Sci.* **2012**, *54*, 1135–1144. [CrossRef]
51. Mohan, P.; Padmanabhan, V.N.; Ramjee, R. Nericell: Rich monitoring of road and traffic conditions using mobile smartphones. In Proceedings of the 6th ACM Conference on Embedded Network Sensor Systems, Raleigh, NC, USA, 5–7 November 2008.
52. Festa, D.C.; Mongelli, D.W.E.; Astarita, V.; Giorgi, P. First results of a new methodology for the identification of road surface anomalies. In Proceedings of the 2013 IEEE International Conference on Service Operations and Logistics, and Informatics, SOLI, Dongguan, China, 28–30 July 2013.
53. Bierlaire, M.; Chen, J.; Newman, J.P. *Modeling Route Choice Behavior From Smartphone GPS Data*; Report TRANSP-OR 101016; Ecole Polytechnique Fédérale de Lausanne: Lausanne, Switzerland, 2010.
54. Dailey, D.J.; Cathey, F.W. Virtual speed sensors using transit vehicles as traffic probes. In Proceedings of the IEEE Conference on Intelligent Transportation Systems, Singapore, 6 September 2002.
55. Axer, S.; Friedrich, B. Level of service estimation based on low-frequency floating car data. *Transp. Res. Procedia* **2014**, *3*, 1051–1058. [CrossRef]
56. Biral, F.; Da Lio, M.; Bertolazzi, E. Combining safety margins and user preferences into a driving criterion for optimal control-based computation of reference maneuvers for an ADAS of the next generation. In Proceedings of the IEEE Intelligent Vehicles Symposium, Las Vegas, NV, USA, 6–8 June 2005.
57. Vaiana, R.; Iuele, T.; Astarita, V.; Caruso, M.V.; Tassitani, A.; Zaffino, C.; Giofrè, V.P. Driving behavior and traffic safety: An acceleration-based safety evaluation procedure for smartphones. *Mod. Appl. Sci.* **2014**, *8*, 88. [CrossRef]
58. Barthauer, M.; Friedrich, B. Evaluation of a signal state prediction algorithm for car to infrastructure applications. *Transp. Res. Procedia* **2014**, *3*, 982–991. [CrossRef]
59. Olaverri-Monreal, C.; Errea-Moreno, J.; Díaz-Álvarez, A. Implementation and Evaluation of a Traffic Light Assistance System Based on V2I Communication in a Simulation Framework. *J. Adv. Transp.* **2018**. [CrossRef]
60. Hu, J.; Fontaine, M.D.; Park, B.B.; Ma, J. Field evaluations of an adaptive traffic signal—Using private-sector probe data. *J. Transp. Eng.* **2015**, *142*, 04015033. [CrossRef]
61. Saust, F.; Bley, O.; Kutzner, R.; Wille, J.M.; Friedrich, B.; Maurer, M. Exploitability of vehicle related sensor data in cooperative systems. In Proceedings of the IEEE Conference on Intelligent Transportation Systems, Funchal, Portugal, 19–22 September 2010.
62. Krajzewicz, D.; Heinrich, M.; Milano, M. COLOMBO: Investigating the Potential of V2X for Traffic Management Purposes assuming low penetration Rates. In Proceedings of the 9th ITS European Congress, Dublin, Ireland, 4–7 June 2013.
63. Astarita, V.; Vincenzo Pasquale, G.; GUIDO, G.; Vitale, A. A Single Intersection Cooperative-Competitive Paradigm in Real Time Traffic Signal Settings Based on Floating Car Data. *Energies* **2019**, *12*, 409. [CrossRef]
64. Astarita, V.; Festa, D.C.; Vincenzo, G. Cooperative-competitive paradigm in traffic signal synchronization based on floating car data. In Proceedings of the 2018 IEEE International Conference on Environment and Electrical Engineering and 2018 IEEE Industrial and Commercial Power Systems Europe (EEEIC/I&CPS Europe), Palermo, Italy, 12–15 June 2018.

65. Gradinescu, V.; Gorgorin, C.; Diaconescu, R.; Cristea, V.; Iftode, L. Adaptive traffic lights using car-to-car communication. In Proceedings of the IEEE Vehicular Technology Conference, Dublin, Ireland, 22–25 April 2007.
66. Astarita, V.; Giofrè, V.P.; Guido, G.; Vitale, A. The Use of Adaptive Traffic Signal Systems Based on Floating Car Data. *Wirel. Commun. Mob. Comput.* **2017**, *2017*, 1–13. [CrossRef]
67. Chandan, K.; Seco, A.M.; Silva, A.B. Real-time Traffic Signal Control for Isolated Intersection, using Car-following Logic under Connected Vehicle Environment. *Transp. Res. Procedia* **2017**, *25*, 1610–1625. [CrossRef]
68. Zhang, R.; Ishikawa, A.; Wang, W.; Striner, B.; Tonguz, O. Intelligent Traffic Signal Control: Using Reinforcement Learning with Partial Detection. *arXiv* **2018**, arXiv:1807.01628.
69. Sharma, S.; Lubmann, J.; So, J. Controller Independent Software-in-the-Loop Approach to Evaluate Rule-Based Traffic Signal Retiming Strategy by Utilizing Floating Car Data. *IEEE Trans. Intell. Transp. Syst.* **2019**, *20*, 3585–3594. [CrossRef]
70. Marinelli, M.; Palmisano, G.; Astarita, V.; Ottomanelli, M.; Dell'Orco, M. A Fuzzy set-based method to identify the car position in a road lane at intersections by smartphone GPS data. *Transp. Res. Procedia* **2017**, *27*, 444–451. [CrossRef]
71. Giofre, V.P.; Astarita, V.; Guido, G.; Vitale, A. Localization issues in the use of ITS. In Proceedings of the 5th IEEE International Conference on Models and Technologies for Intelligent Transportation Systems, Naples, Italy, 26–28 June 2017.
72. Giofrè, V.P.; Maciejewski, M.; Merkisz Guranowska, A.; Piątkowski, B.; Astarita, V. Real road network application of a new microsimulation tool TRITONE. *Arch. Transp.* **2013**, *27*, 111–121. [CrossRef]
73. Astarita, V.; Giofrè, V.; Guido, G.; Vitale, A.; Festa, D.; Vaiana, R.; Iuele, T.; Mongelli, D.; Rogano, D.; Gallelli, V. New features of Tritone for the evaluation of traffic safety performances. In *Transport Infrastructure and Systems*; Taylor & Francis Group: London, UK, 2017; pp. 625–632.
74. Astarita, V.; Giofré, V.; Guido, G.; Vitale, A. Calibration of a new microsimulation package for the evaluation of traffic safety performances. *PROCEDIA Soc. Behav. Sci.* **2012**, *54*, 1019–1026. [CrossRef]
75. Astarita, V.; Guido, G.; Vitale, A.; Giofré, V. A new microsimulation model for the evaluation of traffic safety performances. *Eur. Transp. Trasp. Eur.* **2012**, *51*, 1–16.
76. Krajzewicz, D.; Erdmann, J.; Behrisch, M.; Bieker, L. Recent Development and Applications of {SUMO—Simulation of Urban MObility}. *Int. J. Adv. Syst. Meas.* **2012**, *5*, 128–138.
77. Ekeila, W.; Sayed, T.; El Esawey, M. Development of Dynamic Transit Signal Priority Strategy. *Transp. Res. Rec. J. Transp. Res. Board* **2009**. [CrossRef]
78. Tan, L.; Zhao, X.; Hu, D.; Shang, Y.; Wang, R. A study of single intersection traffic signal control based on two-player cooperation game model. In Proceedings of the 2010 WASE International Conference on Information Engineering, ICIE, Beidaihe, China, 14–15 August 2010.
79. Abdelghaffar, H.M.; Yang, H.; Rakha, H.A. Isolated traffic signal control using a game theoretic framework. In Proceedings of the IEEE Conference on Intelligent Transportation Systems, Rio de Janeiro, Brazil, 1–4 November 2016.

Article

A Traffic Flow Prediction Method Based on Road Crossing Vector Coding and a Bidirectional Recursive Neural Network

Shuanfeng Zhao *, Qingqing Zhao, Yunrui Bai and Shijun Li

School of Mechanical Engineering, Xi'an University of Science and Technology, Xi'an 710054, China; zhaoqingqing7@163.com (Q.Z.); rickybe0525@163.com (Y.B.); ajun386@163.com (S.J.)

* Correspondence: zsf@xust.edu.cn; Tel.: +86-029-8558-3159

Received: 17 August 2019; Accepted: 5 September 2019; Published: 8 September 2019

Abstract: Aiming at the problems that current predicting models are incapable of extracting the inner rule of the traffic flow sequence in traffic big data, and unable to make full use of the spatio-temporal relationship of the traffic flow to improve the accuracy of prediction, a Bi-directional Regression Neural Network (BRNN) is proposed in this paper, which can fully apply the context information of road intersections both in the past and the future to predict the traffic volume, and further to make up the deficiency that the current models can only predict the next-moment output according to the time series information in the previous moment. Meanwhile, a vectorized code to screen out the intersections related to the predicting point in the road network and to train and predict through inputting the track data of the selected intersections into BRNN, is designed. In addition, the model is testified through the true traffic data in partial area of Shen Zhen. The results indicate that, compared with current traffic predicting models, the model in this paper is capable of providing the necessary evidence for traffic guidance and control due to its excellent performance in extracting the spatio-temporal feature of the traffic flow series, which can enhance the accuracy by 16.298% on average.

Keywords: deep leaning; intelligent transportation; vectorization coding; BRNN; traffic forecast

1. Introduction

In many cities of China, frequent traffic congestion in the main road intersections, especially during the rush hours, has brought an increasingly severe challenge to the urban road traffic system [1,2]. To improve traffic safety and efficiency, some researchers are connecting vehicles to each other and to the road infrastructure [3]. Nonetheless, the long-studied techniques on traffic systems in many developed countries have already paved the way to the birth of the Intelligent Transportation System (ITS) [4]. In the past ten years, ITS has exerted its significant influence on many aspects such as traffic guidance, supervision on fatigue driving [5,6], monitoring traffic conditions, emergency support, as well as the prediction of traffic flow. Noticeably, for a more accurate and efficient identification on the traffic condition, as well as the more perfect and intelligent traffic system, the difficulties and key points stand on the prediction of the traffic flow, of which the core study point in turn lies on how to predict traffic volume.

The variation pattern of traffic volume is affected by factors which are very complicated, thereby with the characters of nonlinearity and spatial-temporal correlation, etc., which have brought significant difficulties in predicting the traffic flow [7,8]. Of the current predicting methods on traffic flow, the mainstream models contain the parametric model and nonparametric model, as well as the mixed model, etc.

Among those, the Seasonal Autoregressive Integrated Moving Average model (SARIMA), a parametric model advocated by Kumar, taking into account the influence that the various seasons exert upon traffic flow, can predict the short-term traffic flow with just a small amount of data input, which have solved the problem of applicability that the traditional model ARIMA cannot deal with; but the considerable deviation that this model generates during the fluctuation of traffic flow may have a significant impact on the prediction results, and the results of the experiments based on the nonlinear and instable traffic flow data set are lower than expected [9]. In order to minimize the deviation and to further perfect the method related to the nonlinear traffic flow, nonparametric models such as nonparametric regression, wavelet theory, etc., were proposed by scholars, which can better process the data with complicated variations when targeting the traffic time series; but due to the enormous calculation and complex structure, this kind of model is not applicable for practical operation [10,11]. Thereby, many mixed models have been proposed to improve the prediction performance. Li et al. firstly proposed a mixed model combining the Support Vector Regression (SVR) with ARIMA to predict the sample data by preprocessing the traffic flow data set to extract the data representing the traffic features from the complicated and complex samples [12]. Cheng et al integrated the advantages of the Particle Swarm Optimizer (PSO) (aiming at optimizing the weight and threshold of the BP neural network) and the BP neural network (aiming at predicting the traffic flow) into one single model to fulfill the purpose of an accurate prediction by enhancing the accuracy of prediction and the rate of convergence [13]. Liu et al combined the Support Vector Regression (SVR) with the k-nearest Neighbor method to construct the more accurate KNN-SVR model, which is obtained by comparing the three other models—SVR, BPNN and KNN—with the SVR model retrained by the historical traffic flow series that is reorganized according to the KNN algorithm [14]. Chen Xiaobo et al analyzed the urban traffic road network through a model constructed by combining the Least Squares SVR (LSSVR) and the Genetic Algorithm (GA), which can simulate the variation rules to accomplish the prediction of traffic flow [15]. Although those mixed models above have succeeded in integrating the advantages and to some extent optimized the prediction, the enormous difficulty in combination and different results from various combinations have negatively impacted the actual effect.

Nowadays, with the computer becoming increasingly pervasive and powerful, the development of the Artificial Intelligence and deep learning theory have opened a new horizon in the field of traffic flow prediction. The general deep learning algorithm contains Deep Residual Network, Convolutional Neural Network and Recurrent Neural Network, etc. Ma et al. constructed an integrated model algorithm combining the Restricted Boltzmann Machine (RBM) and Recurrent Neural Network (RNN) after fully investigating the rule of traffic series, and the consequence showed that the model had more a accurate prediction and thus was applicable to predict the traffic volume of the urban road network [16]. But this model, like other neural network prediction models, has the flaws of the gradient disappearing and a gradient blow-up, and thus is not applicable for prediction on a long-term time series. Deng Xuankun et al. combined the feature components extracted by the Convolutional Neural Network (CNN) with an LSTM model to solve the problem on predicting the traffic flow series [17]. This model has stronger real-time capability, but was too complicated, due to the countless parameters. Jonathan Mackenzie et al. [18] adopted the Regression Analysis method in the experiment part to compare the expectation value (Y) with the actual traffic figure (Y), and to assess the statistical data from traffic prediction model with the Mean Absolute Percentage Error (MAPE), Root Mean Squared Error (RMSE) and GEH [18,19]. Meanwhile the MAPE and RMSE were recorded for the sake of comparison with previous studies. At last, GEH was considered to be the most useful to assess the traffic prediction. Tian and Pan firstly studied a variant of the Long Short Term Memory (LSTM), one of the Recurrent Neural Network [20]. After realizing a hidden layer containing 5–40 units, they assessed the traffic data with the California Caltrans Performance measuring system, and eventually made a comparison with the methods of Random Walk, Radial Basis Function Support Vector Machine, Single-hidden Layer Feedforward Neural Network and SAEs; the consequence revealed that the MAPE and RMSE deriving from the LSTM were impressive in all testing sections.

In the test, T. Pamula proposed the simplest network structure to realize the function of monitoring the difference of the traffic data series in the network map by predicting through various neural network structures [21,22]. Huang et al developed a deep system structure consisting of the Deep Belief Network (DBN) and a multitasking regression layer to obtain the optimized prediction result, in which the author had tested the DBN in different depth and with various node numbers [23]. K. Halawa et al. proposed the expanding method for the main road intersection in the urban traffic network through introducing an additional hidden layer in the network to realize the reflection of the multi-layer sensor to the complicated relation among different variates [24]. J. Guo et al. developed the filter method based upon the least square, minimum mean square, generalized linear model and Kalman filter by adopting the Kalman filter with the adaptive mechanism to the variance, which revealed an excellent fitting ability to the varying flow characteristic [25]. Min and Wynter presented a multivariate auto-regression model based on Vector-ARMA, which contains the dependency relationship between the adjacent testing points, and the consequence showed an outstanding prediction accuracy to the different vehicle speed and traffic volume [26].

As known from the studies above on the prediction of the traffic flow, the deviation of the traffic flow time series in a single road intersection is usually significant, and the deviation reduces as the flow curve or such is taken account into the prediction input data of the road intersection. However, the similar flow curve between the road intersections does not indicate a necessary relationship among them, neither the influence relation among different road intersections. Accordingly, to explore the spatio-temporal relationship among the road intersections, this paper designs a vectorized code which extracts the spatial feature of the road intersections by vectorizing them and meanwhile screens out those with the closest relation to the predicting point; eventually, the track vector matrix will be input into BRNN model for training and predicating, and be evaluated and compared with other predicting models. As a result, the consequence shows that the predicting method in this paper has higher accuracy.

2. Theory

2.1. Model Introduction

Recurrent Neural Network (RNN) is specialized for processing the series data. Compared with the classical feedforward neural network, the structure is relatively simple, which just reconnects the output of the hidden layer (or the output layer) back to the hidden layer forming a closed loop. It can also be considered as to add a memory unit in the feedforward neural network. When the neure transmits forward, the hidden layer does not only send messages to the front end, but also stores the message in the memory unit. When the next message performs, it will be sent to the hidden layer with the previous message stored in the memory units together to extract the features, and those processed messages will then be stored in the memory units again. The typical RNN model structure is as shown in Figure 1.

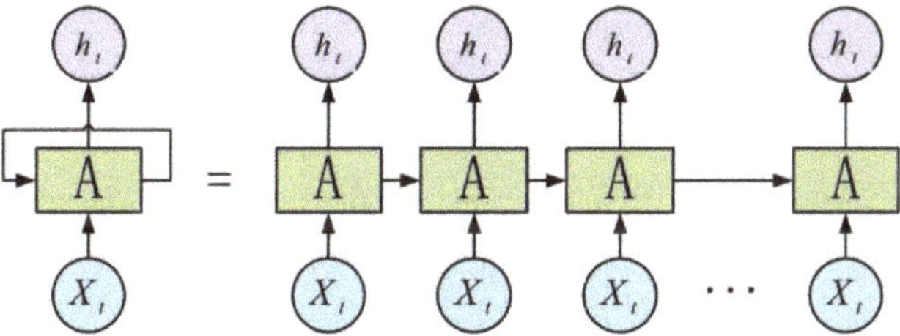

Figure 1. Computation flow of RNN model.

However, as the standard RNN is unidirectional, when processing the time series, the status value in the position i is only related to output from position 0 to position i, but has nothing to do with that from position i + 1 to the end, namely information above only; to obtain the context information, RNN may calculate inversely to allow the status value in every position, and can obtain the information above from the forward RNN and the information below from the inverse RNN in order to make full use of the information in the past and the future. Such RNN is called bidirectional RNN, and according to which this paper proposed the Bi-directional Recurrent Neural Network (BRNN) [27]. The core conception of BRNN is that two RNNs will read the information from opposite directions with one starting from the beginning and the other starting from the end, as the series data are input into the model; the model will store the information respectively in two hidden layers connecting to the same output layer, and such a bidirectional recycle structure can provide every output series point in the output layer with the complete context information of the past and the future; namely the output of the neure in time t depends on both the elements in the past and the future. Thereby, this model is able to visit the context information in the past as well as that in the future, and make full use of the input information during the model training. Whereas, RNN is incapable of providing the context information in the future. Theoretically, the input of BRNN contains three parts including the input from the input layer, as well as the previous-moment and next-moment output from the hidden layer. Consequently, the output layer can integrate perfectly the complete context information in every moment of the input series. The typical structure of BRNN is as shown in Figure 2.

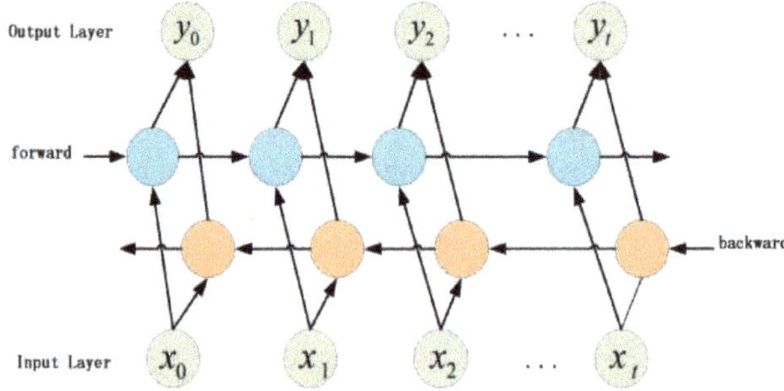

Figure 2. Computation flow of Bi-directional Regression Neural Network (BRNN) model.

$$y^{(l_i)} = softmax(w_{yh_f}h_f{}^{(l_i)} + w_{yh_b}h_b{}^{(l_i)} + b_y) \tag{1}$$

$$h_f{}^{(l_i)} = sigmoid(w_{h_fx}x^{(l_i)} + w_{h_fh_f}h_f{}^{(l_{i-1})} + b_{h_f}) \tag{2}$$

$$h_b{}^{(l_i)} = sigmoid(w_{h_bx}x^{(l_i)} + w_{h_bh_b}h_b{}^{(l_{i-1})} + b_{h_b}) \tag{3}$$

From Figure 2, it is clear that the hidden layer of the network reserves two parts of information, namely the forward information $h_f^{(li)}$, and the inverse information $h_b^{(li)}$, while the final output $y^{(li)}$ contains both parts. b_i is the output bias. Known from the three Equations above, the six unique weights including the weight matrix w_{h_fx}, w_{h_bx} (input layer to hidden layer), w_{yh_f}, w_{yh_b} (hidden layer to input layer) and $w_{h_fh_f}$, $w_{h_bh_b}$ (hidden layer to hidden layer) are applied repeatedly in every moment, but each pair of the values between the forward and inverse is quite different.

2.2. The Vectorization of the Road Intersection

In the intelligent transportation system (ITS), the camera installation applied in recording the passing vehicles is involved. The cameras are installed in each road intersection with a unique serial number represented as l_i (i is the integer between 0 and n, and n is the sum of the road intersections).

The data collected in every road intersection contain the information including the serial number of the road intersection, the vehicle numbers and time, which are the primary data for research. Since each vehicle passes through a different intersection in sequence, vehicle trajectory T_P can be represented by a series of road intersection serial numbers sorted by time.

$$T_P : \{l_0, l_1, l_2, \cdots, l_s\} \tag{4}$$

where p is the number of the vehicle. $l_0, l_1, l_2, \cdots, l_s$ is the road intersection serial number and $\{l_0, l_1, l_2, \cdots, l_s\}$ is arranged according to the time sequence of the vehicles passing through the intersection. s is the length of the vehicle track. In order to get the track data of each vehicle passing through the intersection serial number in a certain period of time, we need to process the original data into track data. Since the original data set is large, this paper adopts the trajectory data set statistical algorithm to generate the trajectory data (See Appendix A for the specific algorithm).

In the neural network training, as the intersection serial numbers of the traffic flow data are not in the vector form, they are unable to be directly input into the neural network, thus for the convenience of calculation and understanding of the neural network, the intersection serial number should be vectorized. This paper will regard each track data as a natural-language document with each intersection serial number as the single word in the document. Firstly, according to the data set the intersection serial numbers will be collected, which will be vectorized into the corresponding serial number after that. Finally, the vector matrix will be obtained depending on the vectorized intersection serial number. The vectorized serial number model can simplify the process of the text content into the vector operation in k-dimensional vector space, and the similarity in the vector space can be used to express the similarity of the text meaning [28]. This model connects all words to the hidden layer casting out the most time-consuming, nonlinear hidden layer, meanwhile inputting the word vector of the context of the testing word and outputting that of the testing word. As the large number of weight matrices need updating during the model training, which results in the excessive calculation, the training speed may possibly slow down. Therefore, to reduce the burden of training, this model integrates the negative sampling method which updates only the input and element weight in the negative sample set, and tends to considerably reduce the calculation amount during the gradient descent, and thereby accelerate the training speed [29].

Negative Sampling may generate a random negative example in the source code (the variate can be set in the code, and if the original word is generated during the random generation of the negative example, the number may possibly be less), where the original word is positive example with the label as 1, the other random generated label is 0, so the input f is:

$$f = \sigma\left(neu1^T \cdot \mathrm{syn}1\right) \tag{5}$$

Loss is the negative Log likelihood (due to the random gradient descent method, here represents only a single layer in a word), namely:

$$Loss = -LogLikelihood = -label \cdot \log f - (1 - f) \tag{6}$$

The gradient can be expressed as follows:

$$\begin{aligned} &Gradient_{neu1} = \tfrac{\partial Loss}{\partial neu1} \\ &= -label \cdot (1-f) \cdot syn1 + (1-label) \cdot f \cdot syn1 \\ &= -(label - f) \cdot syn1 \end{aligned} \tag{7}$$

$$\begin{aligned} &Gradient_{syn1} = \tfrac{\partial Loss}{\partial syn1} \\ &= -label \cdot (1-f) \cdot neu1 + (1-label) \cdot f \cdot neu1 \\ &= -(label - f) \cdot neu1 \end{aligned} \tag{8}$$

The gradient transfer from the hidden layer to the output layer is even simpler, because the hidden layer is the sum of the variates in the input layer; thereby, the gradient of the input layer is namely that of the hidden layer.

As shown in the Figure 3, the sentence position is the subscript of the current word in the sentence. Set the specific sentence A B C D as an example, when the code below is entered for the first time, the current word is A, its sentence position is 0. *b* is the word generated randomly between 0 and window −1, and the size of entire window is (2 × window + 1 − 2 × b), which means to read window −b words on both the left and right side. It is obvious that with the window slipping from left to right the size is random, and when the random variation is from 3 (when b = window −1) to 2 (when b = 0), the random value b decides the size of the current window. In the code, nue1 is the vector of the hidden layer, namely the sum of the corresponding vector of the context (the words in the window except the current word.)

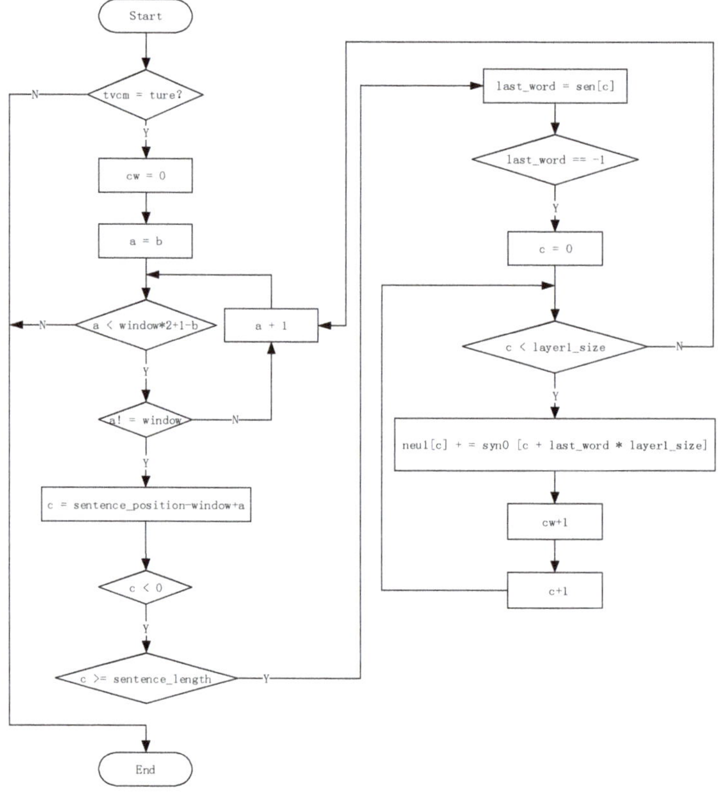

Figure 3. Part of the source code flow chart of the vectorization coding model.

As shown in Figure 4, the prediction of the vectorized code model $P\ (w_t/w_{t-k}, w_{t-(k-1)}, \ldots\ w_{t-1}, w_{t+1}, w_{t+2}, \ldots, w_{t+k})$ can be obtained. The operation from input layer to the hidden layer is actually the sum of the context vector. The structure of the vectorized code model consists of four layers including the input layer, forward hidden layer, inverse hidden layer and the output layer, which is a discriminative model.

The core concept is the probability that the appearance of the current word will be maximum when the number of words appearing around the current word in the context is c, namely conditional probability maximization. When the words (the number of words is c) around the current two words tends to appear frequently, the vector of the two words will be very close, and the vector distance of this word will shrink.

Based on such a theory, it can be considered that in the date set constituted by the road intersection serial numbers, when the current vehicle is passing the testing road intersection, if the similar situation happens frequently in the adjacent road intersections, the relevance of those road intersections is high. Thereby, the spatial-tempo feature of the traffic flow can be fully manifested by the corresponding vector of the road intersection.

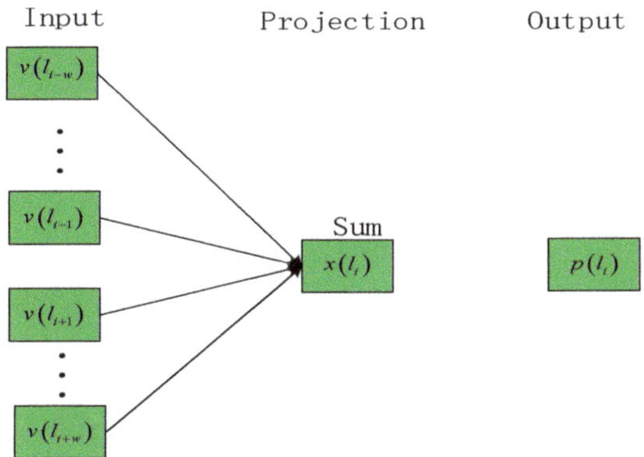

Figure 4. Calculation process the vectorization coding model.

Where, $v(l_i)$ is the vector of road intersection l_i. w is the step length. The Equation for the output of the projection layer and output layer is as follows:

$$X(l_i) = \sum_{x=i-w}^{i-1} v(l_x) + \sum_{x=i+1}^{i+w} v(l_x) \tag{9}$$

The adjust equation of the vectorized code parameter based on the negative sampling is:

$$\theta(l_x) = \theta(l_x) + \eta\left[L^{l_i}(l_x) - \frac{1}{1+e^{-X^T(l_i)\theta(l_i)}}\right]X(l_i), l_x \in NEG(l_i) \tag{10}$$

$$L^{l_i}(l_x) = \begin{cases} 0 \quad , \quad l_x = l_i \quad ; \\ 1 \quad , \quad l_x \neq l_i \quad ; \end{cases} \tag{11}$$

where, η is learning rate. *NEG*(l_i) is the negative sampling set. The update Equation of the vector of the input road intersection is:

$$v(l_x) = v(l_x) + \eta\sum_{l_x \in NEG(l_i)\cup\{l_i\}}\left[L^{l_i}(l_x) - \frac{1}{1+e^{-X^T(l_i)\theta(l_i)}}\right]\theta(l_x) \tag{12}$$

$$l_x \in \{l_{i-c}, \cdots, l_{i-1}, l_{i+1}, \cdots, l_{i+c}\}$$

This paper considers all track $\{T_p, p \in V, V\ is\ the\ set\ of\ all\ vehicle\ Numbers\}$ as the input of vectorized code model for the road intersection. The corresponding vector matrix for each track data is obtain by obtaining the corresponding vector of each road intersection serial number through training.

2.3. Calculation of Spatial Relations

The spatial relationship between intersections can be expressed by calculating the distance between the corresponding vectors of each intersection. The spatial closeness of any two intersections can be expressed by Equation (13).

$$\text{distance}(l_i, l_j) = \|v(l_i) - v(l_j)\| \tag{13}$$

For each intersection serial number l_a, we calculated the Euclidean distance between l_a and the corresponding vectors of other intersection serial numbers and sorted the results from smallest to largest. After this, we can get the serial number of K intersections that are closest to it. Their serial number is $\{l_b, l_c, l_d, \cdots\}$. Combine them with l_a, and the result is $G_a : \{l_a, l_b, l_c, l_d, \cdots\}$. The elements in this group are ordered from smallest to largest distance from the vector of the first intersection serial number (See Algorithm A2 in the Appendix A). Finally, the vehicle trajectory matrix of these intersections is input into the BRNN model for training and prediction. Such prediction results can not only reflect the close relationship between adjacent road intersections in traffic flow, but also reflect the spatial traffic flow relationship.

3. Experimental Results and Discussion

3.1. Data Description

This paper adopts the monitoring data of road intersections in partial sections in Shen Zhen, of which the list information of the intersection serial number includes six aspects—intersection serial number, ID of the monitoring point, name of the monitoring point, direction, the involved road and date. The passing vehicle record data in the intersections includes the plate number, vehicle color, time, the name of the monitoring point and the ID of the vehicle lane. Setting the intersection with the serial number 101000206 as the predicting point, nine intersection serial numbers with the closest relation to the predicting point are obtained by calculating the spatial relationship, as shown in Figure 5. Those ten intersections contain 3,469,741 primary data which were collected during 12 March 2018 to 1 April 2018. Through the trajectory data set statistic algorithm we obtained 725, 570 track data which were low-quality primary data with a different step length S for each track owing to the deviation during the data collection, such as equipment defect on certain intersections, identification error of the moving vehicle, failure of the information collection and position error of the moving vehicle, and so on. In order to solve those questions above, this paper input the corresponding vector matrix of the track data into the BRNN for training and prediction. BRNN will make full use of the historical and future information of the spatio-temporal series to predict the intersection serial number that the track actually passed. After several trainings, the model can obtain the high-quality track data, perform the counting statistics on the time of fixed step length, and thereby predict the traffic volume in the predicting intersection. Comparing the actual traffic volume on 1 April with the predicting value:

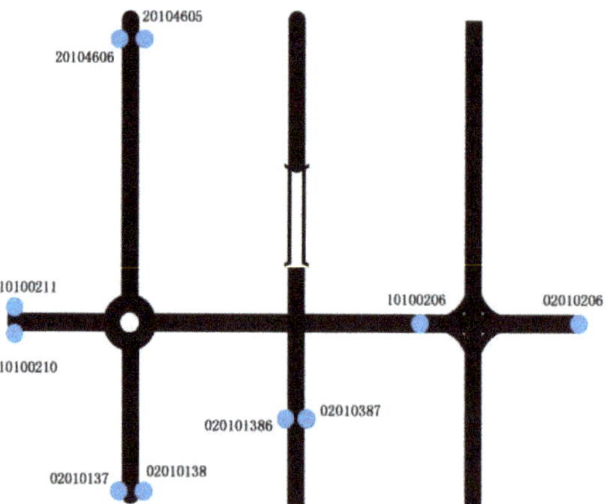

Figure 5. Distribution and ID of intersections in the testing road network.

3.2. Error Evaluation Index

This paper uses mean square error (MSE) and accuracy (ACC) as evaluation indicators. The specific definition is as follows:

$$MSE = \frac{1}{N}\sqrt{\sum_{i=1}^{N}(y_i - \hat{y}_i)^2} \tag{14}$$

$$ACC = \left(1 - \frac{1}{N}\sum_{i=1}^{N}\left|\frac{\hat{y}_i - y_i}{y_i}\right|\right) \times 100\% \tag{15}$$

where: y_i is the true value of traffic flow; $\hat{y}_i$ is the predicted result; N is the number of predicted samples.

3.3. Result of the Intersection Vector

During the training of the vectorized mode model, the negative sampling set needed setting, which will be set as 10 here. In order to improve the training speed and the vector quality of the obtained word, the loss value of the vectorized code model with the different step length w needs analyzing as well for a more accurate intersection vector. To analyze the variation rule of the loss value under a different step length, with the step length w between 1 and 8, after training experiments, the variation pattern of the loss value under different conditions can be summarized as shown in Figure 6.

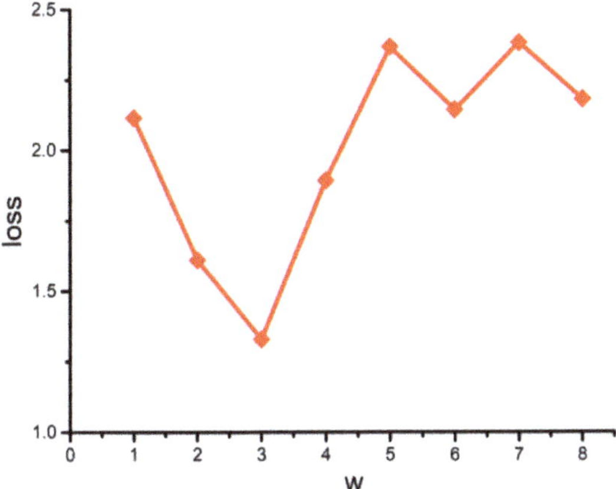

Figure 6. Linear graph of loss variation.

As known from the graph, the loss value meets a decrease before increasing, and slightly fluctuates after that. When w is 3, this loss value reaches the bottom, and thereby the intersection vector is obtained by setting the step length w as 3. Accordingly, the variation curve of the loss value with the iterations under the situation of negative sampling set being 10 and the step length w being 3, is obtained as shown in Figure 7.

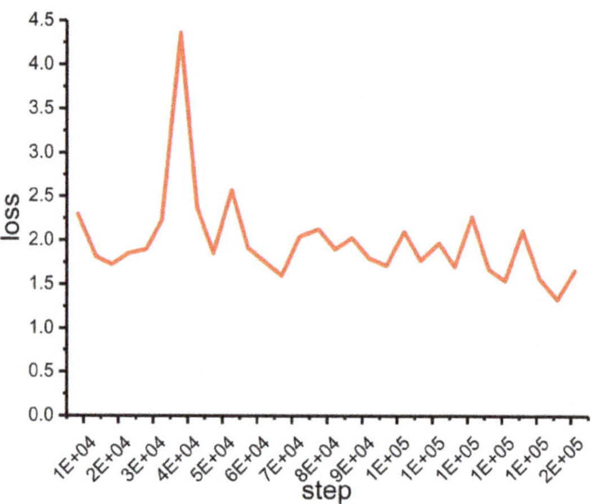

Figure 7. Curve of loss value varying with iterations.

3.4. Result of BRNN Experiment

After a great deal of training of BRNN, the traffic volume on the predicting point (10100206) has been predicted during 25 March to 31 March, as shown in Figure 8.

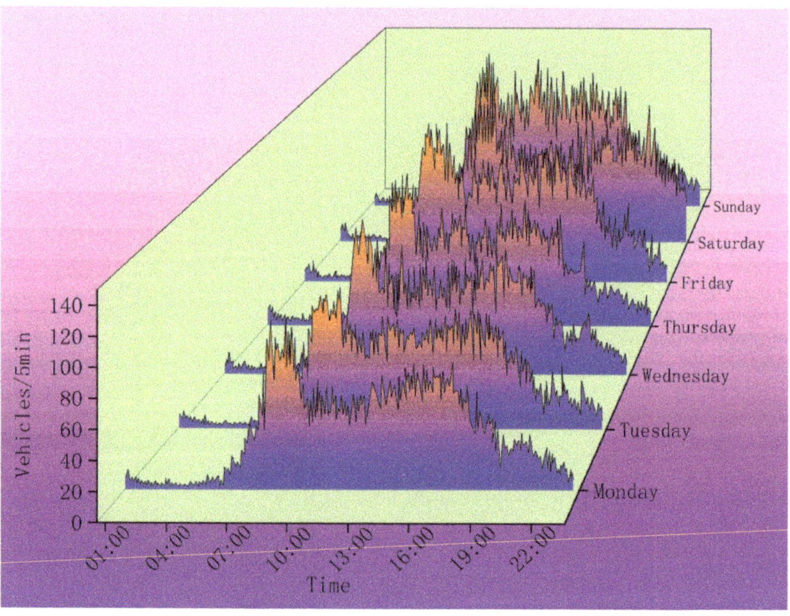

Figure 8. The predicted value in a week.

As shown in Figure 8, the rush hour shows during 7 a.m. to 10 a.m. and 4 p.m. to 7 p.m. when the traffic is intense and the congestion is severe. According to the peak value and traffic intensity, traffic volume on Saturday and Sunday is considerably large. Accordingly, it is evident that the variation curve accords with the actually daily traffic situation.

As shown in Figure 9, the predicting training results of the intersection (10100206) in one week reveal the deviation existing between the predicting results of BRNN and the true value. In order to more thoroughly analyze the predicting performance of BRNN, the RNN is trained as well during the training of the BRNN, and the result of comparison of the two models is as follows:

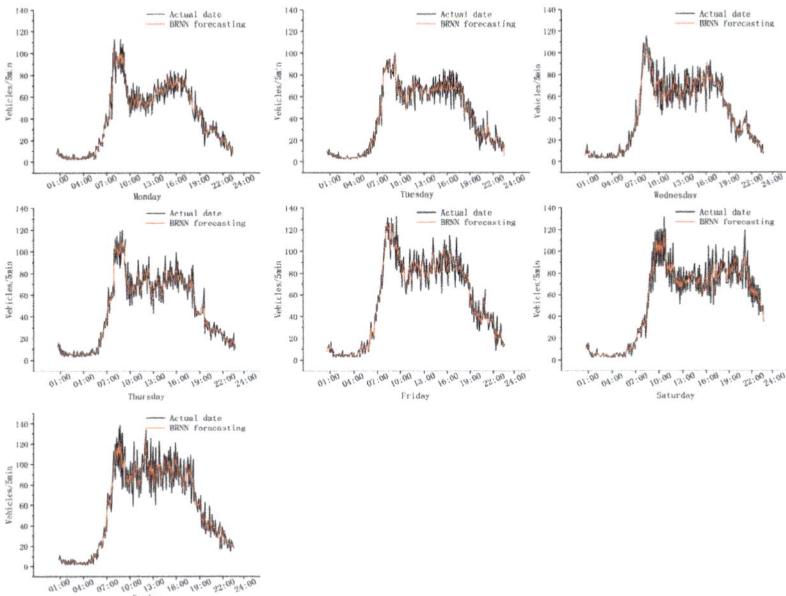

Figure 9. The comparison results of predicted and real traffic flow in a week's time.

As shown in Figure 10, the result of BRNN is better than that of RNN. From the result of RNN, it is evident that RNN has higher similarity with short-term data series, but lower accuracy on long-term data series. This is because RNN cannot make use of the information in the past and the future of any specific moment, and thus will lose certain characteristics as the time runs longer, which tends to generate the predicting deviation.

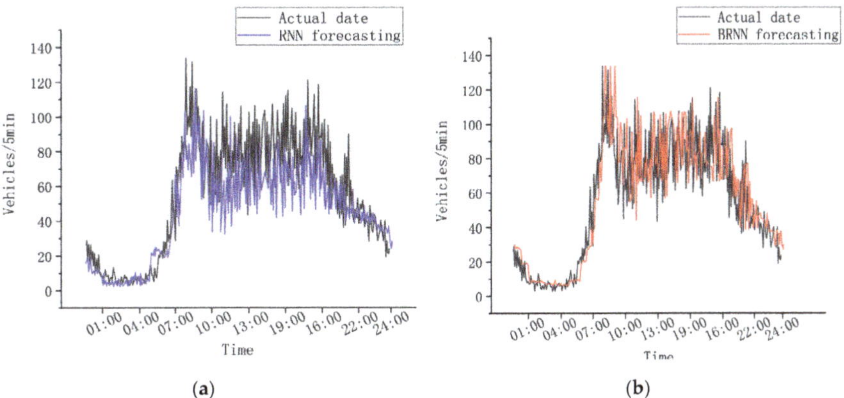

Figure 10. This is the model prediction curves of RNN and BRNN. This first graph (**a**) is the comparison diagram of real traffic flow and predicted traffic flow by RNN, while the second (**b**) is the comparison diagram of real traffic flow and predicted traffic flow by BRNN.

3.5. Data Fitting and Residual Analysis

As shown in Figure 11, the discrete figure consists of three types of data on one coordination, and we fit those data by a nonlinear curve due to the nonlinear characteristic of the traffic flow, with GaussAmp as the fitting function during fitting to construct 95% of the confident belt, the Levenberg-Marquardt optimizing algorithm and the root of Reduced Chi-square to shrink the standard deviation. The result is as follows:

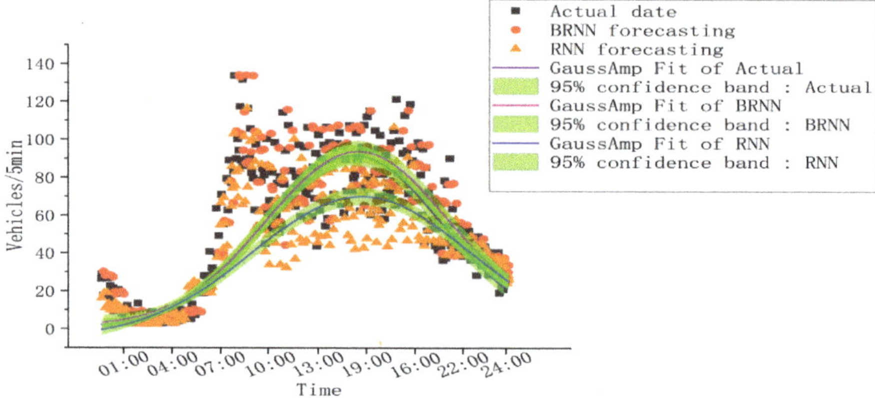

Figure 11. Fitting curve of the scatter value.

The fitting residual graphs of true value, BRNN and RNN, are shown as Figures 12–14. The graphs show the result of fitting the Y value in the traffic flow prediction in the standard form, proportion form and column form. Comparing the intensity of the value near degree 0 in the three graphs, it is obvious that the density of BRNN is higher than RNN, thus with a better fitting result. In Figure 12, the fitting result of RNN is lower than expected due to the RNN being incapable to use the future information and losing part of data with input data becoming increasingly dense during training. However, the fitting result of BRNN is satisfied due to the capability of making full use of information in the past and the future, which makes the predicting result close to the true value. In Figure 13, all data points are distributed around the diagonal line, and the data point from BRNN is closer to the line with high density, and thus less deviation. In Figure 14 the column graphs of three predicting results reveal that the prediction of BRNN is closer to the true traffic volume.

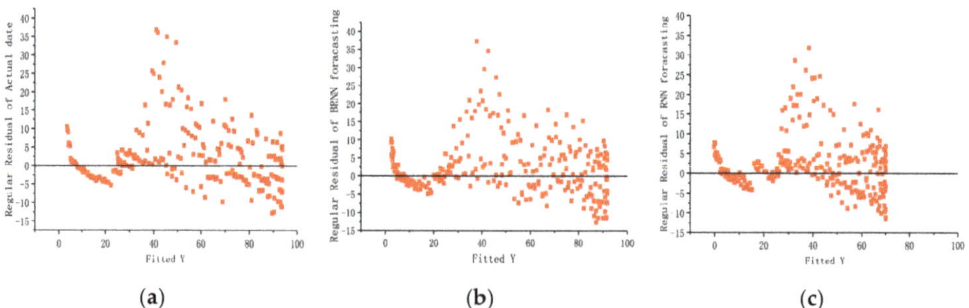

Figure 12. This is the residual diagram of the predicted results for each model. The graph (**a**) is the regular residual of actual data under the fitted Y value. That of (**b**) is the regular residual of BRNN model under the fitted Y value. The one of (**c**) is the regular residual of RNN model under the fitted Y value.

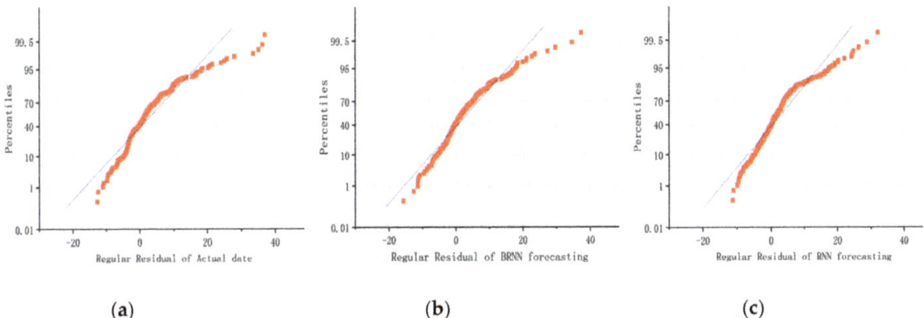

Figure 13. The graph designated (**a**) is the regular residual of actual data under the percentiles form, (**b**) is the regular residual of BRNN forecasting under the percentiles form, and (**c**) is the regular residual of RNN forecasting under the percentiles form.

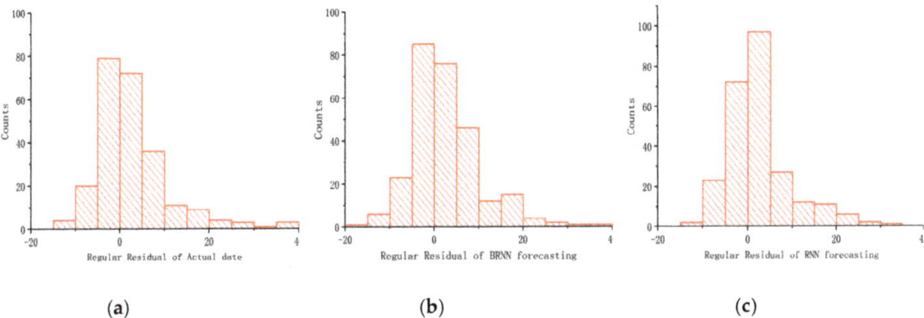

Figure 14. This is a histogram of regular residual of traffic flow prediction results. Graph (**a**) is a histogram regular residual of the actual data. Graph (**b**) is a histogram regular residual of the BRNN model. Graph (**c**) is a histogram regular residual of the RNN model.

The Equations used in the data fitting are shown in Table 1. The fitting results of the data are shown in Table 2.

Table 1. Models and equations used in the fitting.

Model	Equation
GaussAmp	$y = y_0 + A * \exp\left(-0.5 * ((x - x_c)/w)^2\right)$

Table 2. Data table of fitting results.

Plot	Actual Date	BRNN Forecasting	RNN Forecasting
y_0	0.8452 ± 2.02181	1.70427 ± 2.6873	−4.33261 ± 2.886
x_c	9.08585 ± 0.0815	9.229 ± 0.08364	0.34728 ± 0.1072
w	2.7974 ± 0.12182	2.87558 ± 0.1420	3.42083 ± 0.2021
A	90.87526 ± 2.808	92.04544 ± 3.179	74.42464 ± 2.888
Reduced Chi-Square	8.64135	7.40129	6.03239
R-square(COD)	0.80646	0.79486	0.79342
ADJ.R-square	0.8043	0.79227	0.79093
Residual sum of squares	96066.57313	90180.34987	68932.96234

From the squared value of the correlation coefficient R (COD) of the BRNN and RNN, it is apparent that the closer the squared value of R is to 0, the better the fitting degree. When the squared value of R of the BRNN and RNN are 7.40129 and 6.03239, respectively, that of BRNN is much closer to 1, and thus the deviation between the curve after fitting and the primary data point is smaller, and the prediction is more accurate. As to the residual sum of squares, that of BRNN is obviously larger than that of RNN, which means BRNN is better.

3.6. The Effect of Track Length on the Model

In data collection, due to the error of the acquisition equipment (or other error factors), we get the wrong original data. The original data can be processed to obtain the track data, and the vector matrix corresponding to the track data can be used as the input of the model for training and prediction. Therefore, when errors occur in the collected data, the most significant impact on the model is the accuracy of prediction results. Let us discuss how error data affects model accuracy. In other words, explore the influence of trajectory length s con model accuracy.

As shown in Section 2.2, this s is the number of intersections, so when the data is wrong, we get that the length of the trajectory will change or stay the same. In this case, it reduces the track collection density, affects the description of moving vehicle tracks, and also gets the wrong track data. Therefore, it is equivalent to discussing the influence of s on model accuracy.

There are two cases of the most common trajectory missing (we only discuss the most common cases). For example, if the HD camera fails, it will not work properly. At this time, the serial number of the road intersection of the trajectory data will be missing, and the track length s will become smaller. In another case, when the camera recognizes an error in the license plate number, the serial number of the intersection does not change, but the identified vehicle is no longer a real vehicle.

When s changes. We select the trajectory of $s \geq 5$. Because when the trajectory length is less than 5, the trajectory does not reflect the moving characteristics of the vehicle between road intersections (If there are too many such trajectories, it can be inferred that the damage of the collection equipment is too serious. It is recommended that the equipment at the intersection be inspected and repaired). Set the default length of the track length to 7 (value range is 5, 7, 8, 9, 10, 11, 12, 13, 14, 15, 16, 17) to compare the accuracy of the model at different track lengths. The result is shown in the Figure 15.

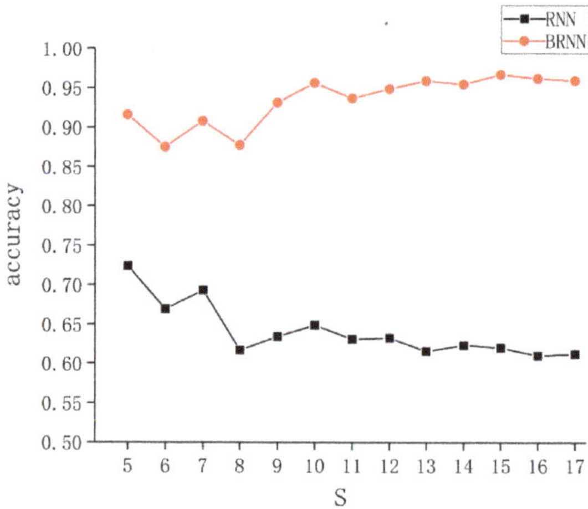

Figure 15. Accuracy of the model for different s values.

As shown in the above figure, the accuracy of the BRNN model is relatively low at the beginning and fluctuates strongly, and then becomes a slow upward trend, then slowly stabilizes, while RNN is reversed. When $s = 6$, the BRNN model has the lowest accuracy rate of 87.53%. When $s = 15$, the BRNN model has the highest accuracy rate of 96.75%.

In summary, when there are more errors in data collection, s will become smaller. As a result, the accuracy of the BRNN model is reduced, and the magnitude of the change is large, which is an unstable state.

When S does not change. Due to camera recognition error. For example, the real license plate number AT6351 is incorrectly identified as AT63S1. The track length S thus acquired is constant, and the prediction accuracy obtained after the model training is higher than the true accuracy. Therefore, the BRNN model has to do more work. This work is to repair the trajectory. This will result in an increase in the computation time of the model (After the track data is repaired, it has little influence on the prediction accuracy of the model and can be ignored).

In order to further assess the predicting performance of BRNN, this paper calculates the predicting results of MSE and ACC, respectively, and makes comparison with RNN, ARIMA, SVR, DBN-SVR, LSTM and KNN-LSTM [30]. The results are shown as in Table 3 that BRNN processes the best predicting performance with an accuracy of 96.75%, which is 24.34%, 25.06, 8.58%, 7.15% and 6.36% higher than RNN, ARIMA, SVR, DNB-SVR and LSTM, respectively. Meanwhile compared with KNN-LSTM, which takes into account the spatio-temporal relationship, the accuracy improves 2.16% as well. Consequently, this experiment illustrates that BRNN has excellent predicting performance.

Table 3. Prediction performance of different models.

Prediction Model	The Evaluation Index	
	MSE	ACC/%
ARIMA	1.5134	61.09
RNN	0.7365	72.36
SVR	0.3266	88.17
DBN-SVR	0.3053	89.60
LSTM	0.1262	90.39
KNN-LSTM	0.0987	94.59
BRNN	0.0831	96.75

4. Conclusions

Based on the deep learning theory, this paper proposed the Bi-directional Regression Neural Network, which can make full use of the intersection information in the past and the future to make up the deficiency of most other models that they are incapable to reflect the spatio-temporal relation of the traffic flow. Meanwhile, this paper, according to the TensorFlow platform, designed the vectorized code to extract the spatial feature of the intersections and to screen out the intersections related closely to the predicting point, as well as to train and predict through inputting the track vector matrix of those intersections into the BRNN. This paper also carried out the predicting data fitting of BRNN and RNN, and the residual analysis. The results indicated that the BRNN can make full use of the information in the past and future to process better performance, whose accuracy is up to 96.75%. In addition, this paper made the compassion of the deviation between BRNN and RNN, ARIMA, SVR, DBN-SVR and LSTM, as well as KNN-LSTM, and the consequences show that the accuracy of BRNN is 16.298% higher on average, which further proved the excellent predicting performance of BRNN.

Author Contributions: Conceptualization, S.Z. and Q.Z.; Data curation, S.Z. and Q.Z.; Formal analysis, S.Z. and Q.Z.; Funding acquisition, S.Z. and S.L.; Investigation, Q.Z. and Y.B.; Methodology, S.Z. and Q.Z.; Project administration, S.Z.; Resources, Q.Z. and Y.B.; Software, S.Z., Q.Z. and S.L.; Supervision, S.Z.; Validation, S.Z., Q.Z. and S.L.; Writing – original draft, Q.Z.; Writing – review & editing, Y.B.

Funding: This research was supported by the Shaanxi Provincial Education Department serves Local Scientific Research Plan in 2019 (Project NO.19JC028) and Shaanxi Provincial Key Research and Development Program (Project NO.2018ZDCXL-G-13-9) and Shaanxi province special project of technological innovation guidance (fund) (Program No.2019QYPY-055).

Conflicts of Interest: The authors declare no conflict of interest.

Appendix A

Algorithm A1. Trajectory data set statistical algorithm.

```
map:
map (key, value, context):
   passRecord = value. split();
   for pr in passRecord:
         context. write(pr. vehicleNo, pr. locationId);
reduce:
reduce (key, values, context):
   result = temp;
   for value in values;
         result.append (value);
   context.write(key, result);
//The final output is < vehicle ID, intersection serial number list >
```

Algorithm A2. The algorithm of spatial closeness.

```
G = Empty collection;
for la in L: //L is the set of all intersection serial number
   list<lx, distance(la, lx)>1;
   for lx in L:
      if(lx ≠ la):
       d = distance(la, lx);
       1.put(lx, d);
   1.sortByValues();//Sort by distance
   Ga = 1.top(k) ∪ la;//Banked the nearest k intersection serial number collection
   G = G ∪ Ga;
return G;
```

References

1. Zambrano-Martinez, J.L.; Calafate, C.T.; Soler, D.; Lemus-Zúñiga, L.-G.; Cano, J.-C.; Manzoni, P.; Gayraud, T. A Centralized Route-Management Solution for Autonomous Vehicles in Urban Areas. *Electronics* **2019**, *8*, 722. [CrossRef]
2. Shaikh, R.A.; Thayananthan, V. Risk-Based Decision Methods for Vehicular Networks. *Electronics* **2019**, *8*, 627. [CrossRef]
3. Botte, M.; Pariota, L.; D'Acierno, L.; Bifulco, G.N. An Overview of Cooperative Driving in the European Union: Policies and Practices. *Electronics* **2019**, *8*, 616. [CrossRef]
4. Kalamaras, I.; Zamichos, A.; Salamanis, A.; Drosou, A.; Kehagias, D.D.; Papadopoulos, S.; Tzovaras, D. An interactive visual analytics platform for smart intelligent transportation systems management. *IEEE Trans. Intell. Transp. Syst.* **2017**, *19*, 1–10. [CrossRef]
5. Shuanfeng, Z.; Wei, G.; Chuanwei, Z. Extraction Method of Driver's Mental Component Based on Empirical Mode Decomposition and Approximate Entropy Statistic Characteristic in Vehicle Running State. *J. Adv. Transp.* **2017**, *2017*, 1–14.
6. Zhao, S. The implementation of driver model based on the attention transfer process. *Math. Probl. Eng.* **2017**, *2017*, 15. [CrossRef]
7. Gang, Y.; Le, W.; Lizhen, D.; Fangping, X. Traffic Flow Prediction Based on Adaptive Particle Swarm Optimization of Least Square Support Vector Machine. *Control Eng. China* **2017**, *24*, 1838–1843.
8. Sheng, N.; Tang, U.W. Spatial Analysis of Urban Form and Pedestrian Exposure to Traffic Noise. *Int. J. Environ. Res. Public Health* **2011**, *8*, 1977–1990. [CrossRef] [PubMed]
9. Kumar, S.V.; Vanajakshi, L. Short-term traffic flow prediction using seasonal ARIMA model with limited input data. *Eur. Transp. Res. Rev.* **2015**, *7*, 21. [CrossRef]
10. Dihua, S.; Chao, L.; Xiaoyong, L. An Improved Nonparametric Regression Algorithm for Short-term Expressway Traffic Flow Forecasting. *J. Highw. Transp. Res. Dev.* **2013**, *30*, 112–118.

11. Yu, W.; Su, J.; Zhang, W. Research on Short-Time Traffic Flow Prediction Based on Wavelet de-Noising Preprocessing. In Proceedings of the IEEE Ninth international Conference on Natural Computation, Shenyang, China, 23–25 July 2013; pp. 252–256.
12. Li, L.; He, S.; Zhang, J.; Ran, B. Short-term highway traffic flow prediction based on a hybrid strategy considering temporal-spatial information. *J. Adv. Transp.* **2016**, *50*, 2029–2040. [CrossRef]
13. Mikolov, T.; Sutskever, I.; Chen, K.; Corrado, G.S.; Dean, J. Distributed Representations of Word and their Compositionality. In Proceedings of the Advances in Neural Information Processing Systems, Stateline, NV, USA, 5–10 December 2013; pp. 3111–3119.
14. Lee, C.-H.; Chih-Hung, W. A Novel Big Data Modeling Method for Improving Driving Range Estimation of EVs. *IEEE Access* **2015**, *3*, 1980–1993. [CrossRef]
15. Xiaobo, C.; Xiang, L.; Zhongjie, W.; Jun, L.; Yingfeng, C.; Long, C. Short-term Traffic Flow Forecasting of Road Network Based on GA-LSSVR Model. *J. Transp. Syst. Eng. Inf. Technol.* **2017**, *17*, 60–66.
16. Ma, X.; Yu, H.; Wang, Y.; Wang, Y. Large-scale transportation network congestion evolution prediction using deep learning theory. *PLoS ONE* **2015**, *10*, e0119044. [CrossRef] [PubMed]
17. Xuankun, D.; Liang, W.; Hongwei, D.; Zhuang, X. Traffic Flow Prediction Based on Deep Neural Networks. *Comput. Eng. Appl.* **2019**, *55*, 228–235.
18. Mackenzie, J.; Roddick, J.F.; Zito, R. An Evaluation of HTM and LSTM for Short-Term Arterial Traffic Flow Prediction. *IEEE Trans. Intell. Transp. Syst.* **2019**, *20*, 1847–1857. [CrossRef]
19. Feldman, O. *The GEH Measure and Quality of the Highway Assignment Models*; Association European Transport: Glasgow, Scotland, UK, 2012.
20. Tian, Y.; Pan, L. Predicting short-term traffic flow by long short-term memory recurrent neural network. In Proceedings of the IEEE international conference on smart city/SocialCom/SustainCom (SmartCity), Chengdu, China, 19–21 December 2015; pp. 153–158.
21. Pamuła, T. Classification and prediction of traffic flow based on real data using neural networks. *Arch. Transp.* **2012**, *24*, 519–529. [CrossRef]
22. Pamuła, T. Short-term traffic flow forecasting method based on the data from video detectors using a neural network. In *Activities of Transport Telematics (Communications in Computer and Information Science)*; Springer: Berlin, Germany, 2013; pp. 147–154.
23. Huang, W.; Song, G.; Hong, H.; Xie, K. Deep architecture for traffic flow prediction: Deep belief networks with multitask learning. *IEEE Trans. Intell. Transp. Syst.* **2014**, *15*, 2191–2201. [CrossRef]
24. Halawa, K.; Bazan, M.; Ciskowski, P.; Janiczek, T.; Kozaczewski, P.; Rusiecki, P. Road traffic predictions across major city intersections using multilaye perceptrons and data from multiple intersections located in various places. *IEEE Trans. Intell. Transp. Syst.* **2016**, *10*, 469–475. [CrossRef]
25. Guo, J.; Huang, W.; Williams, B.M. Adaptive Kalman filter approach for stochastic short-term traffic flow rate prediction and uncertainty quantification. *Transp. Res. C Emerg. Technol.* **2014**, *43*, 50–64. [CrossRef]
26. Min, W.; Wynter, L. Real-time road traffic prediction with spatiotemporal correlations. *Transp. Res. C Emerg. Technol.* **2011**, *19*, 606–616. [CrossRef]
27. Schuster, M.; Paliwal, K.K. Bidirectional Recurrent Neural Networks. *IEEE Trans. Signal Process.* **1997**, *11*, 2673–2681. [CrossRef]
28. Mikolov, T.; Chen, K.; Corrado, G.; Dean, J. Efficient Estimation of Word Representations in Vector Space. *arXiv* **2013**, arXiv:1301.3781.
29. Mikolov, T.; Sutskever, I.; Chen, K.; Corrado, G.; Dean, J. Distributed Representations of Word and Phrases and their Compositionality. In Proceedings of the 26th International Conference on Neural Information Processing Systems, Lake Tahoe, NV, USA, 5–10 December 2013.
30. Luo, X.; Li, D.; Yu, Y.; Zhang, S. Short-term traffic flow prediction based on KNN-LSTM. *J. Beijing Univ. Technol.* **2018**, *44*, 1521–1527.

Article

Masivo: Parallel Simulation Model Based on OpenCL for Massive Public Transportation Systems' Routes

Juan Ruiz-Rosero [1,*], Gustavo Ramirez-Gonzalez [1] and Rahul Khanna [2]

[1] Departamento de Telemática Universidad del Cauca, Calle 5, No. 4-70, Popayan, Cauca 190002, Colombia; gramirez@unicauca.edu.co

[2] Intel Corporation, 2111 NE 25th Ave., Hillsboro, OR 97124, USA; rahul.khanna@intel.com

* Correspondence: jpabloruiz@unicauca.edu.co; Tel.: +57-(2)-820-9900

Received: 19 November 2019; Accepted: 3 December 2019; Published: 8 December 2019

Abstract: There is a large number of tools for the simulation of traffic and routes in public transport systems. These use different simulation models (macroscopic, microscopic, and mesoscopic). Unfortunately, these simulation tools are limited when simulating a complete public transport system, which includes all its buses and routes (up to 270 for the London Underground). The processing times for these type of simulations increase in an unmanageable way since all the relevant variables that are required to simulate consistently and reliably the system behavior must be included. In this paper, we present a new simulation model for public transport routes' simulation called Masivo. It runs the public transport stops' operations in OpenCL work items concurrently, using a multi-core high performance platform. The performance results of Masivo show a speed-up factor of 10.2 compared with the simulator model running with one compute unit and a speed-up factor of 278 times faster than the validation simulator. The real-time factor achieved was 3050 times faster than the 10 h simulated duration, for a public transport system of 300 stops, 2400 buses, and 456,997 passengers.

Keywords: simulation; parallel; multi-core; public transport; mass transit; OpenCL

1. Introduction

Millions of more people are expected to migrate to urban centers in the following years. The United Nations predicts that for 2050, 66% of the population will live in urban areas [1]. Hence, transportation planners are embracing new ways of tackling the old problem of congestion. Public Transport Systems (PTS) move a large number of people while consuming fewer resources (lower energy requirements, lower accident cost, and air quality improvements). Currently, one of the top applications of the Internet of Things (IoT) is the development of smart cities [2]. Smart cities target the increasing capacity to collect a large quantity of data from a PTS and process them in real-time (big data processing). This capacity has become the basis for the development of more effective transit modeling and simulation [3]. An efficient transit simulation powers the optimization agents for public transport systems. These optimization agents improve the route path, route frequency, number of transfers, the waiting time, bus or train sizes, ticket price, and others. With these optimizations, different studies reduced the wait time by 75% [4], the travel time by 14% [5], and operational cost by 11% [5].

The basis of a PTS simulation is system modeling. This modeling is divided into three different approaches: macroscopic, mesoscopic, and microscopic [6]. The macroscopic models aggregate traffic streams without analyzing the internal elements. The mesoscopic ones model homogeneous groups of actors and the interactions with little detail. Finally, the microscopic ones model in detail the dynamics of each actor (vehicles or passengers) and their interactions with others [6]. The public transport system components have been simulated inside these three different models. The macroscopic modeling has been used for simulating the vehicle accumulation of public transport vehicles along with private cars [7]; furthermore, for the modeling of the public transport vehicles' speed [8] and for the road safety impacts of public transport [9]. Similarly, the mesoscopic models have been used for simulating the crowding levels on-board the public transport vehicles [10], emergency evacuation methodologies [11], traffic safety policy evaluation [12], and a multi-modal mobility simulator for large scale scenarios [13]. The microscopic models are the most used for public transport simulation. These models are included in the simulation of the operations at public transport stops [14,15], the public transport system's reliability [16,17], bus priority methods [18,19], exclusive bus lanes [20,21], bus routes [22], transfer organization of public transportation terminals [23], multimodal systems [24], and traffic networks [25].

Similarly, we performed a literature review supported by the scientometric tool ScientoPy [26] to find the most simulated and modeled components in the PTS. For this, we downloaded a dataset from the Clarivate Web of Science (WoS) and Scopus, with the search string ("public trans*" OR "mass transit") AND ("simulation" OR "model*") applied to the documents' title. Figure 1 shows the top seven most simulated or modeled components for PTS. In-door air quality had the most documents, with papers that included diseases transmission [27] and air quality prediction [28–33]. Second, accessibility included modeling the potential effect of shared bicycles on public transport travel times [34] and modeling access to PTS [35–37]. Line planning simulation and modeling covered models and methods for line planing [38], passenger routing decision [39,40], and PTS connection simulation [41].

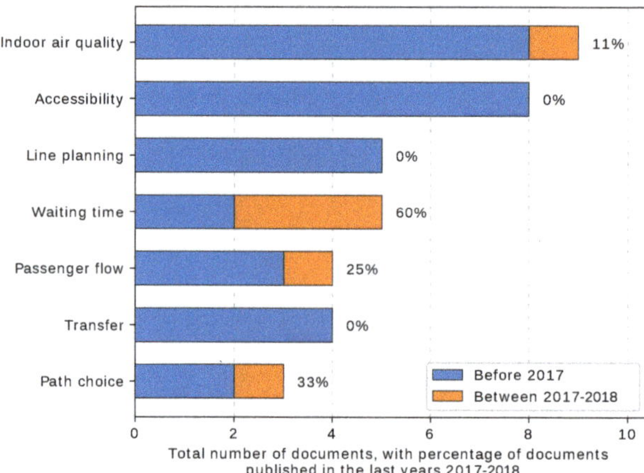

Figure 1. Number of documents related to Public Transport Systems' (PTS) components simulated or modeled reported by the scientific literature in the WoS and Scopus databases.

Waiting time was the topic with the highest growth in PTS simulation and modeling, with 60% of the documents published in the last two years. It includes impacts on passengers' waiting time uncertainty [42], timetable optimization models for minimizing passenger waiting [43], and a stochastic model for the passenger arrival process [44]. Finally, other simulation and modeling components are passenger flow [45–48], transfers [49–52], and path choice [53,54].

These kinds of implementations suffer from two critical issues. First is scalability, because most of them only offer single machine solutions that are not adequate to process large scale data. Second is granularity, due them not considering microscopic traffic situations such as individual passengers/bus/train modeling [55]. For these reasons, novel works have developed new models and simulators that employ parallel architectures to get the simulation results more efficiently. Currently, we find mesoscopic [13,56–60] and microscopic [61] traffic simulators based in GPUs (Graphics Processing Units), HPC (High Performance Computing) [62–64], and microscopic simulation based on Apache Spark [55,65]. Nevertheless, these modeling and simulation approaches do not consider each passenger as a unique entity in the simulation model.

In this paper, we present Masivo. Masivo is a new parallel simulation model based in OpenCL using a multi-core high performance platform for massive public transportation systems. Masivo works with predefined public transport system conditions, which include the stops' total number, the stops' capacity, and the Origin-Destination matrix (OD). This OD matrix and routes' information are updated to this model via CSV files. Masivo gets the simulation results for total alighted passengers and average commute time. Similarly, it shows the performance indicators.

We organized this paper as follows. First, "Section 2. Public Transport Systems' Basis " describes the public transport system, the simulation variables, and output indicators. Then, in "Section 3. Masivo Public Transport Routes' Simulation", we present the components, the input parameters, the output results, and the algorithms for the Masivo simulator. In "Section 4. Results", we summarize the validation and performance results of Masivo. Later, in "Section 5. Discussion", we discuss Masivo's performance against other parallel traffic and public transport simulators. Finally in "Section 6. Conclusions", we summarize all the concluding remarks and future work.

2. Public Transport Systems' Basis

In a public transport system, the buses, trains, or other forms of transport are used by people to travel from one place to another [66]. Figure 2 shows a small public transport system example, which consists of three stops (S_i) separated 2 km and two routes (R_1 and R_2). In this system, we have, for instance, passengers from the origin stop S_1 that travel to the destination stops S_2 and S_3. The origin-destination matrix (OD) denoted as D represents the number of passengers that travel from each stop to another. For example, if we have 10 passengers that travel from S_1 to S_2, the value $D_{1,2}$ is 10.

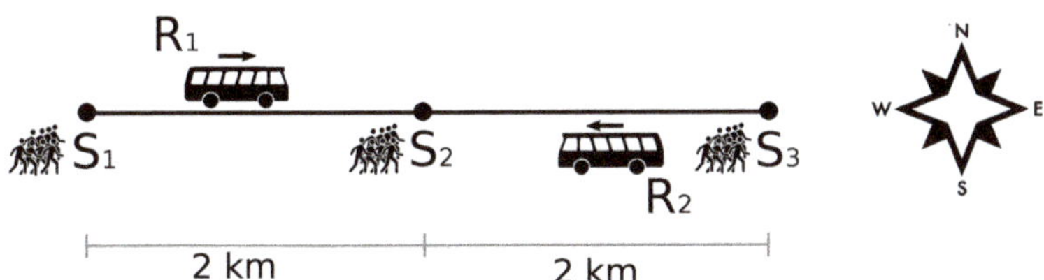

Figure 2. Small example of a public transport systems with three stops and two routes.

In this way, the OD matrix can describe the full system demand, as shown in Equation (1). Here, we see a diagonal of zero, which indicates that there are no passengers that travel from the origin station to the same origin station.

$$D = \begin{bmatrix} 0 & 10 & 20 \\ 10 & 0 & 10 \\ 20 & 10 & 0 \end{bmatrix} \tag{1}$$

The routes are designed to transport passengers from one stop to another. Figure 2 shows two routes (R_1 and R_2). R_1 has the direction west to east, and R_2 has the direction east to west. The RT_r value defines the total trip time (in seconds) for route r, that is the time needed by the bus to complete the full route. The value f_r defines the frequency (in seconds^{-1}) of the route r, and it represents how often the bus is dispatched for each route. Therefore, the number of buses needed for each route is defined by Equation (2) as BR_r, where the symbols "$\lceil$" and "$\rceil$" represent a ceiling function that returns the nearest integer that is greater.

$$BR_r = \lceil RT_r f_r \rceil \tag{2}$$

For instance, we need an R_1 frequency of 5 min ($f_r = 1/300\,\text{s}^{-1}$), and the total trip is 570 s, defined by Equation (3), where N is the number of stops:

$$
\begin{aligned}
RT_1 &= \frac{\text{Total distance}}{\text{Bus avg. speed}} + (\text{Stop time}) \times N \\
RT_1 &= \frac{4\ \text{km}}{30\ \text{km/h}} + (30\ \text{s}) \times 3 \\
RT_1 &= 570\ \text{s}
\end{aligned} \tag{3}
$$

As a result, the number of buses needed for R_1 for this case is defined by Equation (4) as:

$$
\begin{aligned}
BR_r &= \lceil 570\ \text{s} \times \frac{1}{300\ \text{s}} \rceil \\
BR_r &= 2
\end{aligned} \tag{4}
$$

As we can see in Equation (2), if we increase the frequency of a route, we will need to have more buses to meet the demand. Nevertheless, the number of buses is limited by the total fleet size or even by the roads' capacity. Then, to find the route's frequency, we have an optimization problem described by Equation (5), where the objective is to minimize the total travel time, restricted by the total fleet size [67].

$$
\begin{aligned}
&\text{minimize} \sum_{i \in N} \sum_{j \in N} D_{ij} T_{ij}(\mathbf{f}) \\
&\text{subject to:} \sum_{r \in R} \lceil RT_r f_r \rceil \leq \text{total fleet size}
\end{aligned} \tag{5}
$$

where:

N	=	total stops;
R	=	total routes;
D_{ij}	=	OD (Origin-Destination) demand;
T_{ij}	=	passenger travel time;
RT_r	=	round-trip time on route r;
f_r	=	frequency on route r;

The passenger travel time T_{ij} includes the expected waiting time, as a function of the routes' frequencies $\mathbf{f}$. Unfortunately, this travel time cannot be calculated directly, because it depends on other variables such as the buses' capacity, the stops' capacity, and the routes' stops' table. Furthermore, the buses' free seats depend on the buses' capacity and the number of passengers that have boarded the buses at the previous stops. This correlation between the buses' free seats with other variables of the systems converts this into an NP-hard problem, where the approach to this problem relies on heuristic techniques for systems of a reasonable size [67].

2.1. Passenger Demand in Public Transport Systems

The Origin-Destination matrices (OD) specifies the passengers' travel demands between the origin and destination nodes in the network. Geographic Information Systems (GIS) have been used effectively for OD data analysis, supporting geocoding OD survey data to point locations [68]. The OD matrix is dynamic because it is time varying. Inside a city's public transport system, the OD matrix varies on weekdays and weekends and even between different hours of the day. A high quality OD matrix is a fundamental prerequisite for any serious transport system analysis. Surveys have been done to define the OD matrices [68,69]; nevertheless, more sophisticated methods are used today for this purpose, such as passive smart card fare collection system data [70,71] or mobile network data [72].

2.2. Public Transport Vehicle Routes

In public transport systems, the vehicles (trains or buses) follow a route to meet the passengers' demand. The route is defined by a set of stops, where the bus picks up or alights passengers.

Furthermore, the route's frequency is defined as how often the bus is dispatched for each route. That also means how frequent a bus route passes one stop. Due to the dynamic nature of the OD, one of the essential components in determining the public transport service is the selection of the most suitable routes' frequency by time-of-day and day-of-week. The routes' design also includes the development of timetables, scheduling buses, and scheduling drivers. The route layout design depends on the passenger flow because routes are configured to provide a direct or indirect connection between origin-destination demand for passengers.

2.3. *Traffic Simulation Software*

There is much literature that deals with topics about the simulation of traffic networks especially focused on traffic simulation of private vehicles [73–81]. This literature presents simulation systems based on the different models already explained (macroscopic, microscopic, and mesoscopic). With the help of these models, different computational tools have been designed to help to simulate traffic environments. For macroscopic models, there are for example tools such as: Strada [82], Metacor [83], Visum [84], and Emme [85], among others. On the other hand, SUMO [86], PTV VISSIM [87], and Aimsun [88] are some of the tools available that allow traffic simulation with microscopic models. For mixed or mesoscopic models, some of the existing tools are: OmniTRANS [89], TransModeler [90], and the Aimsun [88].

For public transport systems, in Bogotá, Colombia, the Bus Rapid Transit (BRT) system Transmilenio uses two of the most recognized tools for strategic route planning. These tools are PVT Vissim for microsimulation (traffic simulation at intersections, stop crowds simulation, etc.) and Emme for route planning (macrosimulation). With PTV Vissim, it is possible to carry out the specific traffic congestion simulation at an intersection; furthermore, traffic light crossing or lane sections, without these unit simulations interacting with each other to predict a global behavior. On the contrary, Emme allows planning routes based on predictions of passenger demand.

Optimizing the routes' frequencies has been widely studied because this type of optimization can reduce the operating costs and travel times of passengers [5,91,92]. Unfortunately, frequency programming optimization for buses and public transport trains is known to be in the class of NP-hard [93] problems. These kinds of problems cannot be solved in polynomial time [94], which indicates that an algorithm may take an indeterminate time to find the solution to this problem.

2.4. *Traffic Simulation Parameters' Specification*

For this research work, we define the public transport system parameters' specification with the following conditions and variables. First, we have the system conditions that are defined in Table 1. These conditions define the total number of stops, a full OD matrix of size $N \times N$, the max passenger capacity of each stop, and the stop position.

Table 1. System conditions' definitions and symbols.

Variable Name	Symbol
Total number of stops	N
Origin-destination matrix	$\mathbf{OD}$
OD value for stop i to j	D_{ij}
Stops max capacity	SC_n
Stops X position	SPX_n
Stops Y position	SPY_n

On the other hand, Table 2 defines the variables related to the routes and buses. Here, we find the total number of routes, the maximum number of buses available per route, the frequency of each route, the maximum passenger bus capacity, and the route stops' table array. The stops' table array represents the stations (or transit stops) where the bus route stops.

Table 2. Routes variables' definitions and symbols.

Variable Name	Symbol
Total number of routes	R
Max buses per routes	BR_r
Routes frequency	f_r
Max bus capacity	BC
Route stop table array	$stops\_table_r$

Next, Table 3 presents the simulation output variables, which indicate the performance of the public transport system with the actual routes' configuration. These output variables consist of the total departed passengers per stop, the alighted passengers at the destination stop, the total average passengers' commute time, and the average passengers' commute time for the alight stop.

Table 3. Simulation output variables' definitions and symbols.

Variable Name	Symbol
Departed passengers from stop	DPS_n
Alighted passengers at stop	APS_n
Average passengers commute time	ACT
ACT at alight stop n	$ACTS_n$

2.5. Simulation Performance Indicators

Table 4 shows the simulation performance variables. These variables indicate the simulation execution time and how fast the parallel simulation is relative to the real system and the sequential execution.

Table 4. Simulation performance variables' definitions and symbols.

Variable Name	Symbol
Total simulation time	tst
Total parallel simulation execution time	pst
Total sequential simulation execution time	sst
Real-time factor	rtf
Speed-up factor	suf
Efficiency	ef
Number of processing threads	p

Simulation performance can be measured using two measures of effectiveness, namely the real-time factor and the speed-up factor. The real-time factor is a measure of the relative speed of simulation concerning real time [95]. In this work, we measure the real time factor of the parallel implementation by using Equation (6). For example, if we are simulating 1 h (3600 s) of the public transport systems' routes in 3 s of execution time by the parallel model, the real-time factor will be 1200.

$$rtf = \frac{tst}{pst} \tag{6}$$

On the other hand, the speed-up factor (suf) is a measure of parallel simulation performance, which reveals how much faster the simulation is executed in parallel versus the sequential one [95]. Equation (7) describes how to calculate it. For instance, if the simulation execution time in parallel is 3 s and in the sequential one 30 s, the speed-up factor is 10.

$$suf = \frac{sst}{pst} \tag{7}$$

The efficiency (ef) describes where the improvements would be useful [96]. Equation (8) describes the efficiency as a function of the number of processing threads (p). It also represents the relationship between the minimum theoretical execution time for a parallel implementation (sst/p) and the measured processing time for the parallel run $pst(p)$.

$$ef(p) = \frac{sst/p}{pst(p)} \tag{8}$$

3. Masivo Public Transport Routes' Simulation

In this section, we describe the components and the operational description of the developed public transport routes' simulation called Masivo. Masivo is a public transport simulation software built in this research project to evaluate the new parallel simulation model proposed. This software is divided into three main components, as described in Figure 3.

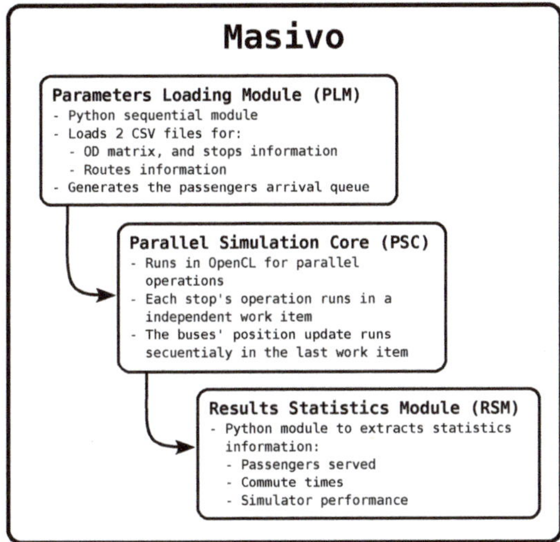

Figure 3. Masivo main components.

First, we have the Parameters' Loading Module (PLM), which is a Python sequential module that loads two CSV files with the information related to the OD matrix, stops, and routes. Then, we have the Parallel Simulation Core (PSC), which runs in OpenCL. The PSC runs the public transport route simulation model, by running each stop's operation in an independent work-item. Finally, we have the Results' Statistics Module (RSM), which is a Python module that extracts the statistical information related to passengers served, commute times, and simulation performance. The following subsections describe in-depth these three components. The full Masivo source code is available on GitHub (`masivo` codebase: https://github.com/jpruiz84/masivo), including installation instructions.

3.1. Parameters Loading Module

The Parameters Loading Modules (PLM) is a Python module that executes sequential operations to load and prepare the passenger arrival queue for the PSC. The PLM opens two CSV files, one with the information needed for the stop operations and others with the information needed for the bus operations, as described in Figure 4. The content of the two input files is described below.

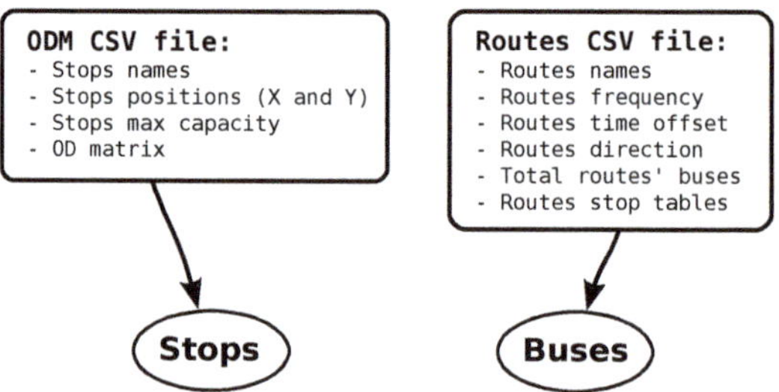

Figure 4. Masivo input parameters.

3.1.1. ODM CSV File

The ODM (OD Matrix) CSV file contains the information related to the operation of the stops. This information includes the names of the stops, stops' position (X and Y), stops' max capacity, and the OD matrix. Table 5 shows an example of an ODM file for a three stop system. The first column (stop_number) has the following list of numbers starting at zero to identify each stop. The second column (stop_name) presents a text string with the stop name due in real transport systems. Here, a name, instead of a number, is used to identify a stop or station. Then, the third and fourth columns indicate the X and Y position in meters. The fifth column shows the maximum passenger capacity at each stop. Finally, the last three columns indicate the OD matrix, where the first row of these columns represents the names of the destination stops.

Table 5. Origin-Destination Matrix (ODM) CSV example file with 3 stops' information.

Stop_Number	Stop_Name	x_Pos	y_Pos	Max_Capacity	Stop_00	Stop_01	Stop_02
0	Stop_00	400	1000	200	0	10	20
1	Stop_01	800	1000	200	10	0	10
2	Stop_02	1200	1000	200	20	10	0

3.1.2. Routes' CSV File

The routes' CSV file contains the information related to the route's operation. This information includes the routes' set, names, routes' frequencies, routes' time offset, routes' direction, routes' notes, and stops' table. Table 6 shows an example of a routes' file with four routes' information. The first column (number) has the following list of numbers starting at zero to identify each route. The second column (name) presents the text string with the route name, as in a real transport system, names composed of letters and numbers are used to identify a route. The third column (freq.) indicates the route frequency in seconds, which represents how often a bus is dispatched for each route. The fourth column (offset) represents the route time offset in seconds, and it is the time when the first bus of each route is sent. The fifth column (buses) indicates the maximum number of buses available for each route. The sixth column (dir.) indicates the direction of this route, where W-E is west to east and E-W is east to west. The seventh column (notes) is used only for information proposes (not used by the PSC) and indicates, in this case, the type of route. Finally, the eighth column (stops_table) contains the stops' table, which indicates the stop stations where the bus has to stop to pick up or alight passengers.

Table 6. Routes CSV example file with 4 routes' information. Freq., frequency; Dir., direction.

Number	Name	Freq.	Offset	Buses	Dir.	Notes	Stops_Table
0	Route_00	300	100	50	W-E	all stops	Stop_00, Stop_01, Stop_02
1	Route_01	300	100	50	E-W	all stops	Stop_02, Stop_01, Stop_00
2	Route_02	900	100	50	W-E	express	Stop_00, Stop_02
3	Route_03	900	100	50	E-W	express	Stop_02, Stop_00

3.1.3. Passenger Arrival Queue Generation

After loading the ODM and routes' CSV files, the PLM generates the passenger arrival queue (*paq*) for each stop from the OD matrix defined as D. The paq_i is an array that contains the passenger that will arrive at the stop during the simulation. The passengers are sorted by the arrival time, starting with the first passenger that arrives at the stop. Algorithm 1 describes the full process to generate the *paq*. First, in Line 2, we define the PASS_TOTAL_ARRIVAL_TIME as 3600 s. This means that the passengers will arrive at the stops from Time 0 to 3600 s during the simulation. Then, in Line 3, we declare a for loop that will run in each of the stops i and then will run the rows of the OD matrix D. The passengers' queues are statics list, with a fixed size of STOP_MAX_PASS (by default, 10,000). Then, we initialize these lists, with a status value in each field that indicates the end of the list (*PASS_STATUS_END_LIST*) and an *arrival_time* that is an unsigned short (UINT16) variable to the maximum possible value. This initialization is done by the operation performed in the loop from Lines 4 to 8.

Algorithm 1 Passenger arrival queue generation algorithm.

```
 1: procedure GENERATEPAQ
 2:     PASS_TOTAL_ARRIVAL_TIME ← 3600;
 3:     for i ← 0 to total_stops do
 4:         for j = 0 to STOP_MAX_PASS do
 5:             pwq_i[j].status = PASS_STATUS_END_LIST;
 6:             paq_i[j].status = PASS_STATUS_END_LIST;
 7:             paq_i[j].arrival_time = UINT16_MAX;
 8:         end for
 9:
10:         for j = 0 to total_stops do
11:             for k = 0 to D[i,j] do
12:                 paq_i[j].orig_stop = i;
13:                 paq_i[j].dest_stop = j;
14:                 paq_i[j].arrival_time = PASS_TOTAL_ARRIVAL_TIME × k/D[i,j];
15:             end for
16:         end for
17:         Sort paq_i by arrival_time;
18:     end for
19: end procedure
```

Later, in Line 10, we define a for loop to cover all destination stops. This will run the columns of the OD matrix D. In Line 11, we set another for loop that will run for each passenger that will travel from stop index i to stop index j. In Lines 12 to 14, we set the passenger's origin stop $paq_i[j].orig\_stop$, the destination stop $paq_i[j].dest\_stop$, and the arrival time $paq_i[j].arrival\_time$. Finally, after the full list of passengers is generated for the stop i, we sort the passengers of paq_i by the arrival time. With this, the PSC can process *paq* efficiently, as we will discuss later.

3.2. Parallel Simulation Core

The Parallel Simulation Core (PSC) runs in an OpenCL kernel the passenger operations and the buses' position update. The simulation tick is 1 s (this means that all simulation components are updated every 1 s). Each passenger has a data structure that contains the data needed for his operations. Figure 5 shows the passengers' operations and the passengers data information. Each passenger has an ID that identifies him/her in the system; furthermore, the origin and destination stop numbers that indicate at which stop the passenger has arrived and at which stop he/she will alight. The arrival time is the time in the simulation that indicates when the passenger has arrived, and the alight time indicates when the passenger has alighted. The status points out the actual status of the passenger inside the simulation (to arrive, arrived, on the bus, alighted, or empty). The empty status means that this space is free for another passenger, such as a seat on the bus. This passenger structure uses 13 bytes of data per passenger.

The stops are defined in the simulator by an array of structures. As described in Figure 5, each element inside the stops' array has a stop number. Furthermore, they include the stop position in the map, total passengers, the last empty index, which indicates which was the last space empty in the waiting queue, and the passengers' queues. Second, we include the arrival queue for passengers that will arrive at the stop. Third is the waiting queue for passengers that have arrived at the stop and are waiting for the bus. Finally, there is the alight queue for passengers that have finished the trip and have alighted at the stop. Each stop structure uses 130 KB of data.

Next, we have an array of structures that describes the buses in the simulator. As shown in Figure 5, each element inside the buses' array has a bus number. Besides, we include the current stop index, which indicates at which stop the bus is, and the last stop index, which indicates which was the last stop where the bus stopped. Similarly, we incorporate here the stops' arrays that have the stops' index where the bus will stop to pick up or alight passengers. Finally, we include the total number of passengers on the bus and the passengers' array that has the data of the passengers that are on the bus.

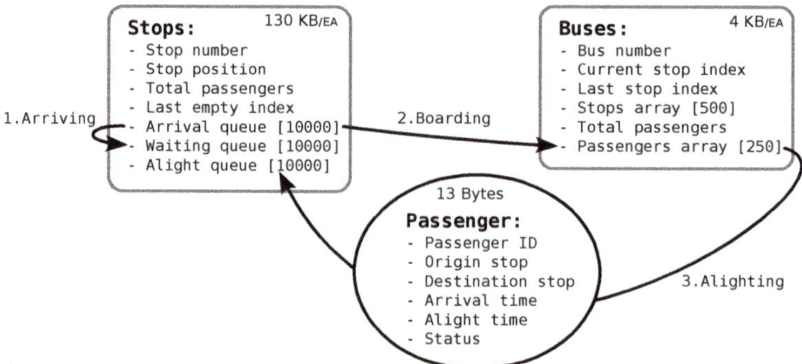

Figure 5. Masivo simulation passenger operations.

Moreover, in Figure 5, we see the three passengers' operations (1, arriving; 2, boarding; 3, alighting). During the arriving, the passenger passes from the arrival queue to the waiting queue at the origin stop. During the boarding, the passenger moves from the origin stop waiting queue to the bus passenger array. Finally, during the alighting, the passengers move to form the bus's passenger array to the alight queue at the destination stop. These passengers' move operations are optimized to run efficiently in OpenCL and are described in the following subsections.

3.2.1. Passenger Arriving

In the passenger arriving operation, the passengers arrive at the stop at the arrival time. Algorithm 2 describes this process. First, we have an infinite while loop that breaks when there are no more passengers that will arrive at the stop in the given simulated time. In Line 2, we check if the total number of passengers remaining in *paq* is zero. If yes, we break the while loop. In Line 6, we extract the working index (*w_index*) of the passengers' arrival queue (*paq*) in w. This working index represents the last passenger's index that has been removed from *paq*. Then, in Line 9, if the arrival time of a passenger that we are processing $paq.p[w].arrival_time$ is greater than the actual simulation time, we do not process this passenger yet. Hence, we break the while loop. Due to the passengers in *paq* being sorted by $arrival_time$, we do not need to process the next passengers, because we already know that $arrival_time$ will be greater than sim_time.

Algorithm 2 Passengers' arriving operation at stop n, where passenger arrival queue paq and passenger waiting queue pwq are specific for stop n.

```
1: while TRUE do
2:     if paq.p[l].total = 0 then
3:         break;
4:     end if
5:
6:     w ← paq.w_index;
7:     if paq.p[w].arrival_time > sim_time then
8:         break;
9:     end if
10:
11:    l ← pwq.last_empty;
12:    pwq.p[l] ← paq.p[w];
13:    pwq.p[l].status ← ARRIVED;
14:    pwq.last_empty ← pwq.last_empty + 1;
15:    pwq.last_total ← pwq.total + 1;
16:
17:    paq.p[l].status ← PASS_STATUS_EMPTY;
18:    paq.p[l].w_index ← paq.p[l].w_index + 1;
19:    paq.p[l].total ← paq.p[l].total − 1;
20:
21: end while
```

Thus, we move the passenger from the passenger arrival queue (paq) to the passenger waiting queue (pwq) only if the conditions of Lines 2 and 7 are not fulfilled. If we have a valid passenger with $arrival_{t}ime$ equal to or less than the simulation time, we move it. For this, in Line 11, we extract the pwq last empty index, which indicates the index of the last empty space in pwq. Then, we assign the actual processing passenger in the arrival queue $paq.p[w]$ to the last empty space in the waiting queue $pwq.p[l]$. Latter, in Lines 13–15, we update the passenger status, and we increment the last empty index and the total counter for pwq. In Lines 17–19, we update the paq passenger's status to PASS_STATUS_EMPTY, we update the w_index of paq, to process the next passenger in the next iteration, and we decrement the total counter of passengers in paq.

3.2.2. Passenger Boarding

The passenger boarding operations run concurrently at each stop. Algorithm 3 describes the boarding operations for the stop n. For this, in Line 1, we start with a main for loop, then run for all the buses. Then, in Line 2, we have an if that only lets us process the bus passenger array (bpa) that is at the current stop. In Line 3, if the total number passengers in the waiting queue is zero ($pwq.total$), we brake, and we do not process more buses, because we do not have passengers at this stop waiting for boarding. In Line 6, we check if the total number of passengers of bus $bpa[j]$ has reached the maximum BUS_MAX_PASS (the bus is full). If yes, we continue to process the next bus.

At this point, in Line 9, we know that passengers are waiting at the stop, and the bus is not full. Therefore, we check for each passenger at the stop to check which of them needs to board the actual bus. For this, in Line 9, we set a temporal variable $last\_empty\_seat\_in\_bus$ to zero. This variable indicates which is the last seat on the bus that is empty. In this way, if there is more than one passenger in the stop to board the bus, we will not start from the beginning of the seats to check which is empty for each passenger.

To check if a passenger is at the stop, in Line 10, we have a for loop, then it runs over passenger waiting queue until STOP_MAX_PASS, that is the last position of the queue (in this case, 10,000). The queue is a static list with a size of 10,000. We know that there will be no more than 10,000 passengers in this queue; the total number of passengers in this queue is always less than this size. To know if we have passed the last passenger that has arrived in the queue, we check that the passenger's status is PASS_STATUS_END_LIST. If yes, we break, and we finish checking for passengers to board at this stop. In Line 15, we check if the status of the passenger ($pwq[k]$) is PASS_STATUS_ARRIVED, which indicates that the passenger has arrived at the stop and is waiting for the bus. Then, in Line 16, we check if the destination stop of this passenger is in the stops' table of the bus that we are processing. If yes, in Line 17, we have a for loop that checks the next empty seat on the bus, with the if condition in Line 18. If the position is empty on the bus, we move the passenger from the waiting list to the bus empty seat in Line 19. In Lines 20 and 21, we update the passenger's status to PASS_STATUS_IN_BUS, and we increment the total number of passengers on the bus. In Lines 22 to 23, we update the status of the empty space in the pwq, and we decrement the total number of passengers at the stop. Finally, in Line 24, we update the temporal variable *last_empty_seat_in_bus* to l, so that when we process the next passenger at this stop in this simulation time iteration, we do not start from the beginning of the bus seats, which we already know are not empty.

Algorithm 3 Passengers' boarding operation at stop n, where pwq is the passenger waiting queue specific for the stop n.

```
1: for j ← 0 to total_buses do
2:     if stop_num = bpa[j].curr_stop then
3:         if pwq.total = 0 then
4:             break;
5:         end if
6:         if bpa[j].total ≥ BUS_MAX_PASS then
7:             continue;
8:         end if
9:         last_empty_seat_in_bus ← 0;
10:        for k ← 0 to STOP_MAX_PASS do
11:            if pwq[k].status = PASS_STATUS_END_LIST then
12:                break;
13:            end if
14:
15:            if pwq[k].status = PASS_STATUS_ARRIVED then
16:                if pwq[k].dest_stop in bpa[j].stops_table then
17:                    for l ← last_empty_seat_in_bus to BUS_MAX_PASS do
18:                        if bpa[j].bpl[l].status = PASS_STATUS_EMPTY then
19:                            bpa[j].bpl[l] ← pwq[k]
20:                            bpa[j].bpl[l].status ← PASS_STATUS_IN_BUS
21:                            bpa[j].total ← bpa[j].total + 1
22:                            pwq[k].status ← PASS_STATUS_EMPTY
23:                            pwq.total ← pwq.total − 1
24:                            last_empty_seat_in_bus ← l
25:                            break;
26:                        end if
27:                    end for
28:                end if
29:            end if
30:
31:        end for
32:    end if
33: end for
```

3.2.3. Passenger Alighting

The passenger alighting operations also run concurrently at each stop. Algorithm 4 describes the boarding operations for the stop n. For this, in Line 1, we start with a main for loop that runs for all the buses. Then, in Line 2, we have an if that only lets us process the bus passenger array (bpa) that is in the current stop. In Line 3, we have an if that only lets us process buses that have more than zero passengers. In this way, at this point, we know that the bus is at the stop, and we have at least one passenger there. Then, in Line 5, we have another for loop that runs for all the bus's seats. Later, the if condition in Line 6 checks if there is a passenger in the actual processing seat. Next, the if condition in Line 7 check if the passenger's destination stop is the same as the actual stop. If yes, we know that the passenger must alight from the bus.

Algorithm 4 Passenger alighting operation at stop n, where passenger waiting queue pwq and passenger alight queue plq are specific for stop n.

```
1:  for j ← 0 to total_buses do
2:      if stop_num = bpa[j].curr_stop then
3:          if bpa[j].total > 0 then
4:
5:              for k ← 0 to BUS_MAX_PASS do
6:                  if bpa[j].bpl[k].status = PASS_STATUS_IN_BUS then
7:                      if bpa[j].bpl[k].dest_stop = pwq.stop_num then
8:                          l ← plq.total;
9:                          plq[l] ← bpa[j].bpl[k];
10:                         plq[l].status ← PASS_STATUS_ALIGHTED;
11:                         plq[l].alight_time ← sim_time;
12:                         plq.total ← plq.total + 1;
13:
14:                         bpa[j].bpl[k].status ← PASS_STATUS_EMPTY;
15:                         bpa[j].total ← bpa[j].total − 1;
16:                     end if
17:                 end if
18:             end for
19:
20:         end if
21:     end if
22: end for
```

To alight the passenger from the bus to the stop, first in Line 8, we get the last empty index of the passenger alight queue (plq) for the temporal variable l. Next, in Line 9, we move the passenger from the bus $bpa[j].bpl[k]$ to the alighting stop $pql[l]$. Later, in Lines 10 to 11, we update the passenger status to PASS_STATUS_ALIGHTED, we set the alight time, and we increment the total number of passengers in the alight queue. Finally, in Lines 14 and 15, we set the status of the empty seat on the bus to PASS_STATUS_EMPTY, and we decrement the passengers' total number for this bus.

3.2.4. Buses' Position Update

The buses' position update runs sequentially per bus in an independent OpenCL work item. This position update consists of moving the bus across the stops and holding the bus at the stops inside the stop table for a fixed time in seconds. Algorithm 5 shows the operations for the bus position update. First, in Line 1, we have a for loop to run for all the buses. Then, in Line 2, we check if we have to start a new bus according to the route frequency and time offset. In Line 4, we check if the bus is already at a stop. If yes, we decrement $in\_the\_stop\_counter$, which indicates how many seconds are remaining for the bus to be held at the stop. In Line 6, we check that *in_the_stop_counter* is zero. If yes, we set *in_the_stop* to FALSE, indicating that the bus has finished the holding time at the stop.

Algorithm 5 Bus update position operations.

```
1: for i ← 0 to total_buses do
2:     check_if_start_the_bus(bus[i]);
3:
4:     if bus[i].in_the_stop = TRUE then
5:         bus[i].in_the_stop_counter ← bus[i].in_the_stop_counter − 1;
6:         if bus[i].in_the_stop_counter = 0 then
7:             bus[i].in_the_stop = FALSE;
8:         end if
9:     end if
10:
11:    if bus[i].in_the_stop = FALSE then
12:        bus[i].curr_pos ← bus[i].curr_pos + TRAVEL_SPEED;
13:    end if
14:
15:    if bus[i].curr_pos = next stop window then
16:        bus[i].curr_stop ← next_stop_i;
17:        bus[i].in_the_stop ← TRUE;
18:        bus[i].in_the_stop_counter ← BUS_STOPPING_TIME;
19:    end if
20: end for
```

Then, in Line 11, we check if the bus is not at the stop. If it is not held at the stop, we increment the current position (*curr_stop*) according to TRAVEL_SPEED. Finally, we check in Line 15 if the bus is in the stop window of the next stop. If yes, we update the *curr_stop* and *in_the_stop* variables accordingly; also, we set *in_the_stop_counter* to BUS_STOPPING_TIME, which by default is 20 s.

3.3. Results Statistics Module

The Results Statistics Module (RSM) is a Python module that extracts the statics information results from the executed simulation. It extracts the number of passengers served, the passengers' commute time, and the simulation performance. Masivo saves this information in the result folder, including CSV files and statistics graphs.

3.3.1. Passengers Served Information

When the simulation is done, each of the stops has a passenger alight queue, which contains the full passenger structure information (passenger ID, origin and destination stops, arrival and alight time, and status), as described in Figure 5. Masivo saves this information and statics about this information in the following output files.

- **total_passengers_results.csv:** contains all the information (passenger ID, origin and destination stops, arrival and alight time, and status) of all passengers inside the simulation.
- **served_passengers_per_stop.csv:** describes per stop the total number of passengers waiting for a bus, the expected total input passengers, the total alight passengers, the total expected alight passengers, and the alighted passengers' percentage.
- **served_passengers_per_stop.eps:** graphs the total alighted passengers contrasted with the remaining passengers expected per stop.
- **commute_time_per_stop.eps:** graphs the average commute time of the alighted passengers per destination stop and per travel direction. This graph is useful to see how the commute time is affected at a specific destination stop depending on traffic congestion on the road.

3.3.2. Performance Information

To know the simulation performance, Masivo saves useful simulation timings and host computer status information. These data are stored in the following files:

- **performance_timeline.csv:** contains the real-time simulation factor over time, with a sampling frequency of 100 s of the simulation. Furthermore, this file contains information related to CPU usage and CPU operation frequency over time.
- **performance_timeline.eps:** graphs the real-time factor over time, the real-time factor low-pass filtered signal, the CPU usage, and the CPU operation frequency.
- **simulation_brief.csv:** contains all simulation configured parameters, total simulation outputs, and total performance indicators. In particular, as inputs, this file contains the buses' average travel speed, bus max passengers, bus stopping time, end simulation time, ODM file, passenger total arrival time, routes' file, and stop to bus windows' distance. As outputs, we find here total alighted expected passengers, total alighted passengers, total alighted passenger percentage, total average commute time, total simulated buses, total simulated passengers, total routes, and total stops. Finally, for the total performance indicators, we include in this file the average CPU usage, average real-time factor, total simulation execution time, OpenCL device name and compute units used, and the limit of the max OpenCL compute units.

3.4. 3D System Visualization Output

For validation purposes, a 3D system visualization output was integrated into Masivo. This visualization system is based on a Python 3D engine called Panda3D. When enabled, a separate thread runs the 3D visualization, with a running interval of 30 fps. In each frame update, the buses' position, buses' occupied seats, stops' alighted passengers, and stops' waiting passengers values are updated from the PSC. Figure 6 shows this 3D output interface. Here, the stops are represented with a blue container, which has a horizontal red bar that indicates the percentage of passengers in the waiting queue relative to the maximum stop passenger capacity. The vertical purple cylinder represents the percentage of alighted passengers relative to the total expected alight passengers. The bus has a horizontal red bar, which represents the percentage of occupied seats relative to the maximum bus passenger capacity.

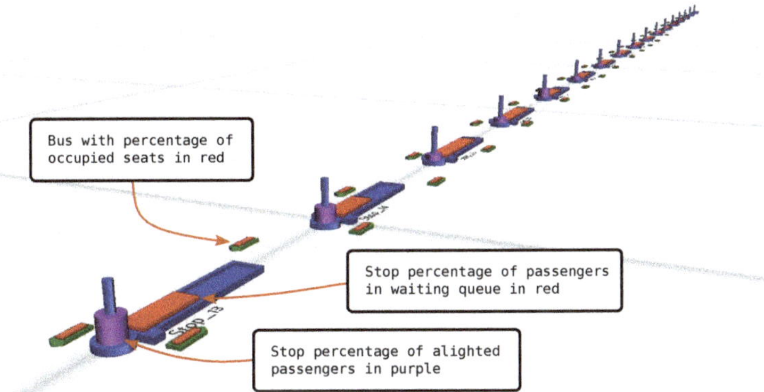

Figure 6. Masivo 3D game engine simulation output.

3.5. Validation Simulator

A validation simulator was built only using Python (called here Pure Python) with dynamic arrays (to emulate passenger's queues). This Pure Python simulator helped to get results that ensured the correct operation of the routes inside an origin-destination demand and then compared these results with the parallel simulator. In this simulator, contrary to the Masivo PSC, the boarding and alighting operations were performed per bus that was held at a stop and not at the stop.

The correct behavior for this simulator was validated with the 3D system visualization output. For this, a three stop small public transport simulation system was simulated. The ODM matrix used here is described by Equation (9), totaling 80 passengers. The passengers arrived at the stops from Time 0 to 3600 s.

$$\text{ODM matrix} = \begin{bmatrix} 0 & 10 & 20 \\ 10 & 0 & 10 \\ 20 & 10 & 0 \end{bmatrix} \tag{9}$$

Two routes, with a 15 min frequency, were used for this system. The routes' files are described by Table 7.

Table 7. Routes used to validate a small 3 stop public transport system.

Number	Name	Freq.	Offset	Buses	Dir.	Notes	Stops_Table
0	Route_00	900	950	50	W-E	all stops	Stop_00, Stop_01, Stop_02
1	Route_01	900	950	50	E-W	all stops	Stop_02, Stop_01, Stop_00

Figure 7 shows the 3D visualization output for three different simulation times. Figure 7a shows the stops at the time 800 s of the simulation. Here, some passengers have arrived at the stops, but the buses have not departed yet. Figure 7a shows the simulation at the time of 1900 s. Here, some passengers have arrived at their destination stops. Furthermore, we see two buses moving passengers. Finally, Figure 7c shows the end of the simulation, where all passengers have arrived at their stops.

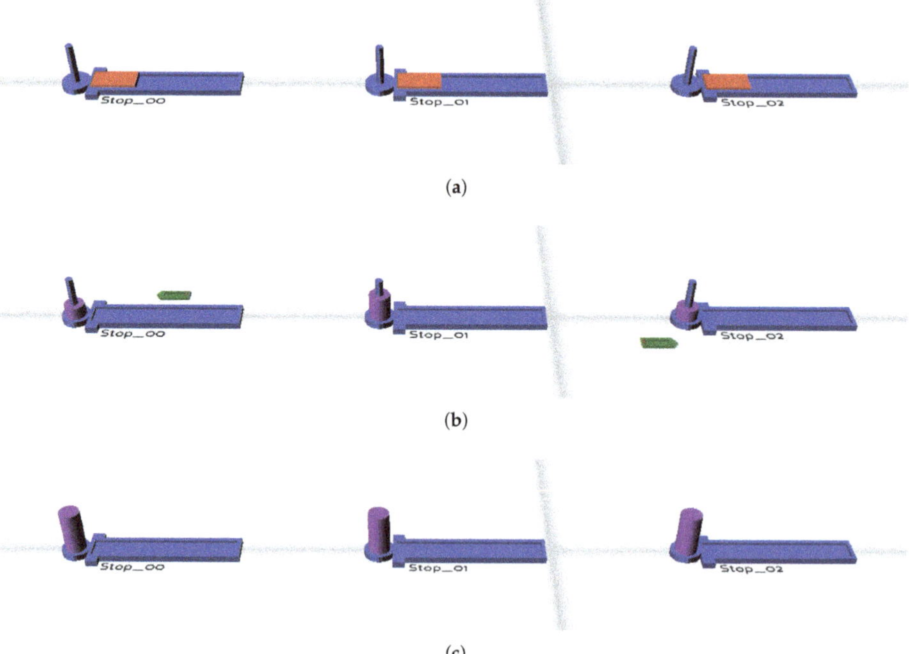

Figure 7. 3D visualization for a three stops small public transport system. (**a**) at 800 s simulation time (**b**) at 1900 s simulation time (**c**) at 7200 s end simulation time.

Furthermore, Figure 8 shows the total alighted and not alighted passengers at each stop. Here, we note that all passengers (80 of 80 expected) alighted at their destination stops.

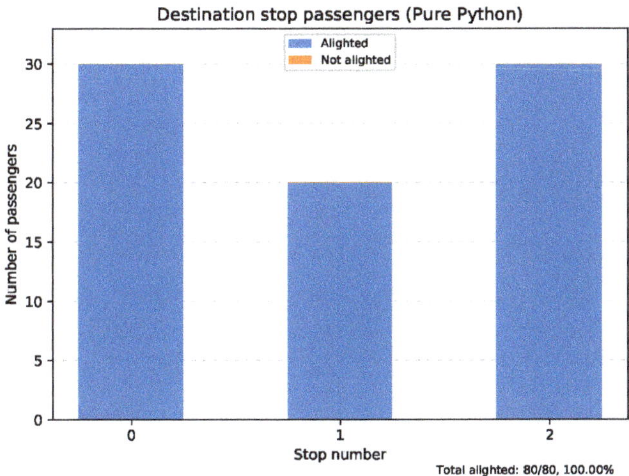

Figure 8. Alighted passengers results for a three stop validation system in Pure Python.

Finally, this validation simulation also calculated the average commute for the passengers per destination stop and travel direction, as shown in Figure 9. Here, we note that the commute time for corner stops (Stop_00 and Stop_02) was greater than the commute time of the center stop (Stop_01). This was because, for instance, Stop_02 received passengers from the closest stop Stop_00 and the furthest stop Stop_01. Meanwhile, Stop_01 received a passenger for the closer stops Stop_00 and Stop_02; hence, the passengers' average commute time was lower.

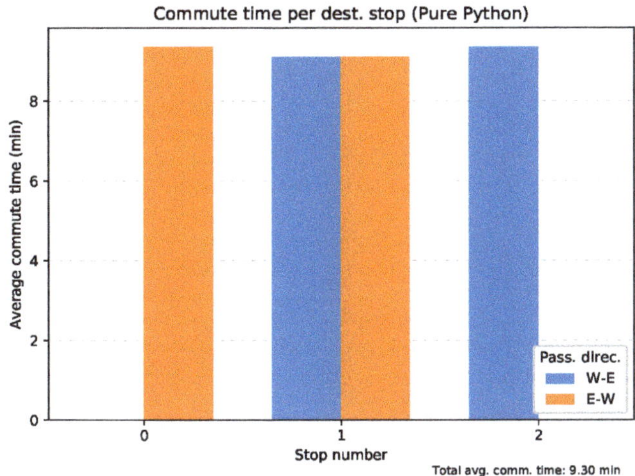

Figure 9. Commute time per stop for a three stop validation system in Pure Python. The abbreviation "dest. stop" refers to destination stop, "Pass. direc." refers to passenger direction, "W-E" refers to West to East direction, and "E-W" refers to East to West direction.

4. Results

In this section, we discuss the simulation model results. First, we present the results for the validation process, where we compare the Masivo PSC outputs with the Pure Python validation simulator. Then, we show the performance results.

4.1. Validation Results

For the validation of the Masivo PSC, four different scenarios were created. These scenarios, with the same conditions, were simulated in the Masivo PSC and the Pure Python validation simulator. Output simulation results related to alighted passengers per stop and the commute time were compared.

4.1.1. Scenario 1, 30 Stops and Two Routes at 54 km/h

The simulation conditions for Scenario 1 are summarized in Table 8. The two routes pick up and alight passengers from all stops, with a bus departing frequency of 1 min. The first route goes from west to east, and the second one goes from east to west. The separation of the stops is 400 m, totaling a line length of 12 km for the 300 stops.

Table 8. Scenario 1 simulation conditions.

Condition	Value
Bus average transit speed (km/h)	54
Bus maximum passengers	250
Bus stopping time (s)	20
End simulation time (s)	7200
Passenger total arrival time (s)	3600
Total passengers	38,721
Total routes	2
Total stops	30

Figure 10 shows the alighted passenger numbers per stop and the total alighted passenger numbers and percentage. The percentage of alighted passengers for the Masivo PSC was 99.85% and 100.00% for the Pure Python validation simulator, giving an error of 0.15%.

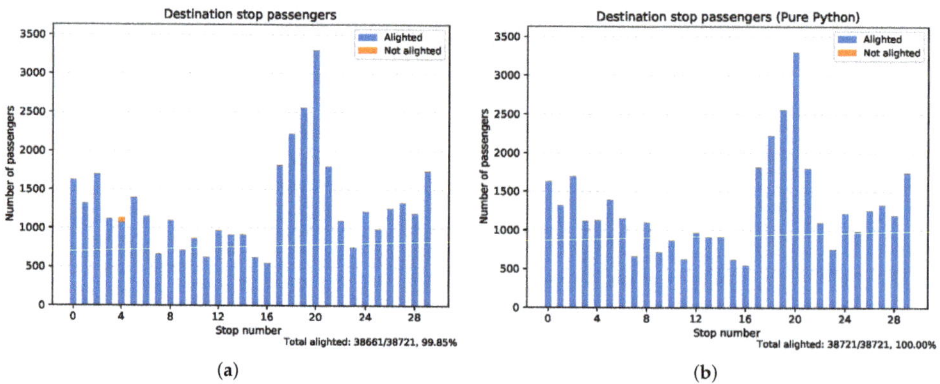

Figure 10. Alighted passengers per stop, 30 stops in Scenario 1. (**a**) Results from Masivo PSC. (**b**) Results from the Pure Python validation simulator.

Figure 11 shows the average passengers' commute time per destination and the total average commute time. Masivo PSC showed a total average commute time of 15.77 min. Meanwhile, the Pure

Python validation simulator showed a total average commute time of 16.44 min, giving us an error of 0.52% relative to the total simulation time.

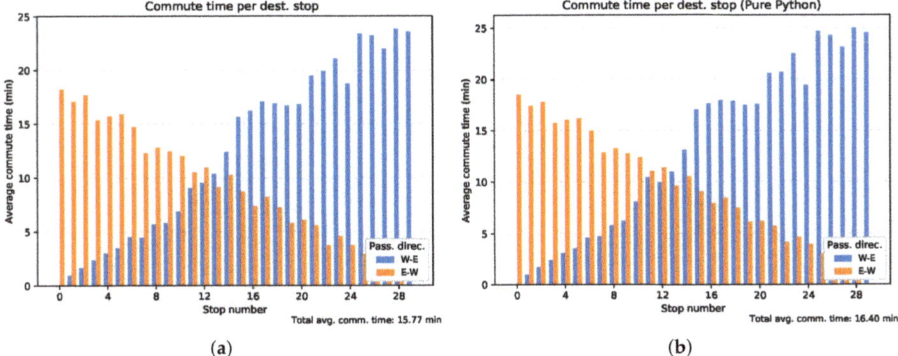

(a) (b)

Figure 11. Passengers' commute time per destination stop, 30 stops in Scenario 1. (**a**) Results from Masivo PSC. (**b**) Results from the Pure Python validation simulator. The abbreviation "dest. stop" refers to destination stop, "Pass. direc." refers to passenger direction, "W-E" refers to West to East direction, and "E-W" refers to East to West direction.

4.1.2. Scenario 2, 30 Stops and Two Routes at 70 km/h

The simulation conditions for Scenario 2 are summarized in Table 9, where the difference between this and Scenario 1 was the transit bus speed, increased from 54 km/h to 70 km/h.

Table 9. Scenario 2 simulation conditions.

Condition	Value
Bus average transit speed (km/h)	70
Bus maximum passengers	250
Bus stopping time (s)	20
End simulation time (s)	7200
Passenger total arrival time (s)	3600
Total passengers	38,721
Total routes	2
Total stops	30

Figure 12 shows the alighted passenger numbers per stop and the total alighted passenger numbers and percentage. The percentage of alighted passengers for the Masivo PSC was 99.62% and 100.00% for the Pure Python validation simulator, giving an error of 0.38%.

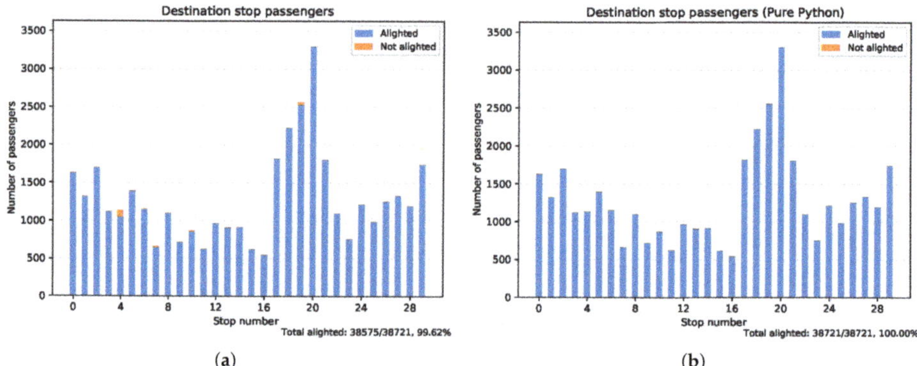

(a) (b)

Figure 12. Alighted passengers per stop, 30 stops in Scenario 2. (**a**) Results from Masivo PSC. (**b**) Results from the Pure Python validation simulator.

Figure 13 shows the average passengers' commute time per destination and the total average commute time. Masivo PSC showed a total average commute time of 14.02 min. Meanwhile, the Pure Python validation simulator showed a total average commute time of 14.47 min, giving us an error of 0.37% relative to the total simulation time.

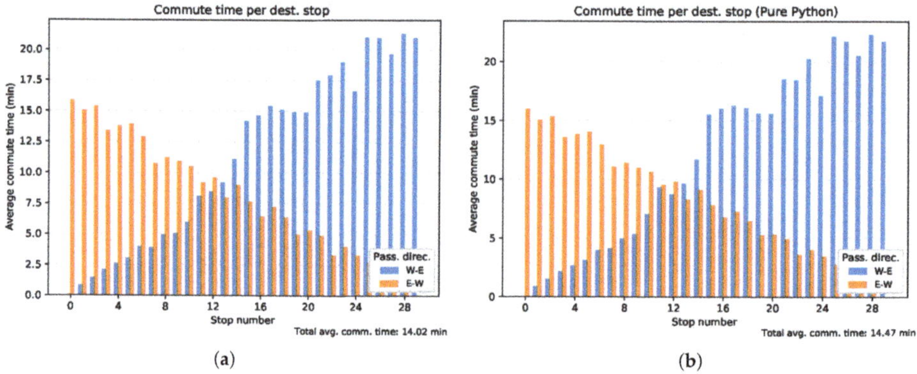

(a) (b)

Figure 13. Passengers' commute time per destination stop, 30 stops in Scenario 2. (**a**) Results from Masivo PSC. (**b**) Results from the Pure Python validation simulator. The abbreviation "dest. stop" refers to destination stop, "Pass. direc." refers to passenger direction, "W-E" refers to West to East direction, and "E-W" refers to East to West direction.

4.1.3. Scenario 3, 30 Stops and Four Routes at 54 km/h

The simulation conditions for Scenario 3 are summarized in Table 10, where the difference between this and Scenario 1 is the number of routes. Here, we have two new routes that do not stop at all stops. These new routes stop at the corners and in the downtown area where the demand is higher than the others. Both new routes have a frequency of 2 min. Precisely, Route 3 stops at Stop_00, Stop_01, Stop_02, Stop_03, Stop_18, Stop_19, Stop_20, Stop_21, and Stop_22. Meanwhile, Route 4 stops at Stop_29, Stop_28, Stop_27, Stop_26, Stop_22, Stop_21, Stop_20, Stop_19, Stop_18.

Table 10. Scenario 3 simulation conditions.

Condition	Value
Bus average transit speed (km/h)	54
Bus maximum passengers	250
Bus stopping time (s)	20
End simulation time (s)	7200
Passenger total arrival time (s)	3600
Total passengers	38,721
Total routes	4
Total stops	30

Figure 14 shows the alighted passenger numbers per stop and the total alighted passenger numbers and percentage. The percentage of alighted passengers for the Masivo PSC was 99.37% and 100.00% for the Pure Python validation simulator, giving an error of 0.63%.

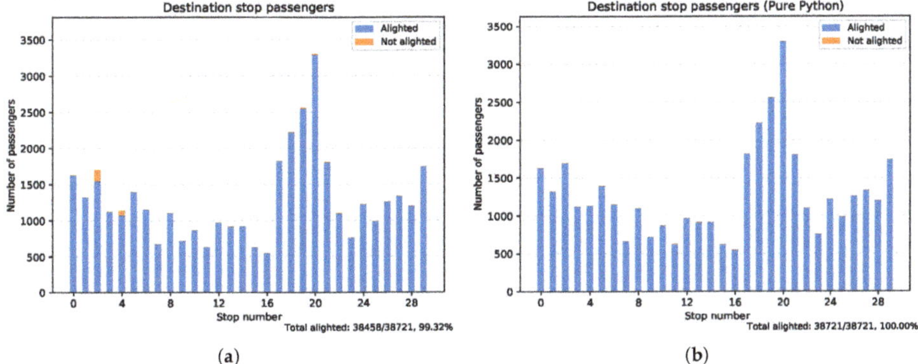

(a) (b)

Figure 14. Alighted passengers per stop, 30 stops in Scenario 3. (**a**) Results from Masivo PSC. (**b**) Results from the Pure Python validation simulator.

Figure 15 shows the average passengers' commute time per destination and the total average commute time. Masivo PSC showed a total average commute time of 15.71 min. Meanwhile, the Pure Python validation simulator showed a total average commute time of 16.33 min, giving us an error of 0.52% relative to total simulation time.

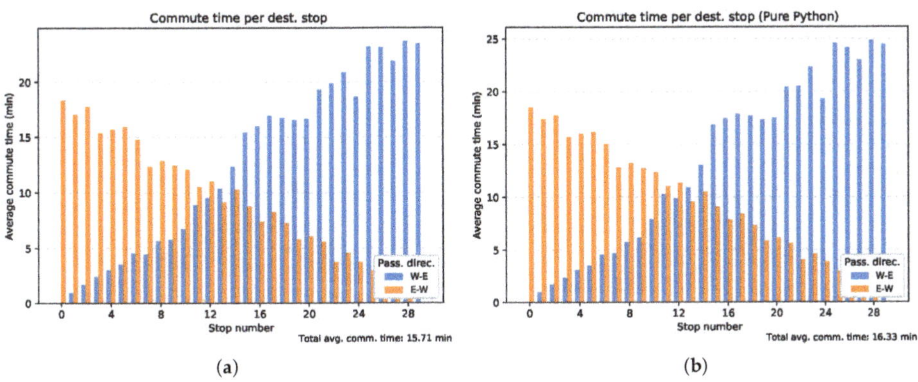

(a) (b)

Figure 15. Passengers' commute time per destination stop, 30 stops in Scenario 3. (**a**) Results from Masivo PSC. (**b**) Results from the Pure Python validation simulator. The abbreviation "dest. stop" refers to destination stop, "Pass. direc." refers to passenger direction, "W-E" refers to West to East direction, and "E-W" refers to East to West direction.

4.1.4. Scenario 4, 300 Stops and Two Routes at 54 km/h

The simulation conditions for Scenario 4 are summarized in Table 11. It consists of 300 stops, which have a normal distribution with more demand for longer distances plus a random number of passengers. Furthermore, corner stops have two times more demand than the others. The two routes pick up and alight passengers from all stops, with a bus departing frequency of 30 s. The first route goes from west to east, and the second one goes from east to west. The separation of the stops is 400 m, totaling a line length of 120 km for the 300 stops. Then, the end simulation time is 10 h, so that all passengers have enough time to finish traveling.

Figure 16 shows the alighted passenger numbers per stop and the total alighted passenger numbers and percentage. The percentage of alighted passengers for the Masivo PSC was 99.52% and 100.00% for the Pure Python validation simulator, giving an error of 0.48%.

Table 11. Scenario 1 simulation conditions.

Condition	Value
Bus average transit speed (km/h)	54
Bus maximum passengers	250
Bus stopping time (s)	20
End simulation time (s)	36,000
Passenger total arrival time (s)	3600
Total passengers	456,997
Total routes	2
Total stops	300

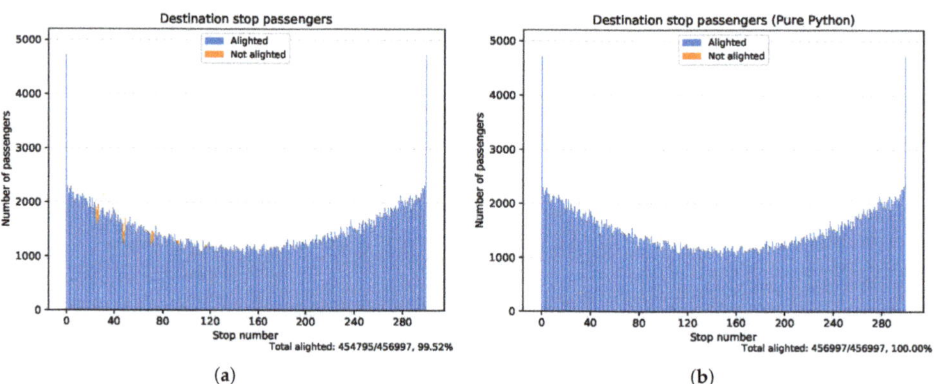

Figure 16. Alighted passengers per stop, 300 stops in Scenario 4. (**a**) Results from Masivo PSC. (**b**) Results from the Pure Python validation simulator.

Figure 17 shows the average passengers' commute time per destination and the total average commute time. Masivo PSC showed a total average commute time of 297.55 min. Meanwhile, the Pure Python validation simulator showed a total average commute time of 298.37 min, giving us an error of 0.13% relative to the total simulation time.

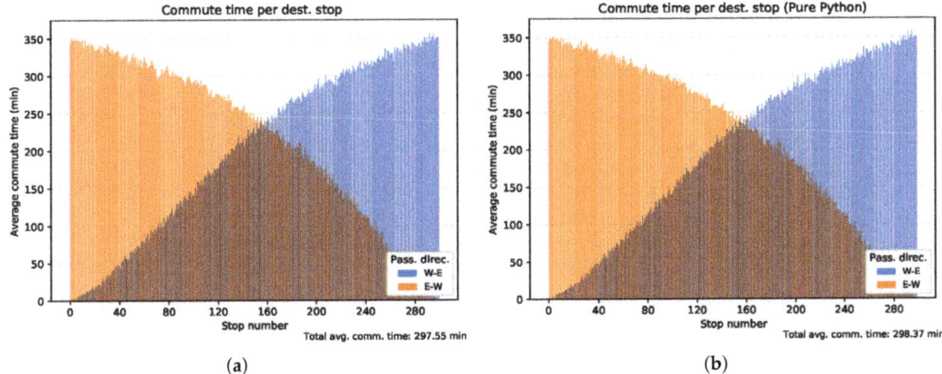

Figure 17. Passengers' commute time per destination stop, 300 stops in Scenario 4. (**a**) Results from Masivo PSC. (**b**) Results from the Pure Python validation simulator.

4.2. Performance

The performance simulation execution was performed in a dedicated server rented from 1&1 IONOS (https://www.ionos.com/servers/dedicated-servers). The specification of the rented server is described in Table 12. Before each performance test, the Linux CPU scaling governors were set in performance mode, to enable the maximum operating frequency.

Table 12. IONOS dedicated server 4XL specifications.

Specification	Value
Type	Dedicated Server 4XL
CPU	Intel Xeon Gold 6210U
# of Cores	20
# of Threads	40
Processor Base Frequency	2.50 GHz
Max Turbo Frequency	3.90 GHz
Cache	27.5 MB
RAM	192 GB
HDD	2 × 4000 GB Hardware RAID 1
OS	Ubuntu 16.04

For the performance tests, the most complex simulation scenario (Scenario 4 previously tested) was used. This is a simulation scenario that has 300 stops and two routes. Table 13 describes the full list of conditions. It consists of 300 stops, which have a normal distribution with more demand for longer distances plus a random number of passengers. Furthermore, corner stops have two times more demand than the others. The two routes pick up and alight passengers from all stops, with a bus departing frequency of 30 s. The first route goes from west to east, and the second one goes from east to west. The separation of the stops is 400 m, totaling a line length of 120 km for the 300 stops. Then, the end simulation time is 10 h, so that all passengers have enough time to finish traveling.

Table 13. Simulation conditions for performance tests, Scenario 4.

Condition	Value
Bus average transit speed (km/h)	54
Bus maximum passengers	250
Bus stopping time (s)	20
End simulation time (s)	36,000
Passenger total arrival time (s)	3600
Total passengers	456,997
Total routes	2
Total buses	2400
Total stops	300

4.2.1. Pure Python Validation Simulator Performance

The performance simulation of Scenario 4 was evaluated in the Pure Python validation simulator. Table 14 shows the performance output values. The total simulation execution time was near one hour (3303 s) for 10 h of simulation time. Then, the average real-time factor was 32.76 times faster than in real-time. Furthermore, we noted that due to this simulator being a sequential implementation, only one of the 40 processing threads was used, then only 2.19% of all the processing power was used.

Table 14. Pure Python simulator performance outputs for Scenario 4.

Performance Indicator	Value
Total simulation execution time (s)	3303.612
Average real-time factor	32.76
Average CPU usage (%)	2.19

Figure 18 shows the real-time factor (RTF) vs. the simulation time. Here, we note that the RTF starts at 780, and then, it has an exponential decline, getting values under 10 times. This decline is because the Pure Python simulator uses dynamic lists to handle passengers and buses. When the system starts, the number of passengers and buses is low. However, then, more passengers arrive at the stops, and more buses are dispatched. Hence, the systems have more elements to process, and the simulation speed decreases.

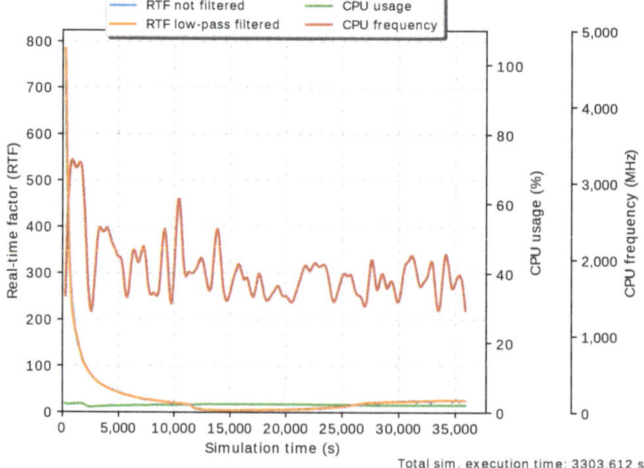

Figure 18. Real-time factor vs. simulation time for the Pure Python validation simulator with CPU usage percentage and CPU operation frequency.

Furthermore, in Figure 18, we see that the CPU usage was always near 2.5%, due to the fact that the sequential implementation will never use more than one processing thread. Finally, the CPU frequency started at 3.2 GHz. However, because the CPU was not fully used, the OS decreased the CPU frequency during the simulation execution.

4.2.2. Masivo PSC Performance

The Masivo Parallel Simulation Core (PSC) performance was also evaluated using Test Scenario 4. Table 15 shows the performance output values. The total simulation execution time was 11.822 s, 278 times faster than the Pure Python simulator. The average real-time factor was 3050 times faster than real-time, and the average CPU usage was 84.21%.

Table 15. Masivo PSC performance outputs for Scenario 4.

Performance Indicator	Value
Total simulation execution time (s)	11.882
Average real-time factor	3050.84
Average CPU usage (%)	84.21

Figure 18 shows the real-time factor vs. the simulation time. Here, we note that the RTF starts in 4300, and then, it has an exponential decline, stabilizing near 2500. This decline is because when the system starts, the number of passengers and buses is low, and the internal for loop that runs the statics list does not need to go to the end of the lists. Later in the simulation, more passengers arrive at the stops, and more buses are dispatched. Hence, the systems have more elements to process, and the simulation speed decreases. This decreasing is not as low as in the Pure Python simulator, due to the concurrent processing operations performed per stop. Finally, from the simulation time of 15,000, we see that the RTF is increasing due to the passengers arriving at the final destination, then there are fewer elements to process.

Figure 19 shows that the CPU usage is near 80%, due to all the cores being used most of the time. Finally, the CPU frequency started and sustained at 3.2 GHz, because the OS maintained this due to the high processing demand.

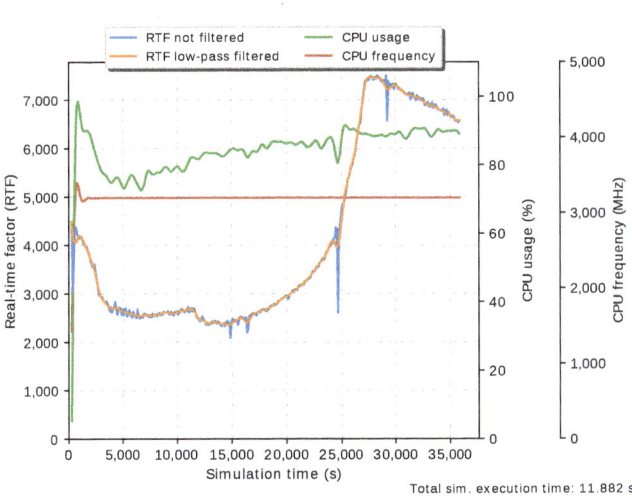

Figure 19. Real-time factor vs. simulation time for Masivo PSC with CPU usage percentage and CPU operation frequency.

As described in Section 2.5, the simulation performance indicators were extracted. Figure 20 shows these indicators for the simulation executed with the compute units limited from one to 40. These results helped us to identify the performance for the different numbers of processing units and the efficiency of the proposed model.

First, Figure 20a shows the simulation execution time vs. the compute units. We see here that the execution time required to run the 10 h simulation was reduced exponentially with more compute units. Nevertheless, we had a low limit, near 10 s, because there were operations in the model that were not parallelized, such as the position update of the buses. Second, Figure 20b shows the real-time factor, which also increased, with the number of cores reaching a final value near to 3000.

Next, Figure 20c has the same shape as the RTF, because it is a function also of the simulation execution time, but in this case, divided by the simulation execution time for one compute unit. We see here that the maximum speed up factor reached was 10.2 times faster than the sequential simulation. Finally, we have the efficiency in Figure 20d, which shows an efficiency of more than 50% for 10 or less compute units.

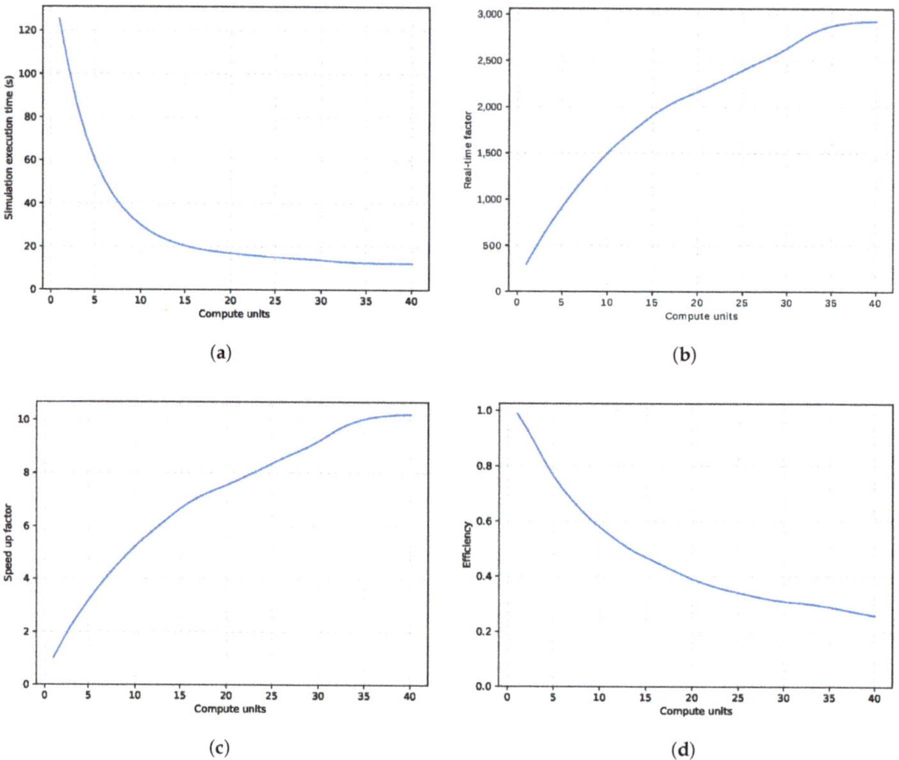

Figure 20. Simulation performance indicators vs. number of compute units. (**a**) Total parallel simulation execution time (pst). (**b**) Real-time factor (rtf). (**c**) Speed-up factor (suf). (**d**) Efficiency (ef).

5. Discussion

In this section, we discuss the performance of Masivo PCS versus other simulation tools. Table 16 summarizes the documents related to parallel simulation applications for traffic and transit systems and also for Masivo PSC. FPGA applications include urban traffic simulation [97], with one document in which the authors in 2005 developed a microscopic simulation of road traffic. They simulated the Portland road network with 124,000 road segments and 100,000 intersections, achieving a speed-up factor of 4.5 [98]. For GPU's parallel architecture, we found five implementations, where four were related to road traffic and one to public transit simulation. In 2012, Wang et al. built a GPU based traffic parallel simulation module, which was able to simulate three hours (10,800 s) of road traffic of a 40 $\times$ 40 lattice road network, with 5000 vehicle agents in 109 s, achieving a speed-up of 105 and a real-time factor of 102 [99]. Later, in 2014, Xu et al. implemented a mesoscopic road traffic simulation on CPU/GPU. This implementation was tested in a road network of 10,201 nodes, 20,100 unidirectional links, and 100,000 vehicles during 1000 simulation ticks (each tick was 1 s). The simulation time for this network was 4720 ms for CPU and 423 ms for GPU, getting a speed-up of 11.2 and a real-time factor of 2364 [60].

In 2017, Song et al. implemented the mesoscopic traffic simulation on GPU developed in [60] for a real-world scenario. The scenario was the Singapore expressway system, which is made up of 3179 nodes, 3388 links, and 9419 lanes with a demand modeled as trips from 4106 OD pairs. In total, 100,000 vehicles loaded into the peak traffic scenario during 1000 simulation ticks (each tick is 1 s) were simulated, and the execution time of this simulation was 1250 ms for CPU against 527 ms for GPU, getting a speed-up 2.37 and a real-time factor of 1897 [59].

Next, in 2019, Saprykin et al. developed GEMSim (GPU-Enhanced Mobility Simulator), a GPU accelerated multi-modal mobility simulator for large scale scenarios. This was a mesoscopic multi-modal, queue based mobility simulator for public transit systems. Here, the large scale scenario of Switzerland with 513,770 nodes, 1,127,775 road links, 27,873 stops, and 21,847 routes was simulated. The simulation ran for a full day (86400 s), and the execution time was 290 s, equivalent to a real-time ratio of 298 for a population of 5.2 million and a real-time ratio of 1300 for 100,000 population size [13]. The same year, Vinh and Tan created a framework for mesoscopic traffic simulation in GPU tested in an updated Singapore expressways network, which has 3186 nodes, 3386 links, and 9437 lanes, with demand modeled as trips from 4103 origin-destination pairs. Up to 300,000 vehicles were simulated, getting a speed-up of five times against the CPU [56].

For multi-core parallel architecture, first in 2012, Potuzak developed a distributed parallel road traffic simulator for clusters of multi-core computers, tested in regular square grids of 64,256 and 1024 crossroads. The simulations ran for 1 h (3600 s) with no number of vehicles specified with an execution time for the best cases of 22 s for 64 crossroads (real-time factor: 163), 86 s for 256 crossroads (real-time factor: 41), and 245 s for 1024 crossroads (real-time factor: 14) [100]. Then, in 2013, Bruegmann et al. developed an online, real-time traffic information system called OLSIMv4 (OnLine Traffic SIMulation version 4), which used microscopic traffic simulations exploiting the thread level parallelism on multi-core machines using a coarse-grained parallel microscopic simulation model. This simulation ran in the highway network of North Rhine-Westphalia (NRW), Germany. The simulated network contained more than 3000 loop detectors bundled into 1300 measurement cross-sections with an average statistical distance of 3.5 km. The simulation ran for 60 min with 120,000 vehicles with an execution time 60 s, resulting in a real-time factor of 60 [101].

In the same year, Fernandes et al. developed a parallel microscopic simulator capable of simulating metropolitan scale traffic using OpenMP. For the simulation, they used a real-world road network of the San Francisco Bay Area, which comprises 145,665 roads with a total of 27,439 km. The simulation ran in 60 s intervals with a step time of 0.1 s and generating a vehicle in each road every 2 s following a negative exponential distribution, totaling 4,349,130 vehicles. With these parameters, they achieved an execution time of 70 s (real-time factor 0.86) and a speed-up factor of 16 times, using 24 processors against the single processor simulation [102].

Table 16. Documents related to traffic and transit parallel simulation.

Document Title	Year	Parallel Architecture	Model Type	Simulation Core	Speedup Factor	Real-Time Factor	Simulation Complexity	Reference
Urban traffic simulation modeling for reconfigurable hardware	2005	FPGA	Microscopic	Road traffic	4.50	Not specified	100,000 intersections	[98]
A GPU based traffic parallel simulation module of artificial transportation systems	2012	GPU	Microscopic	Road traffic	105.00	102.00	1600 intersections and 5000 vehicles	[99]
Mesoscopic traffic simulation on CPU/GPU	2014	GPU	Mesoscopic	Road traffic	11.20	2364.00	10,201 nodes and 100,000 vehicles	[60]
Supporting real-world network oriented mesoscopic traffic simulation on GPU	2017	GPU	Mesoscopic	Road traffic	2.37	1897.00	4106 OD pairs, 100,000 vehicles	[59]
GEMSim: A GPU accelerated multi-modal mobility simulator for large scale scenarios	2019	GPU	Mesoscopic	Public transit system	Not specified	1300.00	27,873 stops and 100,000 passengers	[13]
A framework for mesoscopic traffic simulation in GPU	2019	GPU	Mesoscopic	Road traffic	5.00	Not specified	4103 OD pairs, 300,000 vehicles	[56]
Distributed-parallel road traffic simulator for clusters of multi-core computers	2012	Multi-core	Microscopic	Road traffic	3.32	163.00	1024 Crossroads	[100]
Real-time traffic information system using microscopic traffic simulation	2013	Multi-core	Microscopic	Road traffic	Not specified	60.00	1300 cross-sections and 120,000 vehicles	[101]
Parallel microscopic simulation of metropolitan scale traffic	2013	Multi-core	Microscopic	Road traffic	16.00	0.86	145,665 roads, 4,349,130 vehicles, and 0.1 s step time	[102]
Masivo PSC	2019	Multi-core	Mesoscopic	Public transit system	10.20/278	3000	300 stops, 2400 buses, and 456,997 passengers,	N/A

Masivo PSC uses a mesoscopic simulation model type, due to the buses' congestion in the road not being modeled. It is focused on public transit system simulation. It achieved a speed-up factor of 10.2 compared with the simulator model running with one compute unit and a speed-up factor of 278 times faster than the Pure Python validation simulator. Compared with other models, Masivo PSC focused on the buses' route behavior inside a PTS, letting us define the routes' stops' table and frequency; and giving us; as a result, the passengers' commute time. On the other hand, other implementations for commute or waiting passengers' time simulation and modeling mentioned in the Introduction used a stochastic collective model for waiting time prediction [44] or a model to predict the waiting time based on the passengers' learning process and adaptation with respect to waiting time uncertainty and travel information [42].

6. Conclusions

A new simulation model for routes in public transport systems using parallel computing called Masivo was presented. Masivo works with a predefined public transport system conditions, which include the stops' total number, stops' capacity, and the OD matrix. This OD matrix can be updated in this model via a CSV file. Masivo was built with three main components. The first one is the Parameters Loading Module (PLM), which is a Python module that loads the stops' information (OD matrix, stops' positions, stops' max capacity) and the routes' information. The second module is the Parallel Simulation Core (PSC), which was built in OpenCL and contains the new purposed simulation model. The PCS was designed to run in parallel the operations of each stop, including the passengers' arrival, boarding, and alighting, and to run the buses' displacement process sequentially over the roads. Finally, the Results Statistics Module (RSM) is a Python tool that ran after the PSC to get and analyze the simulation outputs and the performance results.

Furthermore, for Masivo validation, a Pure Python public transport simulator was built for validation purposes. This simulator performed the arrival operations sequentially for each stop. Then, it ran the boarding and alighting operations of each bus that was currently stooped at a station and, finally, executed the buses' update position operations. Furthermore, it had a 3D output graphic interface to visualize the movement of the buses through the different stops, the stops, the bus occupancy, and the alighted passengers at each stop. This visualization tool helped to validate the consistent operations of this simulator and the Masivo PSC. With the Pure Python public transport simulator validated, we used it to validate the Masivo PSC in four different scenarios. We ran each scenario with precisely the same condition in both simulators to compare the simulation output results. The simulation output results for validation consisted of the total alighted passengers per stop and for the whole system, as well as the average commute time per stop and for the whole system. We found that the relative error was no higher than 0.7 % in all the tested scenarios.

The performance of Masivo was evaluated with the speed-up and real-time factor indicators. Masivo achieved a speed-up factor of 10.2 compared with the simulator model running with one compute unit and a speed-up factor of 278 times faster than the Pure Python validation simulator. The real-time factor achieved was 3050 times faster than the 10 h simulated duration. The performance test scenario used consisted of a public transport simulation with 300 stops, 2400 buses, and 456,997 passengers. This performance value was compared with the most recent traffic and public transport model that performed the simulation of systems with similar size. We found that, to date, there are no models reported specifically designed for PTS that can achieve the real-time factor that we have achieved in this research.

Future work with this model could integrate the passengers' interchange between the different public transport modes, such as the train to bus. Furthermore, this model has a fast execution time, making this a high performance solution to use in a public transport optimization resources algorithm.

Author Contributions: Conceptualization, J.R.-R., G.R.-G., and R.K.; formal analysis, J.R.-R.; methodology, G.R.-G. and R.K.; project administration, G.R.-G.; software, J.R.-R. and R.K.; validation, J.R.-R.; visualization, J.R.-R.; writing, review and editing, J.R.-R.

Funding: This work was funded by the Departamento Administrativo de Ciencia, Tecnología e Innovación (647-2014) and by the Universidad del Cauca (501100005682).

Conflicts of Interest: The authors declare no conflict of interest.

References

1. United Nations. *World Urbanization Prospects*; United Nations: New York, NY, USA, 2014; p. 32. [CrossRef]
2. Ruiz-Rosero, J.; Ramirez-Gonzalez, G.; Williams, J.M.; Liu, H.; Khanna, R.; Pisharody, G. Internet of Things: A Scientometric Review. *Symmetry* **2017**, *9*, 301. [CrossRef]
3. Ji, Y.; Mishalani, R.G.; McCord, M.R. Transit passenger origin–destination flow estimation: Efficiently combining onboard survey and large automatic passenger count datasets. Big Data in Transportation and Traffic Engineering. *Transp. Res. Part C Emerg. Technol.* **2015**, *58*, 178–192. [CrossRef]
4. Budiawan, D.; Soenandi, I.A.; Marpaung, B. Optimilisasi Jumlah Armada Transjakarta di Koridor-8 Jurusan Harmoni-Lebak Bulus dengan Menggunakan Metode Goal Programming. *Tek. Dan Ilmu Komput.* **2014**, *3*, 128–137.
5. Ergun, M.; Kesten, A.S. Alteration of bus routes in large-scale networks. *Sci. Res. Essays* **2011**, *6*, 5865–5882.
6. Barceló, J. Models, Traffic Models, Simulation, and Traffic Simulation. In *Fundamentals of Traffic Simulation*; Springer: New York, NY, USA, 2010; pp. 1–62. [CrossRef]
7. Loder, A.; Ambuehl, L.; Menendez, M.; Axhausen, K.W. Empirics of multi-modal traffic networks—Using the 3D macroscopic fundamental diagram. *Transp. Res. Part C Emerg. Technol.* **2017**, *82*, 88–101. [CrossRef]
8. Oskarbski, J.; Birr, K.; Miszewski, M.; Zarski, K. Estimating the Average Speed of Public Transport Vehicles Based on Traffic Control System Data. In Proceedings of the 2015 International Conference on Models and Technologies for Intelligent Transportation Systems (MT-ITS), Budapest, Hungary, 3–5 June 2015.
9. Truong, L.; Currie, G. Macroscopic road safety impacts of public transport: A case study of Melbourne, Australia. *Accid. Anal. Prev.* **2019**, *132*, 105270. [CrossRef]
10. Drabicki, A.; Kucharski, R.; Cats, O.; Fonzone, A. Simulating the effects of real-time crowding information in public transport networks. In Proceedings of the 2017 5th IEEE International Conference on Models and Technologies for Intelligent Transportation Systems (MT-ITS), Naples, Italy, 26–28 June 2017.
11. Johnsen, A.; Gundersen, E.; Liang, X.; Kaisar, E.; Scarlatos, P. Emergency evacuation methodologies utilizing public transit with meso-simulation. In Proceedings of the Simulation Interoperability Standards Organization—Spring Simulation Interoperability Workshop 2009, San Diego, CA, USA, 23–27 March 2009.
12. Goh, Y.M.; Love, P.E.D. Methodological application of system dynamics for evaluating traffic safety policy. *Saf. Sci.* **2012**, *50*, 1594–1605. [CrossRef]
13. Saprykin, A.; Chokani, N.; Abhari, R.S. GEMSim: A GPU-accelerated multi-modal mobility simulator for large-scale scenarios. *Simul. Model. Pract. Theory* **2019**, *94*, 199–214. [CrossRef]
14. Fernandez, R. Modelling public transport stops by microscopic simulation. *Transp. Res. Part C Emerg. Technol.* **2010**, *18*, 856–868. [CrossRef]
15. Fernandez, R.; Tyler, N. Effect of passenger-bus-traffic interactions on bus stop operations. *Transp. Plan. Technol.* **2005**, *28*, 273–292. [CrossRef]
16. Yatskiv, I.; Pticina, I.; Savrasovs, M. Urban public transport system's reliability estimation using microscopic simulation. *Transp. Telecommun.* **2012**, *13*, 219–228. [CrossRef]
17. Yatskiv (Jackiva), I.; Pticina, I.; Romanovska, K. The Riga Public Transport Service Reliability Investigation Based on Traffic Flow Modelling. In *Reliability and Statistics in Transportation and Communication*; Springer: Berlin/Heidelberg, Germany, 2018, Volume 36. [CrossRef]
18. Ahmed, B. Exploring new bus priority methods at isolated vehicle actuated junctions. *Transp. Res. Procedia* **2014**, *14*, 391–406. [CrossRef]
19. Arasan, V.T.; Vedagiri, P. Micro-simulation study of bus priority on roads carrying highly heterogeneous traffic. In Proceedings of the 22nd European Conference on Modeling and Simulation (ECMS), Nicosia, Cyprus, 3–6 June 2008.

20. Thamizh, A.V.; Vedagiri, P. Microsimulation study of the effect of exclusive bus lanes on heterogeneous traffic flow. *J. Urban Plan. Dev.* **2010**, *136*, 50–58. [CrossRef]
21. Chen, X.; Yu, L.; Zhu, L.; Guo, J.; Sun, M. Microscopic traffic simulation approach to the capacity impact analysis of weaving sections for the exclusive bus lanes on an urban expressway. *J. Transp. Eng.* **2010**, *136*, 895–902. [CrossRef]
22. Chandrasekar, R.; Cheu, R.; Chin, H. Simulation evaluation of route-based control of bus operations. *J. Transp. Eng.* **2002**, *128*, 519–527. [CrossRef]
23. Wang, J.; Chen, S.; He, Y.; Gao, L. Simulation of Transfer Organization of Urban Public Transportation Hubs. *J. Transp. Syst. Eng. Inf. Technol.* **2006**, *6*, 96–102. [CrossRef]
24. Zargayouna, M.; Zeddini, B.; Scemama, G.; Othman, A. Agent-Based Simulator for Travelers Multimodal Mobility. In *Advanced Methods and Technologies for Agent and Multi-Agent Systems*; IOS Press: Amsterdam, The Netherlands, 2013; Volume 252. [CrossRef]
25. Papageorgiou, G.; Damianou, P.; Pitsillides, A.; Aphamis, T.; Charalambous, D.; Ioannou, P. Modelling and Simulation of Transportation Systems: A Scenario Planning Approach. *Automatika* **2009**, *50*, 39.
26. Ruiz-Rosero, J.; Ramirez-Gonzalez, G.; Viveros-Delgado, J. Software survey: ScientoPy, a scientometric tool for topics trend analysis in scientific publications. *Scientometrics* **2019**, *121*, 1165–1188. [CrossRef]
27. Andrews, J.R.; Morrow, C.; Wood, R. Modeling the Role of Public Transportation in Sustaining Tuberculosis Transmission in South Africa. *Am. J. Epidemiol.* **2013**, *177*, 556–561. [CrossRef]
28. Kadiyala, A.; Kumar, A. Multivariate Time Series Models for Prediction of Air Quality Inside a Public Transportation Bus Using Available Software. *Environ. Prog. Sustain. Energy* **2014**, *33*, 337–341. [CrossRef]
29. Kadiyala, A.; Kumar, A. Vector time series models for prediction of air quality inside a public transportation bus using available software. *Environ. Prog. Sustain. Energy* **2014**, *33*, 1069–1073. [CrossRef]
30. Kadiyala, A.; Kumar, A. Multivariate time series based back propagation neural network modeling of air quality inside a public transportation bus using available software. *Environ. Prog. Sustain. Energy* **2015**, *34*, 1259–1266. [CrossRef]
31. Kadiyala, A.; Kumar, A. Univariate Time Series Based Back Propagation Neural Network Modeling of Air Quality Inside a Public Transportation Bus Using Available Software. *Environ. Prog. Sustain. Energy* **2015**, *34*, 319–323. [CrossRef]
32. Kadiyala, A.; Kumar, A. Vector-time-series-based back propagation neural network modeling of air quality inside a public transportation bus using available software. *Environ. Prog. Sustain. Energy* **2016**, *35*, 7–13. [CrossRef]
33. Kadiyala, A.; Kumar, A. Univariate time series based radial basis function neural network modeling of air quality inside a public transportation bus using available software. *Environ. Prog. Sustain. Energy* **2016**, *35*, 320–324. [CrossRef]
34. Jappinen, S.; Toivonen, T.; Salonen, M. Modelling the potential effect of shared bicycles on public transport travel times in Greater Helsinki: An open data approach. *Appl. Geogr.* **2013**, *43*, 13–24. [CrossRef]
35. Saghapour, T.; Moridpour, S.; Thompson, R.G. Modeling access to public transport in urban areas. *J. Adv. Transp.* **2016**, *50*, 1785–1801. [CrossRef]
36. Nurlaela, S.; Curtis, C. Modeling household residential location choice and travel behavior and its relationship with public transport accessibility. In Proceedings of the 15th Meeting of the Euro-Working-Group-on-Transportation (EWGT), Cite Descartes, Paris, France, 10–13 September 2012; Elsevier Science BV: Amsterdam, The Netherlands, 2012; Volume 54, pp. 56–64. [CrossRef]
37. Fuglsang, M.; Hansen, H.S.; Munier, B. Accessibility Analysis and Modelling in Public Transport Networks—A Raster Based Approach. In Proceedings of the 11th International Conference on Computational Science and Its Applications (ICCSA), Univ Cantabria, Santander, Spain, 20–23 June 2011; Springer: Berlin, Germany, 2011; Volume 6782, pp. 207–224.
38. Schoebel, A. Line planning in public transportation: Models and methods. *OR Spectr.* **2012**, *34*, 491–510. [CrossRef]
39. Schmidt, M.; Schoebel, A. The Complexity of Integrating Passenger Routing Decisions in Public Transportation Models. *Networks* **2015**, *65*, 228–243. [CrossRef]
40. Schmidt, M.; SchoBel, A. The complexity of integrating routing decisions in public transportation models. In Proceedings of the 10th Workshop on Algorithmic Approaches for Transportation Modelling, Optimization, and Systems, ATMOS 2010, Liverpool, UK, 9–21 September 2010; Volume 14, pp. 156–169. [CrossRef]

41. Abbas-Turki, A.; Grunder, O.; Elmoudni, A. Simulation and optimization of the public transportation connection system. In Proceedings of the 13th European Simulation Symposium, Marseille, France, 18–20 October 2001; pp. 435–439.
42. Cats, O.; Gkioulou, Z. Modeling the impacts of public transport reliability and travel information on passengers' waiting-time uncertainty. *Euro J. Transp. Logist.* **2017**, *6*, 247–270. [CrossRef]
43. Hassannayebi, E.; Zegordi, S.H.; Yaghini, M.; Amin-Naseri, M.R. Timetable optimization models and methods for minimizing passenger waiting time at public transit terminals. *Transp. Plan. Technol.* **2017**, *40*, 278–304. [CrossRef]
44. Kieu, L.M.; Cai, C. Stochastic collective model of public transport passenger arrival process. *IET Intell. Transp. Syst.* **2018**, *12*, 1027–1035. [CrossRef]
45. Uspalyte-Vitkuniene, R.; Grigonis, V.; Paliulis, G. The Extent Of Influence of O-D Matrix on the Results of Public Transport Modeling. *Transport* **2012**, *27*, 165–170. [CrossRef]
46. Kaiyuan, L.; Lifeng, L.; Feigang, T. A scheduling model and its implementation based on intelligent public transportation system. In Proceedings of the 2nd International Conference on Smart City and Systems Engineering (ICSCSE), Changsha, China, 11–12 November 2017; pp. 52–54. [CrossRef]
47. Janoska, Z.; Dvorsky, J. P system based model of passenger flow in public transportation systems: A case study of Prague Metro. In Proceedings of the 13th Annual Workshop on Databases, Texts, Specifications and Objects (DATESO 2013), Pisek, Czech Republic, 17–19 April 2013; Volume 971, pp. 59–69.
48. Yu, H.F.; Qin, Y.; Wang, Z.Y.; Wang, B.; Zhan, M.H. Research on urban mass transit network passenger flow simulation on the basis of multi-agent. In Proceedings of the International Conference on Information Technology and Computer Application Engineering (ITCAE), Hong Kong, China, 27–28 August 2013; CRC Press-Taylor & Francis Group: Boca Raton, FL, USA, 2014; pp. 225–229.
49. Hadas, Y.; Ranjitkar, P. Modeling public-transit connectivity with spatial quality-of-transfer measurements. *J. Transp. Geogr.* **2012**, *22*, 137–147. [CrossRef]
50. Ceder, A.; Chowdhury, S.; Taghipouran, N.; Olsen, J. Modelling public-transport users' behaviour at connection point. *Transp. Policy* **2013**, *27*, 112–122. [CrossRef]
51. He, R.; Li, Y.; Zhang, Z. Models and genetic algorithms for the optimal riding routes with transfer times limited in urban public transportation. In Proceedings of the 5th International Symposium on Operations Research and Its Applications, Tibet, China, 8–13 August 2005; Volume 5, pp. 340–348.
52. Debinska, E.; Cichocinski, P. The application of multimodal network for the modeling of movement in public transport. In Proceedings of the 13th International Multidisciplinary Scientific Geoconference, SGEM 2013, Albena, Bulgaria, 16–22 June 2013; pp. 559–563.
53. Rasmusseni, T.K.; Anderson, M.K.; Nielsen, O.A.; Prato, C.G. Timetable-based simulation method for choice set generation in large-scale public transport networks. *Eur. J. Transp. Infrastruct. Res.* **2016**, *16*, 467–489.
54. Drabicki, A.; Kucharski, R.; Szarata, A. Modelling the public transport capacity constraints' impact on passenger path choices in transit assignment models. *Arch. Transp.* **2017**, *43*, 7–28. [CrossRef]
55. Fu, Z.; Yu, J.; Sarwat, M. Demonstrating geosparksim: A scalable microscopic road network traffic simulator based on Apache spark. In Proceedings of the 16th International Symposium on Spatial and Temporal Databases, Vienna, Austria, 19–21 August 2019; pp. 186–189. [CrossRef]
56. Vu, V.A.; Tan, G. A Framework for Mesoscopic Traffic Simulation in GPU. *IEEE Trans. Parallel Distrib. Syst.* **2019**, *30*, 1691–1703. [CrossRef]
57. Saprykin, A.; Chokani, N.; Abhari, R. Large-scale multi-agent mobility simulations on a GPU: Towards high performance and scalability. *Procedia Comput. Sci.* **2019**, *151*, 733–738. [CrossRef]
58. Vu, V.; Tan, G. High-performance mesoscopic traffic simulation with GPU for large scale networks. In Proceedings of the 2017 IEEE/ACM 21st International Symposium on Distributed Simulation and Real Time Applications (DS-RT), Rome, Italy, 18–20 October 2017; Volume 2017, pp. 1–9. [CrossRef]
59. Song, X.; Xie, Z.; Xu, Y.; Tan, G.; Tang, W.; Bi, J.; Li, X. Supporting real-world network-oriented mesoscopic traffic simulation on GPU. *Simul. Model. Pract. Theory* **2017**, *74*, 46–63. [CrossRef]
60. Xu, Y.; Tan, G.; Li, X.; Song, X. Mesoscopic Traffic Simulation on CPU/GPU. In Proceedings of the SIGSIM-PADS'14: 2014 Acm Conference on Sigsim Principles of Advanced Discrete Simulation, Denver, CO, USA, 18–21 May 2014; Assoc Computing Machinery: New York, NY, USA, 2014.

61. Xiao, J.; Andelfinger, P.; Eckhoff, D.; Cai, W.; Knoll, A. Exploring Execution Schemes for Agent-Based Traffic Simulation on Heterogeneous Hardware. In Proceedings of the 2018 IEEE/ACM 22nd International Symposium on Distributed Simulation and Real Time Applications (DS-RT), Madrid, Spain, 15–17 October 2018; pp. 243–252. [CrossRef]
62. Janczykowski, M.; Turek, W.; Malawski, M.; Byrski, A. Large-scale urban traffic simulation with Scala and high-performance computing system. *J. Comput. Sci.* **2019**, *35*, 91–101. [CrossRef]
63. Turek, W. Erlang-based desynchronized urban traffic simulation for high-performance computing systems. *Future Gener. Comput. Syst.* **2018**, *79*, 645–652. [CrossRef]
64. Turek, W.; Siwik, L.; Byrski, A. Leveraging rapid simulation and analysis of large urban road systems on HPC. *Transp. Res. Part C: Emerg. Technol.* **2018**, *87*, 46–57. [CrossRef]
65. Fu, Z.; Yu, J.; Sarwat, M. Building a large-scale microscopic road network traffic simulator in apache spark. In Proceedings of the 2019 20th IEEE International Conference on Mobile Data Management (MDM), Hong Kong, China, 10–13 June 2019; Voume 2019, pp. 320–328. [CrossRef]
66. Public-Transport Noun—Definition, Pictures, Pronunciation and Usage Notes: Oxford Advanced Learner's Dictionary. Available online: OxfordLearnersDictionaries.com (accessed on 1 December 2019).
67. Desaulniers, G.; Hickman, M.D. Chapter 2 Public Transit. In *Transportation. Handbooks in Operations Research and Management Science*; Barnhart, C., Laporte, G., Eds.; Elsevier: Amsterdam, The Netherlands, 2007; Volume 14, pp. 69–127. [CrossRef]
68. Barua, B.; Boberg, J.; Hsiao, S.; Zhang, X. Integrating Geographic Information Systems with Transit Survey Methodology. *Transp. Res. Rec.* **2001**, *1753*, 29–34. [CrossRef]
69. Kuwahara, M.; Sullivan, E.C. Estimating origin-destination matrices from roadside survey data. *Transp. Res. Part B Methodol.* **1987**, *21*, 233–248. [CrossRef]
70. Munizaga, M.A.; Palma, C. Estimation of a disaggregate multimodal public transport Origin–Destination matrix from passive smart card data from Santiago, Chile. *Transp. Res. Part C Emerg. Technol.* **2012**, *24*, 9–18. [CrossRef]
71. Farzin, J.M. Constructing an Automated Bus Origin–Destination Matrix Using Farecard and Global Positioning System Data in São Paulo, Brazil. *Transp. Res. Rec.* **2008**, *2072*, 30–37. [CrossRef]
72. White, J. Extracting origin destination information from mobile phone data. In Proceedings of the Eleventh International Conference on Road Transport Information and Control, London, UK, 19–21 March 2002; pp. 30–34.
73. Fellendorf, M.; Vortisch, P. Validation of the microscopic traffic flow model VISSIM in different real-world situations. In Proceedings of the Transportation Research Board 80th Annual Meeting, Washington, DC, USA, 7–11 January 2001.
74. Kotusevski, G.; Hawick, K. A Review of Traffic Simulation Software. *Res. Lett. Inf. Math. Sci.* **2009**, *13*, 1–19.
75. Burghout, W.; Wahlstedt, J. Hybrid traffic simulation with adaptive signal control. *Transp. Res. Rec. J. Transp. Res. Board* **2015**, *1999*, 191–197. [CrossRef]
76. Burghout, W. *Mesoscopic Simulation Models For Short-Term Prediction;* Royal Institute of Technology: Melbourne, Australia, 2005.
77. Balakrishna, R.; Antoniou, C.; Ben-Akiva, M.; Koutsopoulos, H.; Wen, Y. Calibration of microscopic traffic simulation models: Methods and application. *Transp. Res. Rec. J. Transp. Res. Board* **2015**, *1999*, 198–207. [CrossRef]
78. Bu, R. *Simulación: Un Enfoque Práctico*; Ingenieria Industrial; Publisher Editorial Limusa: Mexico City, Mexico, 1994.
79. Yang, Q.; Koutsopoulos, H.; Ben-Akiva, M. Simulation laboratory for evaluating dynamic traffic management systems. *Transp. Res. Rec. J. Transp. Res. Board* **2000**, *1710*, 22–130. [CrossRef]
80. Jayakrishnan, R.; Cortes, C.E.; Lavanya, R.; Pagès, L. Simulation of urban transportation networks with multiple vehicle classes and services: Classifications, functional requirements and general-purpose modeling schemes. In Proceedings of the TRB 2003 Annual Meeting, Washington, DC, USA, 12–16 January 2003.
81. Bazzan, A.L.; Klügl, F. A review on agent-based technology for traffic and transportation. *Knowl. Eng. Rev.* **2014**, *29*, 375–403. [CrossRef]
82. Buisson, C.; Lebacque, J.; Lesort, J. STRADA, a discretized macroscopic model of vehicular traffic flow in complex networks based on the Godunov scheme. In Proceedings of the CESA'96 IMACS Multiconference: Computational Engineering in Systems Applications, Lille, France, 9–12 July 1996; pp. 976–981.

83. Haj-Salem, H.; Elloumi, N.; Mammar, S.; Papageorgiou, M.; Chrisoulakis, J.; Middelham, F. Metacor: A macroscopic modelling tool for urban corridor. In *Towards An Intelligent Transport System, Proceedings of the First World Congress on Applications of Transport Telematics and Intelligent Vehicle-Highway Systems, Paris, France, 30 November–3 December 1994*; TRB: Washington, DC, USA, 1994; Volume 3.
84. PTV. *VISUM 11.50 User Manual*; PTV: Karlsruhe, Germany, 2011.
85. Inro Software. *The World's Most Trusted Transportation Forecasting Software*; Emme: Westmount, QC, Canada, 2015.
86. Behrisch, M.; Bieker, L.; Erdmann, J.; Krajzewicz, D. SUMO–Simulation of Urban MObility. In Proceedings of the The Third International Conference on Advances in System Simulation (SIMUL 2011), Barcelona, Spain, 23–29 October 2011.
87. AG, PTV Planug Trasport Verker. *PTV Vissim 6 User Manual*; PTV: Karlsruhe, Germany, 2013.
88. Casas, J.; Ferrer, J.L.; Garcia, D.; Perarnau, J.; Torday, A. Traffic simulation with aimsun. In *Fundamentals of Traffic Simulation*; Springer: Berlin/Heidelberg, Germany, 2010; pp. 173–232.
89. Maerivoet, S. Models in Aid of Traffic Management. In Proceedings of the Seminar slides for 'Transportmodellen ter ondersteuning van het mobiliteits-en vervoersbeleid, Brussels, Belgium, 3 May 2004.
90. Yang, H.; Cheng, W.; Xiao, H.C.; Pan, Y.W.; Zhang, D.M. Study of Traffic Flow Adjustment Methods to Congestion Area Based on TransModeler. *Sci. Technol. Eng.* **2011**, *8*, 022.
91. Suping, L.F.C.X.C. Frequency optimization of bus rapid transit based on cost analysis. *J. Southeast Univ. (Natural Sci. Ed.)* **2009**, *4*, 038.
92. Cortés, C.E.; Sáez, D.; Milla, F.; Núñez, A.; Riquelme, M. Hybrid predictive control for real-time optimization of public transport systems' operations based on evolutionary multi-objective optimization. *Transp. Res. Part C Emerg. Technol.* **2010**, *18*, 757–769. [CrossRef]
93. Kwan, R.S.K. Case studies of successful train crew scheduling optimisation. *J. Sched.* **2011**, *14*, 423–434. [CrossRef]
94. Michael, R.G.; David, S.J. *Computers and Intractability: A Guide to the Theory of NP-Completeness*; WH Freeman & Co.: San Francisco, CA, USA, 1979.
95. Lee, D.H.; Chandrasekar, P. A Framework for Parallel Traffic Simulation Using Multiple Instancing of a Simulation Program. *J. Intell. Transp. Syst.* **2002**, *7*, 279–294. [CrossRef]
96. Nagel, K.; Rickert, M. Parallel implementation of the TRANSIMS micro-simulation. *Parallel Comput.* **2001**, *27*, 1611–1639. [CrossRef]
97. Ruiz-Rosero, J.; Ramirez-Gonzalez, G.; Khanna, R. Field Programmable Gate Array Applications—A Scientometric Review. *Computation* **2019**, *7*, 63. [CrossRef]
98. Hansson, A.; Mortveit, H.; Tripp, J.; Gokhale, M. Urban traffic simulation modeling for reconfigurable hardware. In Proceedings of the 3rd Industrial Simulation Conference 2005, Fraunhofer-IPK, Berlin, Germany, 9–11 June 2005; pp. 291–298.
99. Wang, K.; Shen, Z. A GPU based trafficparallel simulation module of artificial transportation systems. In Proceedings of the 2012 IEEE International Conference on Service Operations and Logistics, and Informatics, SOLI 2012, Suzhou, China, 8–10 July 2012. [CrossRef]
100. Potuzak, T. Distributed-Parallel Road Traffic Simulator for Clusters of Multi-core Computers. In Proceedings of the 2012 IEEE/ACM 16th International Symposium on Distributed Simulation and Real Time Applications (DS-RT), Dublin, Ireland, 25–27 October 2012. [CrossRef]
101. Bruegmann, J.; Schreckenberg, M.; Luther, W. Real-time Traffic Information System Using Microscopic Traffic Simulation. In Proceedings of the 2013 8th Eurosim Congress on Modelling and Simulation (EUROSIM), Wales, UK, 10–13 September 2013. [CrossRef]
102. Fernandes, R.; Vieira, F.; Ferreira, M. Parallel Microscopic Simulation of Metropolitan-scale Traffic. In Proceedings of the 46th Annual Simulation Symposium (ANSS 2013)—2013 Spring Simulation Multiconference (Springsim'13), San Diego, CA, USA, 7–10 April 2013; Volume 45.

Article

Deep, Consistent Behavioral Decision Making with Planning Features for Autonomous Vehicles

Lilin Qian, Xin Xu *, Yujun Zeng and Junwen Huang

College of Intelligence Science and Technology, National University of Defense Technology, Changsha 410073, China; qianlilin15@nudt.edu.cn (L.Q.); yujunzeng@sina.cn (Y.Z.); hjw756517@163.com (J.H.)
* Correspondence: xinxu@nudt.edu.cn; Tel.: +86-175-5873-6841

Received: 4 November 2019; Accepted: 3 December 2019; Published: 6 December 2019

Abstract: Autonomous driving promises to be the main trend in the future intelligent transportation systems due to its potentiality for energy saving, and traffic and safety improvements. However, traditional autonomous vehicles' behavioral decisions face consistency issues between behavioral decision and trajectory planning and shows a strong dependence on the human experience. In this paper, we present a planning-feature-based deep behavior decision method (PFBD) for autonomous driving in complex, dynamic traffic. We used a deep reinforcement learning (DRL) learning framework with the twin delayed deep deterministic policy gradient algorithm (TD3) to exploit the optimal policy. We took into account the features of topological routes in the decision making of autonomous vehicles, through which consistency between decision making and path planning layers can be guaranteed. Specifically, the features of a route extracted from path planning space are shared as the input states for the behavioral decision. The actor-network learns a near-optimal policy from the feasible and safe candidate emulated routes. Simulation tests on three typical scenarios have been performed to demonstrate the performance of the learning policy, including the comparison with a traditional rule-based expert algorithm and the comparison with the policy considering partial information of a contour. The results show that the proposed approach can achieve better decisions. Real-time test on an HQ3 (HongQi the third) autonomous vehicle also validated the effectiveness of PFBD.

Keywords: autonomous driving; behavioral decision; planning features; deep reinforcement learning; trajectory planning

---

1. Introduction

Autonomous driving (AD) has been vastly investigated in different domains for several decades and has a wide variety of applications, especially in intelligent transportation systems. AD provides more economic, comfortable and safer service than human drivers with fewer collision risks. The behavior decision and trajectory planning system is a key component of AD.

Presently, as a mature, rule-based decision and planning method, finite state machines (FSMs) [1,2] have been applied in most autonomous vehicles. Nonetheless, FSMs still have problems with guaranteeing decisions' accuracy, especially in complex scenarios, due to the difficulty in determining the boundary condition of different states and going through all states with limited experience. To make the vehicles more intelligent and more acceptable for human beings, learning methods have been attractive in recent years to enhance decisions in AD. Among them, neural networks related approaches [3–5] have been used to navigate autonomous vehicles due to their strong capacity of mapping inputs to control signals. The raw pixels from cameras [3,4] were used to train neural networks for keeping a vehicle in a lane, which proves that the task of lane following can be learned from the only limited training signal (steering) in a supervised manner. Fusing multiple inputs

(LIDAR (light detection and ranging) raw inputs, global direction and GPS), a full convolutional neural network [5] generated driving paths in a more explainable output. In fact, both rule-based and supervised learning based decision policies have the bottleneck of exploring unknown or complex scenarios.

Reinforcement Learning (RL) focuses on finding a balance between exploration and exploitation, and learns optimal policy by interacting with environments. Thus, RL promises to learn more complete decision policy than rule-based or supervised learning based methods. Deep reinforcement learning (DRL), a combination of (RL) and deep learning (DL), has been the hottest topic in the artificial intelligence (AI) discipline. Especially after the win of AI in the game Go [6], DRL has been applied in intelligent transportation systems and automated vehicles.

One class of DRL methods use limited features as inputs, where features are specially selected with experience. Some researchers have used DRL for autonomous vehicles in subsystems, such as energy management systems (EMSs) [7], route planning [8], and behavioral decisions (lane changes, on-ramp merging, etc.) [9–11]. Deep Q learning (DQN) [11] was used to learn the lateral control policy with only eight states involving information of vehicles and the road. Aradi et al. [9] used the REINFORCE algorithm to learn the driving policy mapping 16 continuous states as inputs to train the steer and acceleration demand. These methods can hardly be practical, since modeling a vehicle's lane change behavior with limited information is impossible in real traffic. Wolf et al. [10] explored a compact semantic state model of lane change with more complete information and trained the policy with DQN. Although the model can adapt to a variety of scenarios, the agent collides with other vehicles at high probability.

To learn a simple and explainable driving policy, either states or actions or both in the discrete semantic form were used [11–14]. States can be "close to the front vehicle" and "far to the front vehicle" and the actions can be overtake, left change, give way and accelerate, respectively. Li et al. [13] used the hierarchical game theory to train a driving policy from low-level to high-level step by step. You et al. [14] used a Q-network to learn discrete optimal actions. For common traffic, an artificial reward function was used. For traffics with little knowledge, the reward function was learned through inverse reinforcement learning (IRL) with the maximum entropy principle.

Some researchers studied the sub-modules of autonomous vehicles, such as longitudinal following and lateral control [15–18]. Zhu et al. [15] used DDPG [19] to teach a human-like autonomous car-following model, which has high trajectory-reproducing accuracy and generalized-capability. Based on parameterized batch actor-critic learning algorithm, Huang et al. [17] obtained a near-optimal lateral control policy for vehicles. The policy can automatically adjust the fuel/brake signals to track target velocity.

Another class of methods focuses on learning driving policy from raw sensor inputs, such as those from cameras, LIDAR and screen shots of a simulator [3–5,20–22]. Sallab [20] et al. proposed a new framework of DRL for autonomous vehicles, which combines LSTM and attention networks to capture the spatial features. The framework was tested on the simulation platform TORCS. Chen [22] proposed an end-to-end path planner for autonomous vehicles based on a theory called parallel planning. A deep Q-network was used to strength their motion planner with fewer mistakes. However, the effect of a deep Q-network was not given in their publication.

As illustrated in Figure 1, the policy does perform better than experts and traditional decision models do in some situations. However, it can hardly be directly used for behavioral decisions or control in reality. One concern is the transferability of the policy learned by end-to-end methods. The screen shots used as input states differ from raw pixels of real cameras. Besides, the same abstract scenarios, in reality, may have quite abundant images due to different weather, seasons, road conditions, etc. Thus, it is questionable to use a simulator's policy trained from images. The other concern is the inconsistency between behavioral decision and trajectory planning. In the autonomous driving system, the low-level planner will check whether space is enough to perform a maneuver from the decision level. Taking a left change command from the behavioral decision system, for example, the motion planner, firstly, evaluates whether the gaps are enough for the ego agent to merge in. If space is wide enough, the maneuver is executed with a path generated; otherwise, the command will be ignored for safety's sake. The final conservative strategy must be a compromise between an optimal policy and a safe one.

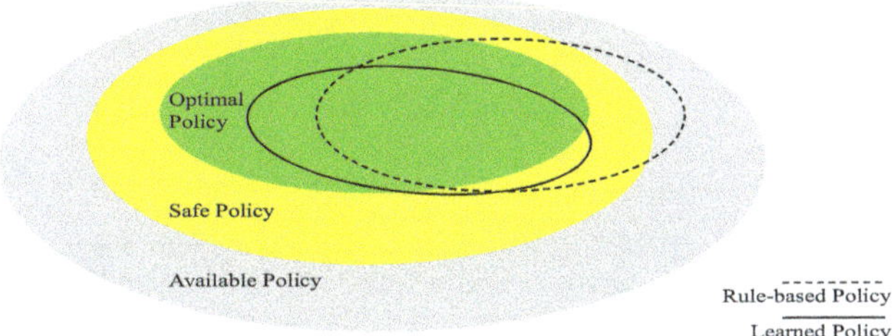

Figure 1. Explanation of polices in decision space. The yellow area represents safe policy. The green area represents the optimal policy. The dotted ellipse is the rule-based or experience-based policy. The solid ellipse is the learned policy.

A policy with safety ensured can be implemented in autonomous vehicles. Generating a policy by combining path planning features can be one feasible choice. If given the traffic participants and environments, a system can provide all possible maneuvers with corresponding reasonable paths; then, the decision policy can be reinforced with learning methods by selecting the best maneuver. Proceeding with previous work, the synchronous maneuver searching and trajectory planning algorithm (SMSTP) [23], we propose a new planning-features-based behavior decision method whose architecture is shown in Figure 2. SMSTP outputs no-collision trajectories which represent the safe policy in yellow of Figure 1, while the target of our proposed PFBD approach was to learn the optimal policy in green.

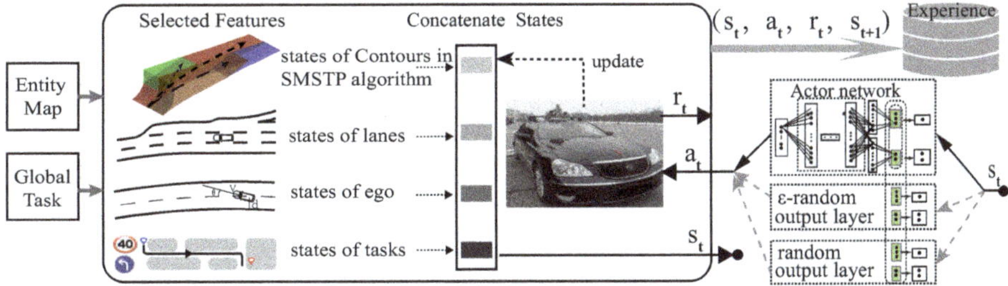

Figure 2. The sample and evaluation architecture of decision and planning for autonomous vehicles. A contour [23] is one transitable spatio-temporal area for autonomous vehicles in the synchronous maneuver searching and trajectory planning algorithm (SMSTP) algorithm.

In the SMSTP algorithm, three key parameters, the contour to drive, the expected pose and velocity at predicting horizon in the contour, are selected empirically. To learn an optimal policy, we creatively propose to use the DRL algorithm to learn the aforementioned three key parameters instead. The SMSTP algorithm provides a basic policy with a safety guarantee, while PFBD aims at optimizing the behavioral decision towards the optimum. Once the three key parameters are decided, a no-collision trajectory can be optimized numerically.

To the authors' knowledge, the deep combination of trajectory planning and behavioral decisions for autonomous vehicles is proposed for the first time here. In brief, we summarize the contributions of the proposed approach in this paper as follows:

- We came up with a compact semantic framework of the feature representation which includes planning space's states, lanes' states, ego vehicle's states and task's states. The states selected from planning space are necessary and representative for making behavioral decisions in various traffic scenarios.
- The proposed PFBD method selects a route generated from the planning space with a safety guarantee. Besides, it learns better policy through exploring the environment instead of struggling with the rules.
- The proposed method learns a policy through deep reinforcement learning. The semantic training data can be sampled from both simulations and real traffic. The simulated samples help the policy generalize the unknown scenarios.
- We applied the method to autonomous driving in situations, such as lane changing, zebra passing and ramp-merging. Both simulation and real-time tests were performed. The PFBD method achieved competitive performance compared to rule-based methods.

1.1. Reinforcement Learning Basics

RL originates from the trial-and-error search with delayed rewards, and it is dedicated to learning the optimal policy, which maximizes a long-term reward in the subsequent decisions. The agent learns a control policy by continuously interacting with its environment, and the policy maps states to actions from interactions.

The agent is in a state $s_t \in S$ of itself and its environment at the time $t \in [0, 1, 2, \dots]$, and it has a possible action $a_t \in A_{s_t} \subseteq A$ on state s_t, where S and A are the sets of all the environment states and available actions, respectively. An agent in a new state s_{t+1} observes an immediate reward $r_t = r(s_t, a_t) \in R$ after executing the action a_t. A new state transition tuple (s_t, a_t, r_t, s_{t+1}) is then sampled to train for a control strategy, which is defined by a policy π mapping the states to the actions $\pi : S \rightarrow P(A)$. The agent aims at maximizing the long-term discounted return defined as:$R_t = \sum_{t=1}^{\inf} \gamma^{t-1} r_t$, where $\gamma \in [0, 1]$ is the discount. On the policy π, the state-action value function:

$$Q^{\pi}(s_t, a_t) = \mathrm{E}_{\pi}[r_t | s_t, a_t], \tag{1}$$

describes how good the policy is after taking an action a_t in state s_t.

The policy π is often acquired by recursively solving the Bellman equations:

$$Q^{\pi}(s_t, a_t) = \mathrm{E}_{s_{t+1} \sim S, a_t \sim \pi}[r(s_t, a_t) + \gamma\, \mathrm{E}_{\pi}[Q^{\pi}(s_{t+1}, a_{t+1})]]. \tag{2}$$

The driving policy in a realistic traffic situation can be viewed as a deterministic one. The policy then can be described using a function $\mu : S \leftarrow A$, and the expectation of action can be avoided:

$$Q^{\mu}(s_t, a_t) = \mathrm{E}_{s_{t+1} \sim S}[r(s_t, a_t) + \gamma\, \mathrm{E}_{\mu}[Q^{\pi}(s_{t+1}, a_{t+1})]]. \tag{3}$$

Deep Reinforcement Learning

For a value-only based RL algorithm, such as deep Q-learning, it adopts a deep neural network (DQN) instead of a table to estimate the value to each state-action pair. DQN uses a greedy $\mu(s) = \arg\max_a Q(s, a)$ in discrete action spaces. Approximated by parameter θ^Q, one agent optimizes the function by minimizing the loss:

$$Loss(\theta^Q) = \mathrm{E}_{\mu}[(Q(s_t, a_t) - y_t)^2], \tag{4}$$

where the term $y_t = r(s_t, a_t) + \gamma Q(s_{t+1}, \mu(s_{t+1}) | \theta^Q)$ is called the target, as the loss tries to make the value function be more like this target. The long-term accumulation of reward leads to high variance. Besides, DQN only works well with few actions; thus, failing in continuous action spaces.

DDPG [19], a model-free off-policy algorithm, can be used for environments with continuous action spaces. DDPG has two pairs of actor-critic networks with the same structure. The critic network approximates the state-action value function $Q(s,a|\theta)$ by parameter θ, which updates by minimizing (4). The actor network approximates the agent's policy $\mu(s|\omega)$ by parameter ω, which deterministically maps the observed state s to an certain action a. The actor network is updated by the derivative of (3) respectively to the actor parameters:

$$\nabla_{\omega}\mu \approx \frac{1}{N}\sum_{i} \nabla_{a} Q(s,a|\theta)|_{s=s_i,a=\mu(s_i)} \nabla_{\omega}\mu(s|\omega)|_{s=s_i}, \tag{5}$$

where N is the number of state transitions.

Two main technologies are employed by DDPG to enable a stable learning process. The first one is the replay buffer, which is the set E of previous experiences in the form of state transitions (s_t, a_t, r_t and s_{t+1}). The learning policy may shift towards and even over-fit to training samples. Therefore, old experiences are indispensable for robustness. To balance the learning process, new experiences are given a higher probability than older ones of being sampled in a minibatch. The other technology is the target network originally used in DQN. The target term in (4) depends on the same parameters to be learned while the network is being trained. This problem makes loss optimization unstable. The solution is to use another pair of networks $Q^{'}(s,a|\theta^{'})$ and $\mu^{'}(s|\omega^{'})$ with a time delay. A second network, called the target network, lags behind the first one, which is called the evaluation network. The target network updates by a smooth average:

$$\begin{aligned} \theta^{'} &= \tau\theta + (1-\tau)\theta^{'} \\ \omega^{'} &= \tau\omega + (1-\tau)\omega^{'} \end{aligned} \tag{6}$$

where $\tau \ll 1$.

To address the over-approximation of actor-critic networks, the twin-delayed deep deterministic policy gradient algorithm (TD3) [24] based on DDPG employs three technologies. The delayed policy update changes the target networks slowly every d steps, and this makes networks steadier by avoiding over accumulating the rewards. The policy smoothing supposes similar states should have similar actions by adding noise to the target actions in target Q networks:

$$y_t = r(s_t, a_t) + \gamma Q(s_{t+1}, \tilde{a}|\theta^{'}) \tag{7}$$

$$\tilde{a} = (\mu(s_{t+1})|\omega^{'}) + \epsilon \tag{8}$$

$$\epsilon \sim \text{clip}(\mathcal{N}(0,\sigma), -c, c),$$

where σ is a parameter for noise, c is the boundary for clipping. Clipped double Q-Learning uses two randomly initialized critic networks as estimates and always takes the minimum one to update the target:

$$y_t = r(s_t, a_t) + \gamma \min_{i=1,2} Q(s_{t+1}, \tilde{a}|\theta_i^{'}). \tag{9}$$

2. Problem Formulation

2.1. Vehicle Models

The vehicle uses a kinematic bicycle model for forward motion and lateral movement:

$$\begin{aligned} x_{t+1} &= x_t + v_t cos(\theta_t)\nabla T \\ y_{t+1} &= y_t + v_t sin(\theta_t)\nabla T \\ v_{t+1} &= v_t + a_t\nabla T \\ \theta_{t+1} &= \theta_t + \frac{\tan(\delta_t)}{L} v_t \nabla T, \end{aligned} \tag{10}$$

where x and y are the vehicle's lateral and longitudinal positions; v and a are the velocity and acceleration; and θ and δ are the vehicle's heading and steering angles, respectively. L is the wheelbase and ∇T is the time step size. Since this paper focuses on the decision level, the dynamic of other traffic vehicles is assumed to be controlled by fixed acceleration speed. To simulate generally, different behavior modes of traffic vehicles are used; namely: aggressive mode, regular mode and modest mode. The behavior modes are characterized by different acceleration speeds, safe distances to follow a front vehicle, safe distances to overtake a front vehicle and safe distances to avoid near back vehicles. Each participant vehicle emerges randomly around the ego agent with random target speed. These simulated vehicles are also enabled to change lanes within a random time interval following general traffic rules in their own behavior modes. These settings enable the vehicle to explore more potential types of traffic scenarios, which are prerequisites to learn the optimum policy.

2.2. Planning Space

In real traffic flows, an experienced driver not only observes the nearest vehicle in each lane but also looks far ahead in a defensive manner. Looking far ahead helps drivers notice potential hazards and drive smoothly. Therefore, the traffic vehicles in front of the ego agent's planning scope should be considered.

In previous work [23], the horizon for decision and planning was fixed as T_h. The planning scope in Figure 3 shows the range between the end pose (X_d) of agent decreasing to zero velocity, and the end pose (X_a) of an agent accelerating to the maximum velocity allowed within T_h. Vehicles inside such a scope influence the agent's behavior and the states for decision.

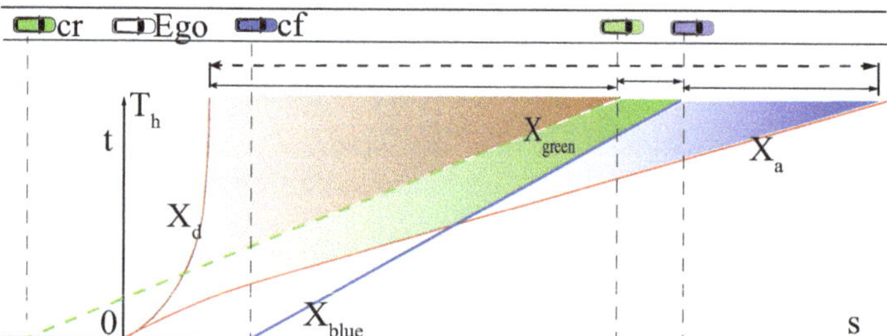

Figure 3. The planning scope (the dotted line with a double arrow) of the ego agent (in white). Two surrounding vehicles are in dim green and in dim blue at time 0. At time T_h, these two vehicles are in light colors. For the one-lane situation, the only available route is in the green contour. The rear-end distance c_{rd} and the front-end distance c_{fd} are the green vehicle's and blue vehicle's distances at time T_h in the future, respectively. The solid lines with double arrows are the planning scopes for three contours.

3. Methodology

3.1. States

The agent's behavioral decision depends on all relevant traffic entities. An entity can be a traffic participant, a pedestrian, a car or a bicycle with pose and velocity information. The lanes and signs, such as stop signs and speed limit signs with particular instructive information, are also entities. However, to generate a safe decision based on the topological routes, entities' information must be translated into planning features with specific meanings, as summarized in Table 1.

Features representing these entities are crucial for the agent's decision, which are pre-processed as following.

Table 1. The feature states for behavioral decision. Five contours exist in each lane. At most three lanes are considered. The '$\circ$' represents that a feature is scaled to $[-1, 1]$ before it is concatenated into states.

States (of)	Features	Meaning
Contour($\mathbf{s_c}$)	c_{rd} ($\circ$)	the planning scope's rear-end distance
	c_{fd} ($\circ$)	the planning scope's front-end distance
	c_{fv} ($\circ$)	the planning scope's front-end velocity
Lane($\mathbf{s_l}$)	l_l ($\circ$)	lane's effective length
	l_a	lane's attribute(accelerating, decelerating,..)
	l_c ($\circ$)	lane's curvature
	l_v ($\circ$)	lane's speed limitation
Ego agent($\mathbf{s_e}$)	e_v ($\circ$)	agent's absolute velocity
	e_a	agent's acceleration speed
	e_h ($\circ$)	agent's heading relative to lane's direction
Task (discrete,($\mathbf{s_c}$))	tk_d	the degree of task (mandatory, discretionary)
	tk_l	the lane-error to target lane of task
	tk_p	the right of way in task
Task (continuous,($\mathbf{s_c}$))	t_p	the position of special task
	t_v	the target velocity to follow at t_p
	–	the preserved states for future use

3.1.1. States of Contours in Planning Space

To make a safe decision, the features for available planning space are extracted from results of SMSTP algorithm, which guarantee safe topological routes in Figure 3. Each contour corresponding one route has three features, as shown in Figure 4. Two of them are the rear-end and front-end poses of a contour, c_{rd} and c_{fd}, respectively. One feature is the recommended velocity c_{fv} at the front end. If the contour is split by a vehicle at the front end, then this feature is related to the velocity of that vehicle; otherwise, it depends on the maximum velocity that ego agent can reach at time T_h.

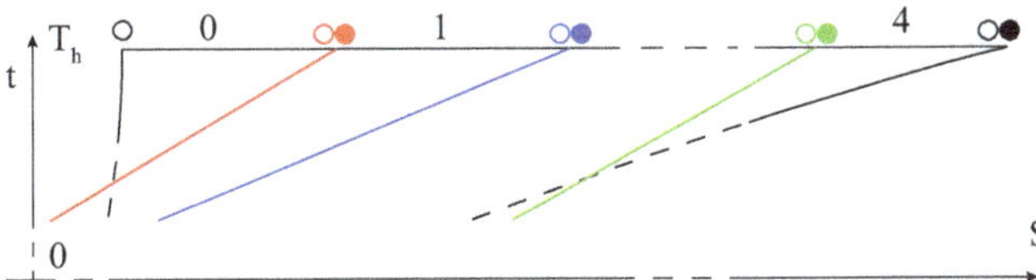

Figure 4. Five contour states in a lane's planning scope. The hollow circles and solid circles represent distance and velocity states of a contour in corresponding colors, respectively. The colored lines stand for trajectories of other traffic participants.

As the input sizes of networks are fixed, the upper number of contours in each lane is limited to five according to a safe following distance, empirically. Besides, two adjacent contours have one shared feature: that the front-end distance c^i_{fd} of i'th contour equals the rear-end distance c^{i+1}_{rf} of $i+1$'th contour (Figure 4). The state for contours in each lane has total 11 features. Therefore, 33 features consist of the state of contours $\mathbf{s_c}$ when three lanes (left lane, current lane and right lane) are considered. The features of distance can be quite enormous compared to some states, such as lane ID $[-1,0,1]$. Hence, the distance $d \in [0,334]$ and velocity $v \in [0,33.4]$ are mapped to $[-1,1]$:

$$out = 2\frac{in - in_{min}}{in_{max} - in_{min}} - 1, in \in (d,v). \tag{11}$$

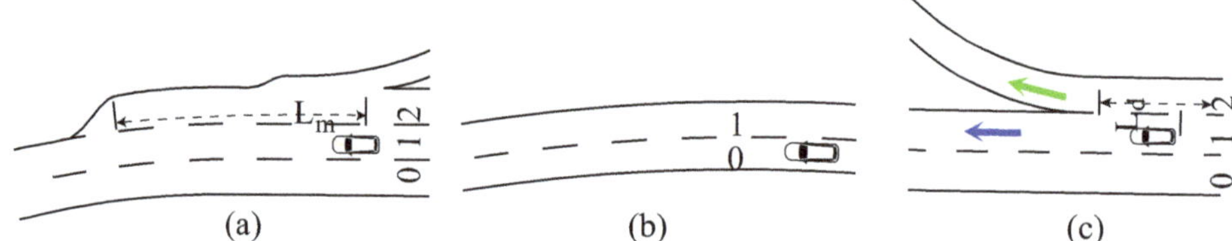

Figure 5. Lane states in different lane structures. (**a**) An in-ramp road. (**b**) A two-lane roa. (**c**) An off-ramp road.

3.1.2. States of Lanes

The state of lanes $\mathbf{s_l}$ contains all structural constrains that decide where a vehicle can go. One key feature is the length of a lane l_l. Taking scenario (a) in Figure 5 for example, when the agent goes in lane **1**, the merging lane's (lane **2**) length is, L_m, from the agent's position to the right lane's end. The left lane does not exist in scenario (b), hence the length is zero. In scenario (c), when the task is going straight along the arterial road, the rightest detached lane's length is, L_d, from the agent's position to the separation point. However, when the task is leaving the arterial road, the lengths of lane **0** and lane **1** are both L_d. So that feature l_l depends on the global task and lane's structure simultaneously.

One inherent feature of lanes is the curvature l_c which limits the maximum speed. The other two features are the lane's attribute l_a and the speed limit l_v of a lane. The rightmost lanes in scenario (a) and scenario (c) are the accelerating lane and decelerating lane in the on-off ramp area, respectively, which means vehicles in these lanes should adjust their speeds to the suggested ones as quickly as possible.

3.1.3. States of the Ego Agent

The state of ego agent $\mathbf{s_e}$ includes absolute velocity e_v, acceleration speed e_a and the angle between vehicle's heading lane's direction e_h. Here, acceleration and velocity are both scaled to $[-1,1]$. e_h is restricted between –0.26 rad and 0.26 rad first, which is then scaled to $[-1,1]$.

3.1.4. States of Tasks

Driving along a road without following traffic rules is relatively simple. However, autonomous vehicles of this kind are not acceptable in realistic situations. An agent's capability includes both going into a target lane and reaching the given position at a specific speed. The former is guided by task features in a discrete form, while the latter is instructed by target features in a continuous form.

Three separate task-level features are employed here. The task-level tk_d means to which degree the agent should follow a given task. When the agent's task is to leave the arterial road in scenario (c) of Figure 5, the mandatory task-level requires the agent to arrive at the rightmost lane before coming to the separation point. The discretionary task-level means that the agent can select any available lane to maintain a high cruise speed in scenario (c). The other one, lane-error tk_l, is the relative distance to the target lane. Lane-error represents how many lanes there are between the current lane and the target lane, which is the main punishment for task requirements. Another key feature is the prioritization tk_p.

When two vehicles' distances to intersections have no significant difference, a vehicle with the right of way is encouraged to go in priority. Otherwise, it should go after vehicles in other roads.

Two continuous target features, the target pose t_p to reach and the target velocity t_v to follow, guide the agent to arrive at the position with given speed. Taking intersection cross, for example, a recommended speed can be 25 km/h for the sake of safety and traffic rules, if the agent traverses at higher speeds, some punishment will be given. These two features also represent speed limits from traffic signs. Including four preserved ones, the state of task $\mathbf{s_{tk}}$ has nine features.

The final state for MDP is $\mathbf{s_t} =< \mathbf{s_c^t}, \mathbf{s_l^t}, \mathbf{s_e^t}, \mathbf{s_{tk}^t} >$ with a total of 57 features.

3.2. Actions

The action $a_t =< a_l^t, a_d^t, a_v^t >$ uses three sub-actions jointly to decide an agent's behavior and optimizing parameters in a select contour. The first discrete action $a_l^t \in [-1, 0, 1]$ selects a lane to go among the left lane, the current lane and the right lane if they exist. The second continuous action $a_d^t \in [-1, 1]$ represents the target position to reach T_h between X_d and X_a. The third continuous action $a_v^t \in [-1, 1]$ stands for the target velocity to follow at time T_h. The inverse transformation of (11) is used to get either the predicted distance $\bar{a}_d^t \in [0, 334]$ or the velocity $\bar{a}_v^t \in [0, 33.4]$ by

$$out = \frac{in + 1}{2}(o_{max} - o_{min}) + o_{mim},\ in \in (a_d, a_v), \tag{12}$$

where o_{max} and o_{min} are the maximal and minimal expected values of inputs. In fact, due to the values of action a_d^t and action, a_v^t can hardly be 1 or -1 through hyperbolic sine function. We enlarge both o_{max} and o_{min} a bit and truncate the outputs to expected ranges.

Action a_l^t selects the target lane, and it has several contours in the corresponding planning scope. The undetermined contour for the trajectory optimizer depends on $\bar{a}_d^t$. If $\bar{a}_d^t$ locates inside one contour of the planning scope in Figure 4, then this contour is selected. Otherwise, either the nearest or the furthest contour is selected and $\bar{a}_d^t$ is adjusted to either c_{rd}^0 or c_{fd}^i (i depends on the number of contours in the lane). In the trajectory optimizer layer of the SMSTP, the chosen contour, adjusted $\bar{a}_d^t$ and $\bar{a}_v^t$, determine the final trajectory in predicted horizon T_h.

3.3. Rewards Design

The ego agent employs an action and gets a reward r_t at time t, which evaluates the current state and action. r_t guides the policy to learn proper actions for higher rewards and reflects to what degree the agent drives as we expect.

A well-designed reward represents the desired policy for complex driving tasks. Therefore, r_t is a compromise among agent's safety, traveling efficiency, and task achievement in linear combination form:

$$r_t = r_s^t + r_e^t + r_t^t \tag{13}$$

$$r_s^t = \begin{cases} -10, & \text{a collision} \\ -2, & \text{a lane change} \\ 0, & \text{else} \end{cases} \tag{14}$$

$$r_e^t = \min(e_v - \frac{\sum_{i=1}^{n}(v_i) + l_v}{n+1}, 0) \tag{15}$$

$$r_t^t + = \begin{cases} \max(-5, -2 * tk_l), & \text{lane-error} \\ \max(-5, -|e_v - t_v), & \text{target speed error } k \\ \ldots, & \text{else,} \end{cases} \tag{16}$$

where r_s^t is the safety penalty, r_e^t is the reward for agent's driving efficiency, r_t^t is the penalty for deviation from task's target and v_i is the velocity of the i's traffic participant. As frequent lane changes can cause danger in traffic flow, a small penalty is set in r_s^t of (14) to discourage unnecessary lane changes. A reasonable agent drives as fast as possible within the speed limit of a lane. However, being a cooperative member of the traffic flow is much safer. Hence, r_e^t in (15) leads the agent to drive at the average speed of traffic, and to pursue the maximal speed of the lane, whether there are traffic participants or not. r_t^t provides an agent the ability to perform different tasks, which depends on the requirements. For example, $-5tk_l$ in (16) punishes the agent for deviating from the target lane when the agent should be in the target lane. $-|e_v - t_v|$ punishes the agent for being unable to follow given speed limitations. To achieve other tasks, r_t^t should adjust the penalty term accordingly. Each term in r_t^t is not less than -5.

3.4. Neural Networks

The actor and critic networks are presented in Figure 6. A state $\mathbf{s_t}$ enters the actor network, which has five neurons as the output layer in green. A softmax function takes the first three neurons as input and outputs the discrete action a_l^t. Two continuous actions a_d^t and a_v^t are the last two neurons adding some noises. The critic network takes a state $\mathbf{s_t}$ as input, and concatenates the first hidden layer with the output layer of actor network as the input of second hidden layer. A scalar value is output as the action-value (Q-value).

The actor network has three mutual hidden layers, and each layer has 256, 512 and 128 neurons, respectively. The private layers connecting discrete and continuous actions both have 64 neurons. The critic network has three hidden layers, and each has 256, 128 and 64 neurons, individually. Rectified linear unit (ReLu) activation functions are used in the hidden layers. Meanwhile the output layer of actor network uses tanh functions to constrain the value of actions.

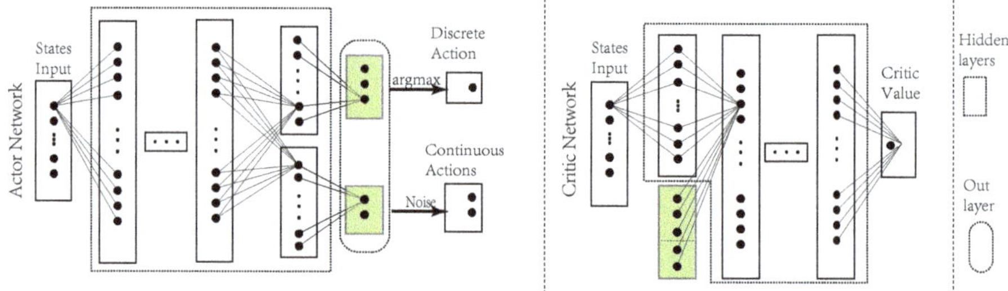

Figure 6. The structure of actor–critic networks. The actions are sampled from the output layer (in green color) of the actor network. The output layer is concatenated with the second layer of critic network. The discrete action is calculated from output layer through a softmax function. The continuous actions are sampled from the other two unions with small noises

3.5. Algorithm Procedure

The procedure of PFBD method for autonomous vehicle safe decision is Algorithm 1. The algorithm includes two main parts. The first part is for sampling and evaluation. Experiences of state transitions are sampled before the policy is trained. Algorithm 1, firstly, initializes the actor-critic networks and sets a task for agent along with the penalty function for different tasks. After initializing the state s_0 in each episode, action a_t is generated according to sampling mode among random sampling and ϵ-random sampling. a_t shifts agent to next state s_{t+1}, and a reward r_t follows. When the number of state transitions $(\mathbf{s_t}, \mathbf{a_t}, \mathbf{r_t}, \mathbf{s_{t+1}})$ reaches N_{max}, the transitions are stored as a group into "Experience" on the local disk. If a policy has been trained, we set sampling mode to "Evaluation" to evaluate its performance.

For the training part, N_{env} groups of samples in Experience are loaded from disk. The newer stored samples are given a higher priority to balance the shifting of policy. Then, the parameters of evaluation and target networks are updated in sequence.

Algorithm 1 Planning features based behavior decision algorithm (PFBD)

Initialization: Initialize evaluation actor-critic networks' parameters $\theta_1, \theta_2, \omega$ randomly
Initialize target actor-critic networks' parameters $\theta_1' \leftarrow \theta_1, \theta_2' \leftarrow \theta_2, \omega' \leftarrow \omega$
for task $k = 1$ to N_k **do**
  Set a **Task** and **sampling mode**, load the traffic map, reset traffic participants
  Initialize the agent's state s_0
  for step $t = 0$ to T **do**
    generate **action** a_0 according to **sampling mode**
    switch (sampling mode)
    case Random sampling: randomize the output layer of the actor network
    case ϵ-random sampling: randomize the output layer of actor network with probability ϵ
    case Evaluation: use the trained policy (forward through actor network)
    end switch
    Employ **action** a_t to next state s_{t+1} based on **SMSTP** algorithm, calculate the reward r_t
    Save transition $(\mathbf{s_t}, \mathbf{a_t}, \mathbf{r_t}, \mathbf{s_{t+1}})$, $++n_s$
    if $n_s == N_{max}$ **then**
      Store N_{max} transitions as a group into **Experience**, $n_s = 0$
    end if
  end for
  for epoch $ep = 1$ to N_{epoch} **do**
    Load N_{env} groups of **Experience** from disk with priority, randomly sample a minibatch of $N_{minibatch}$ transitions
    Update critic in evaluation network by minimizing (4), target is updated by (9)
    Update actor in evaluation network following (5)
    Update the target networks following (6)
  end for
end for

4. Simulation and Experiment

Since the proposed approach focuses on behavioral decision and planning with only abstract entities for autonomous driving, we tested the method with our vehicle simulation platform instead of real traffic. A schematic view of sampling and evaluation architecture is depicted in Figure 2.

4.1. Simulation Setting

4.1.1. Environment

Experiments included three typical scenarios of different traffics, as shown in Figure 7. The first one was a straight six-lane road scenario (Scenario A). The second one was a curve road scenario (Scenario B). Scenario B had seven zebras in the straight part of the road. The last one was a merging road (Scenario C). The lane's width was set to four meters. The maximum sensible ranges of the front and back were set to 200 m and 100 m, respectively.

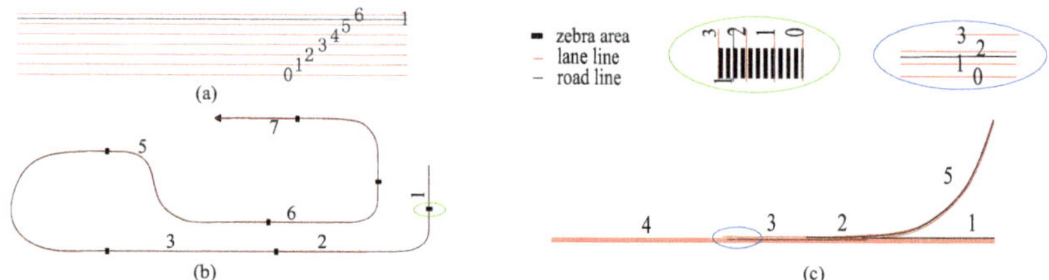

Figure 7. Different traffic scenarios used to train and evaluate the performance agent's policy. (**a**) A straight road with six lanes. (**b**) A curved road with zebras in the straight parts. (**c**) A merging road. The red lines represent lane lines. The black lines are roads.

4.1.2. Traffic Participants

The other traffic participants and their behaviors were set as anticipated in Section 2.1. The participants changed their behavior modes at a probability $p = 0.005$. (If five traffic participants exist, the probability of at least one changing its behavior in one second is 0.2217.) The changing of behavior mode ensures abundant experiences of different scenes. A vehicle was reset when it got out of the agent's range.

The participants employ an expert system to make behavioral decisions and to execute longitudinal control. The expert system considers the distances and velocities of the nearest vehicles in the surrounding lane. The decision system has been well tested in real traffic experiments [25].

4.1.3. Hyper-Parameters for TD3

A minibatch of 64 state transitions was randomly selected from 32 groups of experiences to train the policy every two steps. The optimization method for TD3 is the Adam algorithm [26], and the learning rates for actor and critic networks are 10^{-2} and 10^{-4}, respectively. The discount for future reward was set to 0.99. We used deep learning framework, libtorch (Pytorch for C++) [27], to train the agents.

4.2. Performance Index

To evaluate the capabilities of PFBD method, the following metrics are introduced.

- Collision Rate [%]: the collision rate is defined as

$$R_c = \frac{CR}{V * T * L}, \tag{17}$$

  where CR is the number of crashes that happened, V is the traffic volume in the agent's sensible range, $T \geq 1$ is the time in hours the agent drives and L is the total distance traveled. This is the most critical metric for evaluating the agent's performance.
- Average velocity (m/s): the average velocity represents how fast the agent drives and to what degree the expected velocity is matched.
- Average distance between collisions (km): the distance between two collisions is to avoid the influence of agent's velocity and duration time.
- Rule-violation rate (%): the violation rate indicates the agent's capability of achieving given tasks, such as decreasing when a zebra is coming, keeping in specific lanes and leaving the ramp's accelerating lane before coming to its end.

4.3. Simulation and Evaluation

In the simulation, we trained four different agents with four corresponding polices. Three agents' policies, P_A, P_B and P_C were trained individually on three typical scenarios (see Section 4.1), respectively. To validate the versatility of PFBD, one more agent's policy P_T was trained on all three scenarios. Besides, we compared the performance of PFBD method with the rule-based expert system and an untrained random policy.

An episode is reached in the following three cases:

- Case 1: the agent is made to unavoidably collide with a vehicle in 1.5 s.
- Case 2: the agent has successfully made more than 20 decisions.
- Case 1: the agent updates for 600 steps without collisions.

The last two rules prevent an episode with too many running steps.

To generate the desired behavior, the reward function for scenarios B and C was modified accordingly. For scenario B with zebras on the road, r_t^t in (16) is added with a new term:

$$r^t_{t(zebra)} = \max(-5., -0.5 * \frac{v_{error}}{1 + sqrt(\min(d_{zebra}, 100))}) \qquad (18)$$
$$v_{error} = \sqrt{\max(10^{-4}, v - v_{zebra})},$$

where v_{zebra} is the recommended velocity to pass a zebra area and d_{zebra} is the distance to a zebra. For task entering the main road in scenario **C**, r_t^t in (16) is added with:

$$r^t_{t(ramp)} = -0.5 * \sqrt{-\min(100, d_{ramp}) + 100}, \qquad (19)$$

where d_{ramp} is the distance to ramp's end.

At the first stage, we sampled more than one million state transitions in each scenario randomly. Then, the agents, P_A, P_B and P_C were trained from samples in each scenario according to Algorithm 1, respectively. P_T was trained on samples from all three scenarios. Then, we used the trained agents to sample with a probability ϵ randomly selecting the actions. The agents did not stop being trained until the loss stopped decreasing for 10 epochs.

4.3.1. Performance of PFBD on Different Scenarios

Figure 8 shows the decision and planning result of trained agents in the simulation environment. Figure 8a,b illustrates that the agent takes over the front vehicle (ID = 9) and follows a relatively quicker vehicle (ID = 13). Figure 8c–e shows that the agent merges into the middle lane timely before the ramp(right lane) ends.

The statistical results for four agents are listed in Table 2. Generally, the trained agents achieved the designed maneuvers. The collision rate and average distance without collision show that the agent can learn proper policy to avoid collisions. Scenario A is easier for the agent to drive in since the agent achieves the highest average velocity and lowest collision rate. For scenarios B and C, the agent has to slow down for the zebra or limit its speed in the ramp. $\mathbf{P_T}$ can also learn proper behaviors on all three scenarios without an evident decrease in performance.

Table 2. The evaluation results of training-acquired policy in different scenarios. - represents no consideration.

	$\mathbf{P_A}$	$\mathbf{P_B}$	$\mathbf{P_C}$	$\mathbf{P_T}$ (Scenario A)	$\mathbf{P_T}$ (Scenario B)	$\mathbf{P_T}$ (Scenario C)
Collision Rate	0.23%	0.39%	1.17%	1.3%	1.6%	1.68%
Average Velocity (m/s)	18.1	15.1	14.5	14.9	12.0	11.8
Av. Dist No Col (ks)	18.2	13.6	14.9	11.5	10.8	9.1
Rule Violation	-	3.15%	2.4%	-	5.33%	3.68%

Note that the average distance without collision is relatively short in the simulation. The main cause is that a traffic participant sometimes emerges too close to the agent since it is randomly reset.

For the abundance of sample distribution, the nearest distance to the agent is 20 m. When the agent is much quicker than the simulated vehicle in the front, a collision can be hard to avoid. The rule violation is also difficult to define clearly. In our simulation, for a successful result in the zebra scenario, the duration of the agent's velocity being smaller than the recommended velocity has to be kept up for at least two seconds.

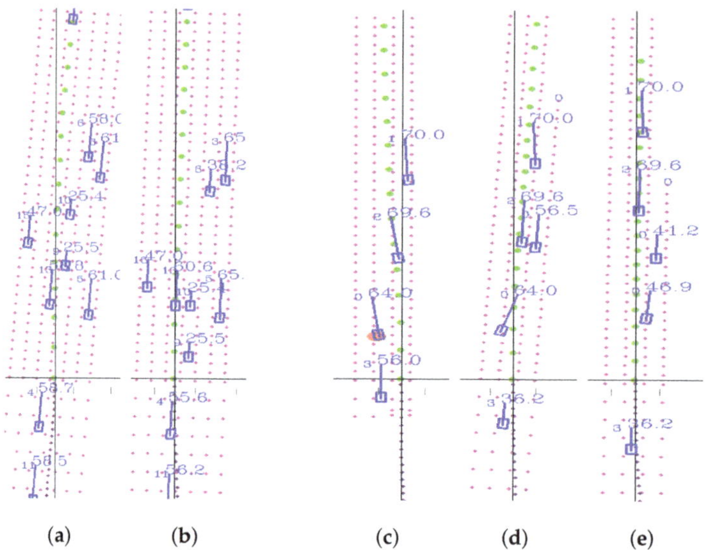

Figure 8. Two typical maneuvers. (**a**,**b**) An overtaking maneuver in a six-lane road with heavy traffic. (**c**–**e**) show a lane merging maneuver at ramp. Each simulated vehicle is labeled with **ID** and speed (km/h) information. The red points are the lane lines. The big green points and small purple ones represent the planned trajectory and the historical trajectory, respectively.

4.3.2. Comparison with the Expert's Policy

DRL-based PFBD method can explore optimal policy via interacting with environments. Nevertheless, the rule-based expert' policy often struggles in dynamic situations with heavy traffic. Therefore, the performance of the trained policy was compared to the expert's policy. The agents P_A, P_B and P_C ran for at least 10^5 iterations and at least 2500 maneuvers executed. The expert's policy described in Section 4.1.2 and a random policy (untrained) were both evaluated.

One critical concern is that the reward of lane-keeping must be recorded in a balanced manner. We did not record a reward in every iteration. On the contrary, either the agent kept in a lane for every 40 steps or a non-lane-keeping maneuver was selected; the reward of lane-keeping was recorded at one time.

Figure 9 depicts the collision times after the agents ran for 10^5 iterations in three testing scenarios. It shows that the learned agents for all three scenarios significantly decreased the collision times after training. Besides, the learned agents achieved competitive performance, as did the conservative traditional expert. In fact, a large number of the collisions derived from a too-close-in-front vehicle after resetting. Although the low-level SMSTP algorithm can plan all non-collision topological paths, whether the trajectory can be well executed is undetermined (the agent cannot decelerate quickly enough). Therefore, collision times depend on the performance of a policy. The simulation results verify that the agents can learn proper policy after training.

For the rewards of the learned agents, as shown in Figure 10, most immediate rewards (in blue points) achieved zero punishments in scenario A and B, while the expert's immediate rewards (in red points) seldom got no punishment. A statistical distribution of immediate rewards is listed in Figure 11. Both agent P_A and agent P_B, more than 40% of the time, received zero punishments. This implies that the agents successfully found an optimal policy in the corresponding states. In contrast, the expert's policy mostly had a small immediate reward, as the expert behaved conservatively and got trapped in dynamic traffic at a relatively low speed most of the time. In the ramp merging scenario, the agent and expert both had a scattered distribution of immediate rewards, which reflected the difficulty in the ramp-merging task. Both policies had to compromise for a successful merging maneuver.

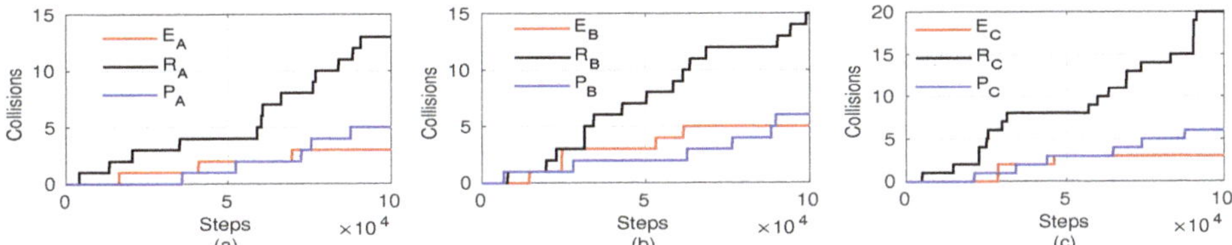

Figure 9. The collision times after 10^5 iterations for three different traffic scenarios. E_*, R_* and P_* represent the policies for expert, random and learned agents, respectively. (**a**) The collision times in scenario **A**. (**b**)The collision times in scenario **B**. (**c**) The collision times in scenario **C**

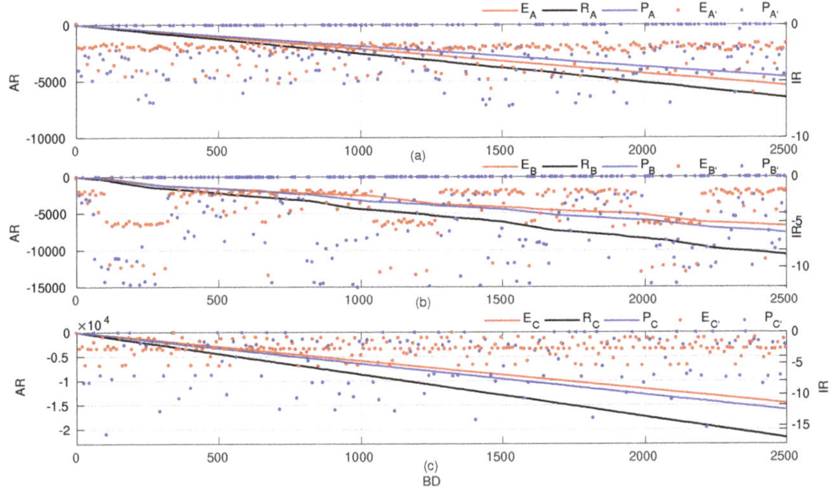

Figure 10. The rewards of reinforcement learning (RL) policy compared to expert's policy and a random policy for 2500 maneuvers. (**a**) Rewards in Scenario **A**. (**b**) Rewards in Scenario **B**. (**c**) Rewards in Scenario **C**. The blue and red points represent the immediate rewards of the trained agent and the expert, respectively. Only one eighth of the points for IR are illustrated for clarity. **AR:** accumulated rewards, **BD:** behavioral decisions, **IR:** immediate rewards.

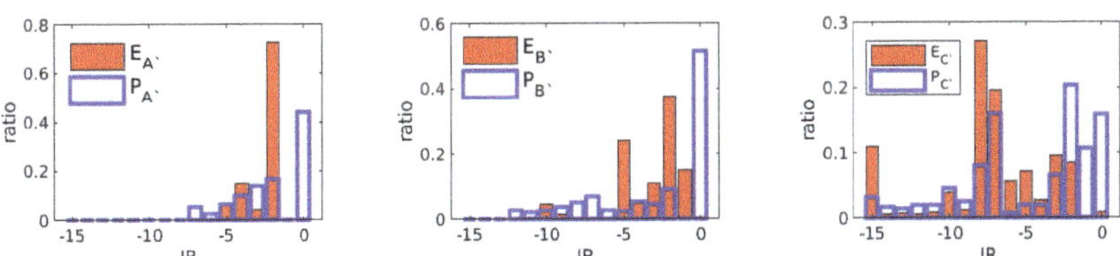

Figure 11. The distribution of immediate rewards for both learned agents and the expert in three testing scenarios.

4.3.3. Comparison with standard DRL

To validate whether the states of contours help to make a better decision, we compared the proposed PBFD with a standard DRL where all the parameters were selected as the same as those in the PBFD except for the dynamic features from the planning layer being not concerned in the standard DRL. To achieve this, we specifically set, in the DRL, the planning scope's front-end distance c_{fd} to zero with small noise. With the above design, we trained a new agent C_A in scenario A, whose behaviors are displayed together with those of the PBFD and the expert in Figure 12.

The results show that agent C_A had almost the same collision times as P_A in 10^5 running steps. It can also be seen in Figure 12a,c that agent C_A had about a 6% chance of achieving zero punishment, which is better than expert's policy (in Figure 11), which could hardly get any zero punishment outcomes. However, agent C_A behaved worse than agent P_A, as 40% of immediate rewards of agent P_A were zero. This means that agent P_A with full knowledge of contour states has larger probabilities of making optimal behavioral decisions. Thus, the creative design of contour is vital to learn better behavior decisions.

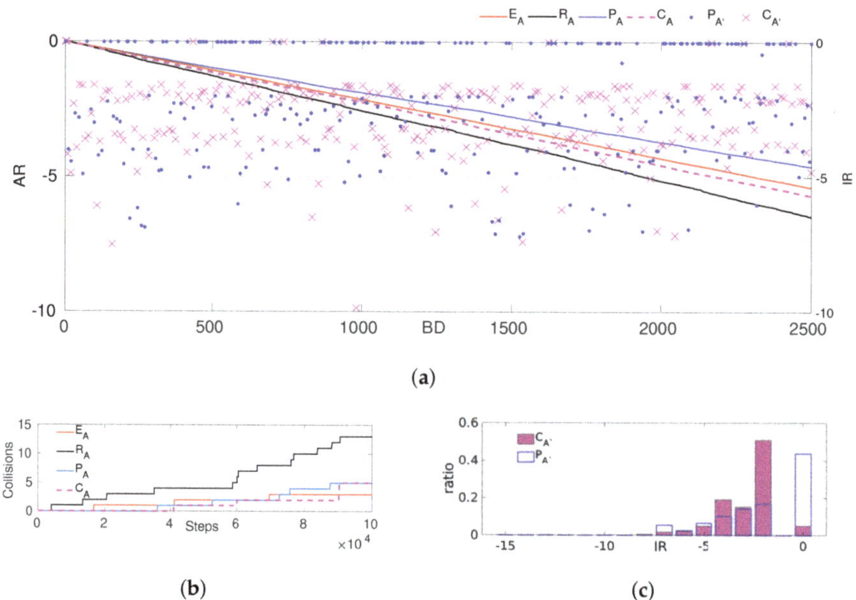

Figure 12. The comparison results of states with or without the profile's information. (**a**) Rewards in Scenario **A**. The blue • and magenta × are the immediate rewards of agent P_A and agent C_A, respectively. (**b**) The collision times in Scenario **A**. (**c**) The distribution of immediate rewards of agent P_A and agent C_A.

4.4. Real-time Experiment On HQ3 Autonomous Vehicle

To validate the capacity of the proposed PFBD algorithm for autonomous vehicle decision making, the real-time experiment on HQ3 autonomous vehicle was conducted in the suburbs with regular traffic. HQ3 [25] was equipped with necessary sensors, including cameras, LIDAR, radars, inertial measurement unit (IMU), etc.

To use the learned agents in real-time test, we kept the same states in both simulation and the real test, and we ignored those states that did not exist in simulations, such as an intersection with turn commands. The two main tests in the suburbs were keeping in the target lane near a zebra (corresponding to Scenario A and B) and merging onto the main road from a side road (corresponding to Scenario C). The agent's maximum speed was limited to 60 km/h. The road network avoided untrained tasks, such left turning or right turning that did not exist in simulation.

Figure 13 shows the agent's behavioral decision and trajectory planning results for the two separate maneuvers. The blue short lines indicate the speeds and directions of the detected surrounding vehicles on the road. In the first test, HQ3 successfully merged to the middle lane. Besides, as the zebra area was coming (see Figure 13c, the planned trajectory led HQ3 to slow down (see green trajectory of Figure 13i). In the second test, our agent tried to merge into the left main road (see Figure 13d–f). HQ3 changed to the left curve lane before the road network of the side road coming to end. As in the real-time test shown, the trajectory planner based on the learned agent can successfully generate a smooth and safe trajectory for HQ3.

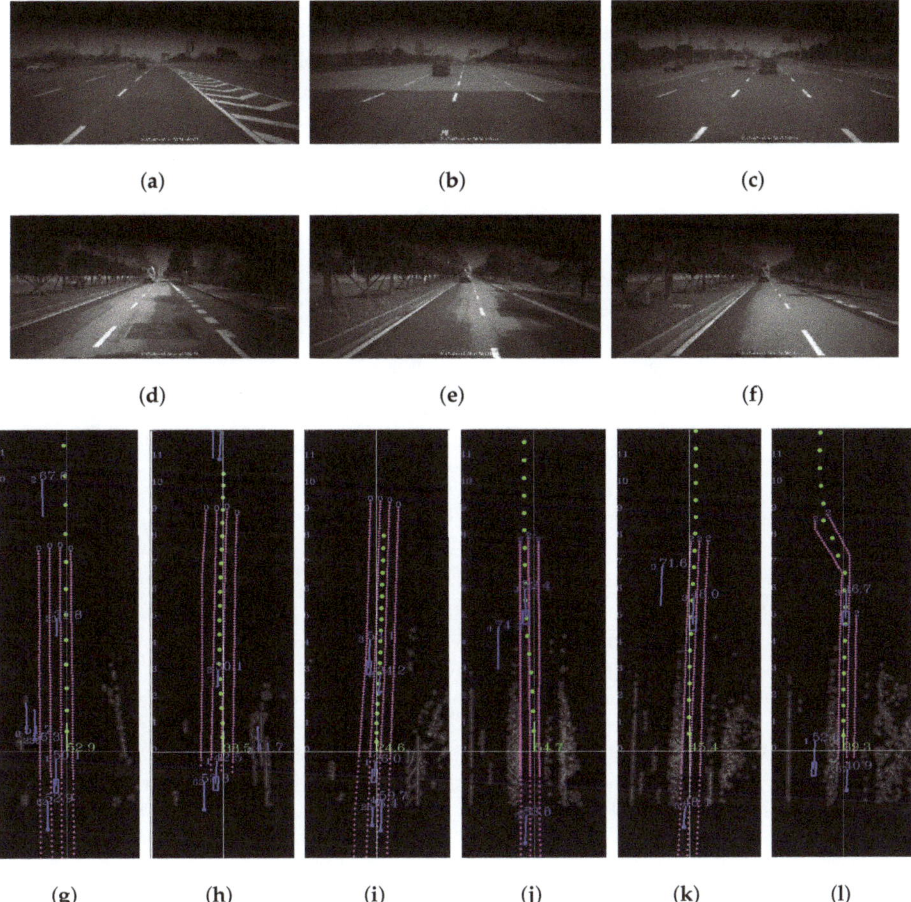

Figure 13. Snaps of two real-time maneuvers in the suburbs. (**a–c** and **d–f**) are the front views of (**g–i** and **j–l**), respectively. The red dotted lines are the lane lines. The green points are the planned trajectory. The gray areas are obstacles.

5. Conclusions

In this paper, a novel planning features-based, deep behavior decision method trained with TD3 was proposed to select an optimal maneuver for autonomous vehicles in dynamic traffic. The PFBD uses rule-based topological trajectories as basic guarantees and learns optimal policy with DRL, which helps to bring learning methods to the autonomous vehicle's core module, behavioral decisions and trajectory planning. The behavioral decision problem was modeled as an MDP process, and the training algorithm was proposed. Due to the creative design of input states and actions, the behavior decision is inherently associated with drivable routes in the topological planning space. Thus, the policy's decision is consistent with trajectory planning, and the path is a safer one. The evaluation results have demonstrated that the proposed method can learn proper behaviors. Besides, comparisons to the expert's policy and the standard DRL validate the effectiveness of PFBD and show its capability to find optimal strategies with larger possibilities. The real-time test also validated the effectiveness of the proposed PFBD.

Since the agent's performance is competitive compared to the rule-based expert's policy, the future work will extend the current discrete behavioral decisions to consecutive ones, which shall evaluate the present behaviors and potential behaviors in every time step. Therefore, some emergent situations could be tackled immediately rather than executing the ongoing behavior to the end. In the meanwhile, more traffic scenarios can be used to test the generalization performance of the proposed method.

Author Contributions: L.Q. came up with the idea and proposed the algorithm; X.X. instructed the main research idea and acquired the funds. L.Q. and J.H. implemented the experiments and analyzed the data; L.Q and Y.Z. accomplished the writing.

Funding: This research was supported by the National Natural Science Foundation of China under Grants 61751311, 61825305 and U1564214.

Conflicts of Interest: The authors declare no conflict of interest.

Abbreviations

The following abbreviations are used in this manuscript:

PFBD	Planning Features based Behavior Decision
DDPG	Deep Deterministic Policy Gradient
DRL	Deep Reinforcement Learning
TD3	Twin Delayed Deep Deterministic policy gradient algorithm
SMSTP	Synchronous Maneuver Searching and Trajectory Planning algorithm

References

1. Montemerlo, M.; Becker, J.; Bhat, S.; Dahlkamp, H.; Dolgov, D.; Ettinger, S.; Haehnel, D.; Hilden, T.; Hoffmann, G.; Huhnke, B.; et al. Junior: The Stanford entry in the Urban Challenge. *J. Field Robot.* **2008**, *25*, 569–597. [CrossRef]
2. Kurt, A.; Özgüner, Ü. Hierarchical finite state machines for autonomous mobile systems. *Control Eng. Pract.* **2013**, *21*, 184–194. [CrossRef]
3. Pomerleau, D.A. ALVINN: An Autonomous Land Vehicle in a Neural Network. In *Advances in Neural Information Processing System 1*; Morgan Kaufmann Publishers Inc.: San Francisco, CA, USA, 1989; pp. 305–313.
4. Bojarski, M.; Testa, D.D.; Dworakowski, D.; Firner, B.; Flepp, B.; Goyal, P.; Jackel, L.D.; Monfort, M.; Muller, U.; Zhang, J.; et al. End to End Learning for Self-Driving Cars. *arXiv* **2016**, arXiv:1604.07316.
5. Caltagirone, L.; Bellone, M.; Svensson, L.; Wahde, M. LIDAR-based driving path generation using fully convolutional neural networks. In Proceedings of the 2017 IEEE 20th International Conference on Intelligent Transportation Systems (ITSC), Yokohama, Japan, 16–19 October 2017; pp. 1–6.
6. Silver, D.; Huang, A.; Maddison, C.J.; Guez, A.; Sifre, L.; Van Den Driessche, G.; Schrittwieser, J.; Antonoglou, I.; Panneershelvam, V.; Lanctot, M.; et al. Mastering the game of Go with deep neural networks and tree search. *Nature* **2016**, *529*, 484. [CrossRef] [PubMed]
7. Qi, X.; Luo, Y.; Wu, G.; Boriboonsomsin, K.; Barth, M. Deep reinforcement learning enabled self-learning control for energy efficient driving. *Transp. Res. Part C Emerg. Technol.* **2019**, *99*, 67–81. [CrossRef]

8. Nazari, M.; Oroojlooy, A.; Snyder, L.; Takac, M. Reinforcement Learning for Solving the Vehicle Routing Problem. In Proceedings of the 32nd Conference on Neural Information Processing Systems (NeurIPS 2018), Montréal, QC, Canada, 3–8 December 2018; pp. 9839–9849.
9. Aradi, S.; Becsi, T.; Gaspar, P. Policy Gradient Based Reinforcement Learning Approach for Autonomous Highway Driving. In Proceedings of the 2018 IEEE Conference on Control Technology and Applications (CCTA), Copenhagen, Denmark, 21–24 August 2018; pp. 670–675.
10. Wolf, P.; Kurzer, K.; Wingert, T.; Kuhnt, F.; Zollner, J.M. Adaptive Behavior Generation for Autonomous Driving using Deep Reinforcement Learning with Compact Semantic States. In Proceedings of the 2018 IEEE Intelligent Vehicles Symposium (IV), Changshu, China, 26–30 June 2018; pp. 993–1000.
11. Wang, P.; Chan, C.-Y.; Fortelle, A.D. A Reinforcement Learning Based Approach for Automated Lane Change Maneuvers. In Proceedings of the 2018 IEEE Intelligent Vehicles Symposium (IV), Changshu, China, 26–30 June 2018; pp. 1379–1384.
12. Shalev-Shwartz, S.; Shammah, S.; Shashua, A. Safe, multi-agent, reinforcement learning for autonomous driving. *arXiv* **2016**, arXiv:1610.03295.
13. Li, N.; Oyler, D.W.; Zhang, M.; Yildiz, Y.; Kolmanovsky, I.; Girard, A.R. Game Theoretic Modeling of Driver and Vehicle Interactions for Verification and Validation of Autonomous Vehicle Control Systems. *IEEE Trans. Contr. Syst. Technol.* **2018**, *26*, 1782–1797. [CrossRef]
14. You, C.; Lu, J.; Filev, D.; Tsiotras, P. Advanced planning for autonomous vehicles using reinforcement learning and deep inverse reinforcement learning. *Robot. Auton. Syst.* **2019**, *114*, 1–18. [CrossRef]
15. Zhu, M.; Wang, X.; Wang, Y. Human-like autonomous car-following model with deep reinforcement learning. *Transp. Res. Part C: Emerg. Technol.* **2018**, *97*, 348–368. [CrossRef]
16. Huang, Z.; Xu, X.; He, H.; Tan, J.; Sun, Z. Parameterized Batch Reinforcement Learning for Longitudinal Control of Autonomous Land Vehicles. *IEEE Trans. Syst. Man, Cybern. Syst.* **2019**, *49*, 730–741. [CrossRef]
17. Xu, C.L.Z.H.X.; Wang, J. Lateral control for Autonomous Land Vehicles via Dual Heuristic Programming. *Int. J. Robot. Autom.* **2016**, *36*. [CrossRef]
18. Lian, C.; Xu, X.; Chen, H.; He, H. Near-Optimal Tracking Control of Mobile Robots Via Receding-Horizon Dual Heuristic Programming. *IEEE Trans. Cybern.* **2016**, *46*, 2484–2496. [CrossRef] [PubMed]
19. Lillicrap, T.P.; Hunt, J.J.; Pritzel, A.; Heess, N.; Erez, T.; Tassa, Y.; Silver, D.; Wierstra, D. Continuous control with deep reinforcement learning. *Comput. Sci.* **2016**, *8*, A187.
20. Sallab, A.; Abdou, M.; Perot, E.; Yogamani, S. Deep Reinforcement Learning framework for Autonomous Driving. *Electron. Imaging* **2017**, *2017*, 70–76. [CrossRef]
21. Bai, Z.; Cai, B.; Wei, S.G.; Chai, L. Deep Learning Based Motion Planning For Autonomous Vehicle Using Spatiotemporal LSTM Network. In Proceedings of the 2018 Chinese Automation Congress (CAC), Xi'an, China, 30 November–2 December 2018; pp. 1610–1614.
22. Chen, L.; Hu, X.; Tian, W.; Wang, H.; Cao, D.; Wang, F. Parallel planning: A new motion planning framework for autonomous driving. *IEEE/CAA J. Autom. Sin.* **2019**, *6*, 236–246. [CrossRef]
23. Qian, L.; Xu, X.; Zeng, Y.; Li, X.; Sun, Z.; Song, H. Synchronous Maneuver Searching and Trajectory Planning for Autonomous Vehicles in Dynamic Traffic Environments. *IEEE Intell. Transp. Syst. Mag.* in press.
24. Fujimoto, S.; van Hoof, H.; Meger, D. Addressing Function Approximation Error in Actor-Critic Methods. In Proceedings of the Thirty-Fifth International Conference on Machine Learning, Stockholm, Sweden, 10–15 July 2018.
25. Qian, L.; Fu, H.; Li, X.; Cheng, B.; Hu, T.; Sun, Z.; Wu, T.; Dai, B.; Xu, X.; Tang, J. Toward Autonomous Driving in Highway and Urban Environment: HQ3 and IVFC 2017. In Proceedings of the 2018 IEEE Intelligent Vehicles Symposium (IV), Changshu, China, 26–30 June 2018; pp. 1854–1859.
26. Kingma, D.P.; Ba, J. Adam: A Method for Stochastic Optimization. In Proceedings of the 3rd International Conference for Learning Representations, San Diego, CA, USA, 7–9 May 2015.
27. Paszke, A.; Gross, S.; Chintala, S.; Chanan, G.; Yang, E.; DeVito, Z.; Lin, Z.; Desmaison, A.; Antiga, L.; Lerer, A. Automatic differentiation in PyTorch. In Proceedings of the NIPS-W 2017, Long Beach, CA, USA, 4–9 December 2017; p. 4.

Article

The Need for Cooperative Automated Driving

Jan Cedric Mertens [1,*], Christian Knies [1], Frank Diermeyer [1], Svenja Escherle [2] and Sven Kraus [2]

[1] Institute of Automotive Technology, Technical University of Munich, 85748 Garching, Germany; knies@ftm.mw.tum.de (C.K.); diermeyer@ftm.mw.tum.de (F.D.)
[2] MAN Truck & Bus SE, 80995 Munich, Germany; svenja.escherle@man.eu (S.E.); sven.kraus@man.eu (S.K.)
* Correspondence: mertens@ftm.mw.tum.de; Tel.: +49-152-51472395

Received: 26 March 2020; Accepted: 29 April 2020; Published: 4 May 2020

Abstract: In this paper we describe cooperation and social dilemmas in multiagent systems, with an analogy applied to road traffic. Cooperative human drivers, based on their perception of trust and fairness, find efficient solutions for such dilemmas. In the development of automated vehicles (AVs) it is therefore important to ensure that this cooperative ability is maintained even without a human driver. Therefore, the topic of cooperative intelligent transport systems (C-ITSs) is discussed in detail and different characteristics of cooperation and their implementation are derived. Further, three planning levels with the corresponding communication techniques are discussed and several methods for maneuver planning are listed. All in all, we hope that this paper will allow us to better classify different cooperative scenarios, develop novel approaches for cooperative AVs (CAVs), and emphasize the need for cooperative driving.

Keywords: cooperative automated vehicles (CAV); cooperation; communication; V2X; C-ITS; ITS-G5; LTE 5G; maneuver planning

---

1. Motivation

Interaction between vehicles in road traffic is inevitable and, to date, these interactions are based on the human driving style, which depends on factors such as the driver´s experience or current emotional situation. With higher automation levels however, the driver will become obsolete, with the advantage being that automated vehicles will drive more rationally and we will no longer see acts of spite in traffic. Yet, there is also a challenge in the replacement of the human driver with an automated system. Until now, the human has been the only cooperative module within the vehicle, and its removal it will also eliminate mutualism and altruism from our traffic, leaving selfish vehicles that strictly follow traffic rules and optimize only their own costs. The German road traffic regulations (StVO) state in the first paragraph that "Use of the road requires constant care and mutual respect" [1]. While this generic statement is understandable for a human driver, an automated system needs a clear formalization of this rule, and this so far does not exist. Without cooperation, traffic will lose efficiency, with conflicts and aggressive behavior [2,3].

Imagine a situation with heterogenous agents such as fast cars and slow heavy trucks, where the trucks want to drive from the on-ramp onto the freeway. Due to their mass, their acceleration and maximal velocity are limited. According to the StVO, the cars on the freeway have the right of way and with no large gap between the cars, the truck has to stop and wait for that gap, thereby shortening the acceleration lane. With more than one truck, this could lead to a full blockage of the acceleration line, not only for the slow trucks but also for fast cars. This problem is a social dilemma, where selfish interactions bring the most benefit for individual agents (cars), until all agents act selfishly and the whole group suffers from increased costs.

In Section 2 we explain the concepts of cooperation and social dilemmas and then reflect on this using several transport scenarios to identify real-life social dilemmas and problems that can arise due to lack of cooperation. In Section 3 we then discuss how cooperation in road transport, i.e., cooperative intelligent transport systems (C-ITS), can be characterized, and we give three examples of cooperative scenarios. In Section 4 we indicate how cooperation can be implemented in maneuver planning and describe how this actually happens in nature by animals and human drivers. Then, in Section 5 we present in detail how cooperation can be implemented for automated vehicles based on three previously presented features, and conclude the paper with a discussion of the upcoming problems and challenges in Section 6 and a conclusion in Section 7.

2. Cooperation and Social Dilemmas in Multi Agent Systems

Road traffic, with its various road users in the shared environment of the road, is a classic example of a multi-agent system (MAS) [4]. In the following we will discuss MASs in general, as well as cooperation among agents and the resulting social dilemmas.

2.1. Multi-Agent System

A multi-agent system consists of different agents which are located in the same environment and try to collectively reach a common goal. Using the example of the Sense–Plan–Act concept of an automated vehicle (AV) [5] and in accordance with the general MAS definition of Dorri et al. [6], we define an agent and its environment as an entity with the following properties:

Agent (e.g., Vehicle)

- **State**: Properties of an agent including the action history. Can be constant or dynamic (e.g., vehicle length = 3 m, velocity = 80 km/h).
- **Perception (Sense)**: Ability to sense the surroundings or just parts of it. Might be noisy (e.g., a radar detecting an object or a communication module receiving a message).
- **Strategy (Plan)**: Decides which action is executed to reach the goal (e.g., to reach the desired speed the first action could be to accelerate).
- **Action (Act)**: The agent's capability to influence its own or other states for a certain cost (e.g., accelerating and breaking or sending a message).
- **Costs**: The sum of all costs generated by the performed actions (e.g., fuel consumption and travel time).
- **Private Goal**: A certain amount of cost to the agent that should not be exceeded (e.g., arrival time within 2 h).

Environment (e.g., Road Network)

- **State:** Properties of the environment (e.g., the weather conditions).
- **Road**: The drivable area on which the agents can move (e.g., two lanes).
- **Infrastructure**: Elements to support the agents with information (e.g., traffic lights).
- **Constraints:** Limitation of the agent´s action scope (e.g., the road traffic regulations.

The core element of an agent is the strategy for the location of decision-making. With several agents this can be combined for distributed problem-solving (DPI) [6], i.e., reaching a greater common goal. Therefore, we define the MAS as the combination of agents working together in an environment towards a common goal, as shown in Figure 1.

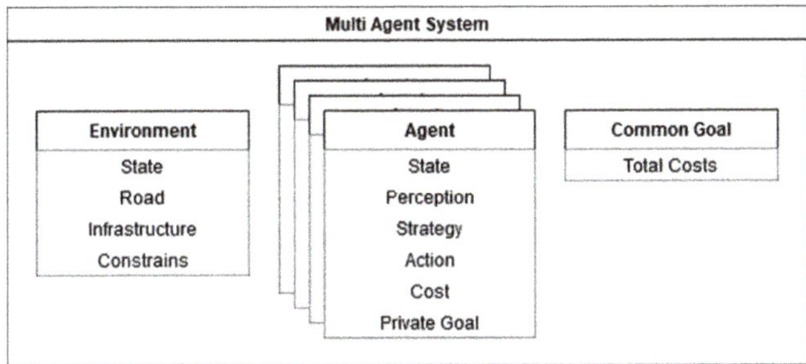

Figure 1. Overview of an multi-agent system with four agents in a shared environment working towards a common goal.

Multi Agent System (e.g., Road Traffic)

- **Agents:** An arbitrary number of entities performing actions and creating costs (e.g., traffic participants such as cars, trucks, and pedestrians).
- **Environment:** One shared environment that allows the agents to interact (e.g., the road network in which the agents are located).
- **Common Goal:** The summed agent's cost within the MAS that should be optimized. Might conflict with the private goals of the agents (e.g., zero fatalities and low CO_2 emissions).

2.2. Cooperation

With multiple agents in one environment, the action of one agent might also influence other agents. If an action intentionally influences at least one other agent it is called an interaction. There are four motivations for interactions which differ in the way the costs of the involved agents are influenced [5,7]:

- **Mutualism**: An agent interacts in such a way that the costs of all involved agents are reduced.
- **Altruism**: An agent interacts in such a way that his/her own costs increase while the costs of other agents are reduced.
- **Selfishness**: An agent interacts in a way that his/her own costs are reduced while the costs of other agents increase.
- **Spite**: An agent interacts such in a way that the costs of all involved agents increase.

Further, an action of an agent is stated to be cooperative if he/she reduces the costs for other or for all involved agents, i.e., through mutualism and altruism. Therefore, cooperation is not necessarily bidirectional, as an altruistic interaction requires a selfish interaction from the counterpart. Acting consequently selfish in a group of mainly cooperative agents is called free-riding and allows one individual agent to achieve very low private costs, while the global cost, i.e., the sum of all private costs, increases. The question of to what extent free-riding can be applied or tolerated by the agents leads us to social dilemmas.

2.3. Social Dilemma

Cooperation in general has the underlying intention of reducing costs; however, if the cooperation is not mutual, this may result in increased costs for some of the involved agents [7]. The question arises as to why an agent should act cooperatively if there is the risk that his/her action will only benefit other agents and increase the agent´s own costs. Such situations are widely discussed within the fields of psychology and game theory, and are called social dilemmas. They are defined by two properties [8]:

- Each individual is sometimes tempted to defect (e.g., for lower costs in the short-term).
- Collective defection leads to higher cost than collective cooperation.

One famous example for a so-called social dilemma is the prisoner's dilemma, as illustrated in Table 1: Two prisoners can choose between betraying their partner or remaining silent (cooperating). If both remain silent, both have to serve one year; if they betray each other, both have to serve two years; and if only one betrays, the one who remained silent has to serve three years while the other one is set free. This example shows that there is no trivial solution to the question of whether an agent should rather cooperate or defect when the intentions of the other agents are not clear [8,9]. Yet, the prisoner's dilemma illustrates only one of many possible social dilemmas. Further famous social dilemmas include the chicken or assurance dilemmas which are, among others, discussed in depth in "The psychology of social dilemmas: A review" [8].

Table 1. The prisoner's dilemma with the actions of Agent A and B with the resulting time in prison.

A \ B	Cooperate	Betray
Cooperate	1 / 1	0 / 3
Betray	3 / 0	2 / 2

The Nash equilibrium, i.e., when no agent can improve his/her situation by decision alone independently of the decisions of others, occurs for the prisoner's dilemma when the agents betray each other [3]. This can lead to zero or two years in prison, while cooperating can lead to one or three years in prison. Simultaneously, we see that the globally optimal solution, i.e., a total prison time of two years, can only be achieved if both agents cooperate. Knowing from the start, however, that an agent is in a group of cooperating agents, he can minimize his private costs by betraying the other cooperating agents. Therefore, an important element in the decision-making process is trust and fairness. If the two agents were to be able to communicate prior to settlement, they could agree on the global optimum instead of the Nash equilibrium with an accumulated prison time of four years.

Imagine a variation of the prisoner's dilemma where a selfish agent can optimize his/her local cost and fulfill its goal by defecting, while a global optimal cost can only be achieved with cooperative behavior. This is similar to the example of the freeway on-ramp with the fast cars and slow heavy trucks. Each agent (car) on the freeway has to decide: "Should I be cooperative and slow down, or should I defect and let the car following me handle the situation?" [10]. With a high amount of defection due to selfish agents, the trucks on the on-ramp cannot find a gap for a lane-change and must come to a full stop, resulting in large costs. If one car on the freeway wants to cooperate after the previous ones have defected, the costs for this altruistic cooperation are already much higher than they would have been for the previous agent (as the truck on the on-ramp has to accelerate from a full stop), leading to even more defection. However, if all agents cooperate and collectively reduce their velocity so that they form a zip-like system (alternating freeway and on-ramp), no further breaking or full stops are required, resulting in a globally optimal solution. A variation of this social dilemma with a fleet of miniature cars resulted in a 35% increased throughput when the cars acted cooperatively as compared to selfishly [11].

The choice for cooperation or defection is mainly based on the cost of interaction, but trust and fairness must also be taken into account [12]. A human agent is less likely to cooperate with an untrustworthy agent and is more likely to cooperate if he/she considers the interaction as fair. In a cooperative environment, cooperation can be exploited when the trustworthiness and fairness of the agents is not tracked. An agent (car) that often asks for cooperation (e.g., to drive onto the freeway) but never grants it is considered a free-rider [13]. As long as the environment is cooperative, the free-rider can achieve very low costs, while the individual costs of the remaining agents and the global costs are not significantly worsened.

Cooperation and social dilemma in road traffic are large research areas, not only for human drivers but also for the development of automated vehicles (AVs) [3,14]. When it comes to developing cooperative AVs, it is important to have a precise definition of the desired cooperation. In road traffic, there are a variety of scenarios in which humans cooperate, although the type of cooperation in each situation can be fundamentally different. It is therefore not enough to say that a cooperative system is being developed, and it is necessary to clearly define the scenarios for which this system is intended. In the following, we give an overview of how different types of cooperation can be described and then we assign three exemplary situations from road traffic to the types of cooperation.

3. Cooperation in Road Traffic

In road traffic, there are many agents (vehicles) in a common environment that follow a set of rules that lead to certain total costs, i.e., the sum of agent costs (local costs), which in most cases is equivalent to the traffic flow. Inefficient driving or accidents reduce the traffic flow; therefore, driving strategies for the vehicles should be found in accordance with the road traffic rules in order to increase efficiency and reduce accidents. A road traffic scenario (hereinafter only referred to as a scenario) is defined by a snapshot of the road traffic, the duration, and the resulting costs, and can either be taken from a real situation or constructed so that there is an unlimited number of scenarios. In each snapshot, the environment, agents, rules, and costs are defined and serve as a starting point. From there, the agent's strategy selects actions over the duration of the scenario that lead to specific costs. The combined actions of all agents form a maneuver that can be described via the agents' trajectories and has the same duration and cost as the scenario. The goal is to find strategies for the agents to produce minimal global costs during a scenario, and one approach is to implement cooperative strategies that lead to cooperative maneuvers, i.e., the agents collectively seek an interaction that lowers global costs and supports fairness. Since there are unlimited snapshots, each of which can lead to unlimited maneuvers, the number of cooperative scenarios is enormous [15,16].

3.1. *Characteristics of Cooperation*

We have developed characteristics of cooperation in order to work out differences and similarities from the variety of cooperative scenarios. The following eight features help to classify cooperative scenarios and maneuvers in road traffic and highlight various aspects to consider when developing cooperative driving systems:

3.1.1. Agents Involved

The number and homogeneity of the vehicles are of interest to the agents involved. Cooperation can take place between two vehicles of the same type, for example between two trucks, but also between a variety of different types of vehicles, such as trucks, cars, and ambulances. Cooperation requires at least two agents, but this number is not limited.

3.1.2. Location

The location can play an important role in the analysis of whether and which cooperative scenario is required. For example, there are static cooperation areas at on-ramps where the type of cooperation required (the creation of a gap for on-ramping vehicles) is already fixed. At these locations, maneuver coordination may also be supported by communication with the infrastructure (roadside units). A platoon, on the other hand, is not tied to a single location but is created at any point on the freeway and extends over long distances.

3.1.3. Urgency and Costs

In order to decide whether a cooperative scenario is necessary, urgency is a crucial factor. Urgency is not symmetrical for all actors involved and can be expressed by the costs that a single agent would avoid by collaborating. For example, if cooperation only brings comfort-related gains, the avoidable costs are very low (low urgency), as compared to cooperation to prevent an accident (high urgency).

3.1.4. Interaction Type

The interaction types can be divided into passive and active cooperation. Only the exchange of information, without adapting one's own maneuver plan to create a mutual or altruistic benefit, is a passive form of cooperation and causes no direct costs. This includes, e.g., wireless communication of messages such as cooperation requests or acknowledgments. If, however, one's own maneuver is adapted, e.g., a lane change is carried out to enable cooperation, we speak of active cooperation.

3.1.5. Duration

The duration of the cooperative scenario has a direct influence on the required planning horizon. For a short duration, such as a lane change on the freeway in order to open a gap for a vehicle on the on-ramp, a forecast of a few seconds is sufficient. However, a truck-overtaking maneuver for example can take up to 45 s, and platooning will require planning over several hours.

3.1.6. Mutuality

In mutual cooperation, all the agents involved benefit (mutuality), but cooperation sometimes requires a subset of the agents involved to accept higher costs in order to reduce global costs (altruism). Therefore, it is necessary to identify which agents are the profiteers and which grant cooperation. In order to increase the motivation for cooperation, an altruistic interaction can be transformed into a mutual interaction if the profiteers compensate the altruistic costs via another channel, e.g., a micropayment.

3.1.7. Preparation Time

The preparation time is the time from the first consideration of the cooperation to the execution of the first cooperative interaction and thus includes the available time for the planning of the cooperative maneuver. In the case of platooning, for example, it can occur that before the maneuver begins it has already been determined which vehicles are to form a platoon at which point. The cooperation is therefore prepared well in advance. In most cases, however, the cooperation takes place spontaneously with only a few seconds of preparation time, and in emergency situations with no preparation time at all.

3.1.8. Initiation

Since several agents are involved in the cooperation, there are also various possible initiators, namely a profiter, a cooperation granter, or a coordinating master. In addition, a distinction can be made between offering, requesting, and announcing a specific interaction for cooperation. For example, the cooperation granter could offer cooperation and the profiter could request cooperation, while a coordinating master could announce certain cooperative interactions and enforce them through sanctions.

3.2. *Examples of Cooperative Situations*

Three exemplary scenarios are presented in order to put the above-mentioned characteristics of cooperative maneuvers into context. These show that cooperation can take many different forms and, in addition, that there is not only one correct way to implement cooperation, but many different possibilities.

3.2.1. Platooning

The general idea of platooning as shown in Figure 2 is to drive with a reduced distance between vehicles in order to use the slipstream of the vehicle in front [17]. The reduced air resistance reduces fuel consumption and the carrier saves money [18]. Therefore, two or more agents are involved, following each other on the motorway. In the short term, the urgency and potential cost savings are very low, but with consistent operation over long distances, fuel savings of up to 10% can be achieved [19]. The type of interaction depends on the position in the platoon; the leading vehicle has a passive role and only exchanges information, especially about its braking maneuvers. This is necessary so that the following vehicles can drive at a reduced distance and react quickly to a breaking situation, thereby performing active maneuvers. This form of cooperation is mutual in the sense that it benefits all agents involved, although the fuel savings for the leading vehicle are significantly lower than for the vehicles in the slipstream. Therefore, although all agents benefit, the benefits are not fairly distributed. A platoon can be formed spontaneously on the motorway or planned well in advance to ensure that two suitable trucks actually meet on the freeway. In both cases, the time available to prepare the cooperation is long and there is no urgency. The biggest profiteer is a following vehicle that wants to enter the platoon, so one can assume that this vehicle should initiate the cooperation and request the transmission of the signals. However, it is also conceivable that a central authority, e.g., a fleet manager of the carrier, initiates the cooperation and forces the vehicles into a platoon.

Figure 2. Three trucks drive in a platoon and use the slipstream to reduce fuel consumption while a fourth truck follows with a normal safety distance.

3.2.2. Lane Merge

In the drive-on scenario or generally in lane-merging scenarios as depicted in Figure 3, the idea is to create fairness by alternately merging vehicles from both lanes, creating a zipper-like pattern [11,20]. Although there may be many vehicles in such a scenario, in most cases only two agents of any type will be involved in the actual cooperation. The location of the cooperation is limited to the drive-up area, and for some vehicles, especially slow trucks, the urgency of the cooperation may be high, as otherwise a braking maneuver must be initiated. When cooperation takes place, it is an active form of cooperation for all agents involved, since the cooperation partner on the freeway adjusts his trajectory so that the profiteer on the driveway can choose a better trajectory. Therefore, it is clear that the cooperation is not mutual, and the grantor accepts costs with an altruistic interaction to enable the profiteer to save (much higher) costs. Since both the duration and the preparation time are short and limited to the time in the drive-up area, all costs incurred can be regarded as one-time expenses. In the long run, however, further costs could arise, e.g., if a very slow vehicle is allowed in front of one's own vehicle. The initialization depends on the vehicle strategy and two possible variants are that the granter proactively offers cooperation or that the profiteer sends a cooperation request.

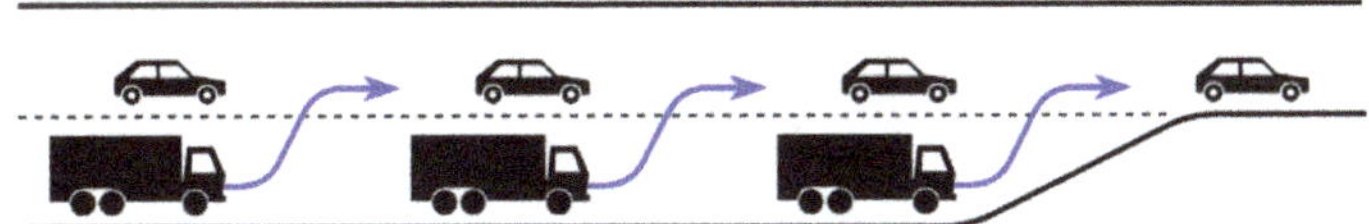

Figure 3. A lane merge on a highway with the cooperative cars creating gaps for the trucks to form a zip-like pattern.

3.2.3. Truck Overtaking

This overtaking scenario is explicitly about a truck overtaking another truck on the freeway while a car approaches on the left lane from behind, as Figure 4 shows. The challenge in this scenario is related to the relative speeds; these are low between the two trucks (1–20 km/h) while being high between the trucks and car (40–180 km/h). This is due to the fact that for example in Germany the speed limit for trucks is 80 km/h, while for passenger cars it is often 120 km/h or there is no limit at all [1]. The number of agents involved in the cooperation is thus three: two trucks and one car. Such a maneuver can take place on any freeway section with at least two lanes and without overtaking prohibition. Further, there is no high urgency for cooperation, because overtaking would only result in a time saving for the overtaking vehicle. For the car, the interaction is active in any case, as the speed must be reduced to allow the overtaking truck to change lanes. The interaction of the front truck can be active if it slows down the speed to allow the overtaking truck to pass faster, or passive if it leaves its speed unchanged and sends only information about its speed. The duration of the overtaking maneuver is regulated by law and may not exceed 45 s in Germany [21]. Since the overtaking truck first has to approach with a low relative speed from behind, there is a lot of preparation time for the cooperation between both trucks, while there is only little time for coordination with the fast-approaching car. The overtaking truck is clearly the profiteer, as cooperation allows it to drive at its desired higher speed, so the maneuver should also be initiated by the overtaking truck. A proactive offer from the other cooperation partners would be a possibility, but in contrast to the drive-on scenario, there is no indication as to when it would make sense here since it is not bound to one location.

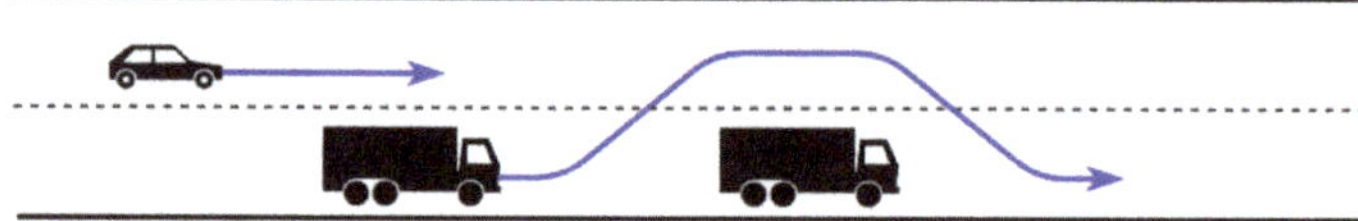

Figure 4. A cooperative truck-overtaking maneuver with the truck in front cooperating to reduce the overtaking time.

4. Implementation of Cooperation

When it comes to implementing cooperation, this means choosing a strategy for the agents that trigger cooperative interactions. With regard to road traffic, strategies are called maneuver planners in the following, and while maneuver planners are not new in automated driving, cooperative maneuver planners are only at the beginning stages of research. However, before going into maneuver planners and automated systems in more detail, we first introduce three characteristics of the implementation of cooperation. Afterwards, we present how cooperative movements have evolved in nature, e.g., with animal swarms, and how cooperation in road traffic is currently applied by humans.

4.1. Characteristics of the Cooperation Implementation

Besides maneuver planning, it is necessary to consider two further fundamental characteristics of cooperation which are partly mutually dependent: the planning level and the communication.

4.1.1. Planning Level

Cooperation can be planned on three levels: Centralized, decentralized with coordination, and decentralized without coordination [22]. "Centralized" means that there is an "all-knowing" master that coordinates the maneuvers of all vehicles, while "decentralized" means that the maneuvers are determined on the vehicles themselves. At the decentralized level with coordination, the vehicles have a communication channel to communicate directly with the vehicles in the neighborhood to plan the maneuver, in contrast to the decentralized level without communication, where the vehicles can be observed in the neighborhood, but there is no possibility of explicit information exchange. In the case of the centralized planning level, the global knowledge is available, theoretically allowing for the global optimum to be found. In the case of the decentralized planning with coordination, only the local knowledge (limited by the communication range) can be accessed, eventually leading only to a local optimum. This is similar at the decentralized planning level without coordination, but further, each vehicle can only plan on its own and must anticipate the behavior of other road users instead of receiving their planed behavior. This means that no negotiation, confirmation, or denial of cooperation is possible.

4.1.2. Communication

A communication channel is required for the centralized and decentralized level with coordination, and we distinguish between direct and indirect communication. Direct communication implies that signals from one vehicle are sent directly to another vehicle, as in the case of the WLAN standard IEEE 802.11p [23] that has been specially developed for vehicle-to-anything communication (V2X). It establishes an ad hoc network between the vehicles and enables a range of approximately 400 m. Indirect communication, on the other hand, uses infrastructure to enable communication over an unlimited range, and a widely-used standard for this is Long-Term Evolution (LTE, c-V2X) [24]. Messages addressed to neighboring vehicles are sent via infrastructure to a backend server, which only forwards the messages to vehicles located in a specific area (geo-messaging). In addition to the communication channel, a standardized communication protocol and message type must also be used, and the development for V2V or V2I is still in its early stages.

4.1.3. Maneuver Planning

The task of maneuver planning is to determine which actions to perform; in the context of automated driving this means that drivable trajectories have to be created. The above-mentioned characteristics of the cooperation, such as the location, the involved vehicles, and the costs have to be considered. For maneuver planning itself, different approaches based on trajectories, atomic maneuvers, state machines, artificial intelligence, and others are available and in development. The special feature of cooperative maneuver planners is that vehicles in the environment are not regarded as un-influenceable obstacles, but as possible partners with whom maneuver can be jointly developed.

4.2. Cooperation in Nature

In nature there are many examples of cooperation, such as the symbiosis of different plants in the jungle or the cleaner fish and sharks. The most interesting example of cooperation in nature, in this context, comes from swarms, herds, and flocks and deals with the collective movement of a large number of individual animals. The biggest motivation behind this cooperation is safety and efficiency: the zebras in a herd are much better protected from the lion than if they are standing isolated and birds flying in a V-formation consume less energy on the long way to wintering near the equator. Cooperation within a swarm follows three basic rules [25]:

- Separation: "avoid crowding neighbors (short-range repulsion)"
- Alignment: "steer towards the average heading of neighbors"
- Cohesion: "steer towards the average position of neighbors (long-range attraction)"

With these elegant rules and without any further communication channels or centralized control, swarms are able to implement complex interactions involving a large number of individuals. A prerequisite for this is an appropriate environment awareness to perceive the distances and speeds of the surrounding animals and to react accordingly. However, the interests of each individual are limited with respect to the global interests of the swarm. A transfer of the swarm to the road traffic is therefore only possible to a very limited extent. While the rules could be applied in a crowd of people walking, a person in a vehicle, especially at high speeds, cannot estimate the distances of other road users and react to them with a necessary small delay [2]. While a small bump among pedestrians is not a problem, such an error can have serious consequences for vehicles at high velocities. Another reason why the swarm behavior is not directly transferable is that in road traffic there is no homogeneous swarm, but rather a combination of many different individual vehicles with different goals and capabilities [25].

4.3. Human Cooperation in Road Traffic

Cooperation in road traffic only occurs if the situation is not clearly regulated by the StVO or an agent voluntarily renounces the rights given by the StVO. At a right-before-left intersection in western countries, there is normally no cooperation, as the situation is clearly determined by the rules of the StVO for example. A driver does not voluntarily stop to give the other vehicle the chance to drive, but because he/she is punished otherwise by law enforcement. Cooperation between humans does not result from a fixed set of rules or an assistance system, but exclusively from the intuition of the human driver. In a situation that is unclear to him/her, the driver might act cautiously and leave others the right of way in order not to provoke an accident, or he/she recognizes that there is a need for cooperation with another vehicle and helps. In the following, the mentioned characteristics of cooperation and its implementation are discussed in order to understand to what kind of cooperation a human driver is capable of.

- Planning level: Cooperation is planned in a decentralized manner, partly with but mainly without coordination. This means that each driver executes what he/she considers as the appropriate action, but usually does not discuss this with the other vehicles.
- Communication: Communication can take place explicitly via horns, light signals, or gestures, but also implicitly via the driver's own behavior. By proactively changing to the left lane in front of an on-ramp, for example, it can be signaled that the resulting gap can be used for the drive on. However, it is not possible to transmit complex plans.
- Maneuver planning: The maneuver is planned in each vehicle itself, based on the experience and intuition of the driver.
- Involved agents: As the complexity increases with the number of participants, usually only two vehicles are involved in the cooperation between human drivers.
- Location: Since no infrastructure is required, the cooperation is largely location-independent.
- Urgency and costs: The urgency and costs can only be roughly estimated based on the driver's experience.
- Interaction type: Due to limited communication, little information can be exchanged and the majority of actions are active and have a direct influence on the traffic situation.
- Duration: Due to limited communication, an agreement for a long time is not possible and cooperative maneuvers are limited to short moments.
- Mutuality: This is strongly dependent on the attitude of the drivers, whereby there are drivers who act altruistically, mutually, or selfishly.
- Preparation time: In most cases, cooperation occurs spontaneously. Far-in-advance planned cooperation can only occur among drivers who have the possibility of prior agreement. An example would be a joint holiday trip involving two vehicles, one of which drives ahead to show the way and the other follows.
- Initiation: Both the profiteer and the grantor can initiate the cooperation. When driving on the freeway, the cooperation can be offered by the grantor on the freeway through a proactive lane change, and the profiteer can request cooperation through the turn signal or force the cooperation by changing lanes on the freeway himself.

This list shows in several points that the limited communication between human drivers is one of the main restricting factors for their cooperation, as was also shown by Fekete et al. [3]. The correct assessment of the situation, especially with respect to the urgency of the cooperation, is very difficult and since there is no suitable opportunity for coordination and knowledge exchange, the situation often remains unclear. Therefore, cooperation might not take place although it would make sense, or, in contrast, cooperation might get started although it is not needed. In these cases, since it is usually an active form of cooperation, costs are incurred from which nobody benefits. A classic example of this is a lane change on the freeway, as shown in Figure 3, where it is not clear for vehicle B whether vehicle A will allow him/her to drive in front, and for B whether A wants to drive in front or behind him/her. If A assumes that B will not let him/her in front but B does, there is a useless deceleration. Such misunderstandings could be avoided by a synchronized level of knowledge among the drivers, but this requires improved communication [14]. Furthermore, due to the limited ability of humans to plan maneuvers, they would only be left with simple cooperative maneuvers. However, due to the human's concept of fairness and emotions, human drivers are often capable of avoiding bad Nash equilibria, leading to lowered common costs due to their cooperation. Although the costs generated by human drivers are not optimal, they are unmatched by cooperative automated vehicles [3].

5. Cooperation for Automated Vehicles

As soon as the human hands over the driving task, it is up to the automated system to cooperate. For this purpose, we discuss the three points on the implementation of cooperation in the context of automated driving.

5.1. Planning Level

The three planning levels of centralized planning (CP), decentralized planning with coordination (DP), and decentralized planning without coordination (DPWC) come with different advantages and disadvantages and are therefore suitable for different scenarios [22,26]. With central planning, all existing knowledge of the vehicles is collected in a central backend and a global optimum is then formed. The prerequisite is therefore a backend with the appropriate computing power and stable communication to the vehicles. As with all centralized systems, this results in a single point of failure, which means that the system loses robustness and further has a high demand for communication. In order to be able to react immediately in safety-critical moments, fast planning is necessary, but communication to the backend and back again can result in a greater delay depending on the communication medium. Furthermore, vehicle control must be transferred from the vehicle itself to the centralized system, which raises concerns about responsibility and legislation.

Communication is also required for decentralized planning with coordination, but since no backend is required, an ad hoc network such as with ITS-G5 can be used. This makes the system more robust, but it is limited to the range of the communication medium. Since only a subset of the knowledge of the central planning level is available for maneuver planning, it can only be optimized locally. The direct communication between the vehicles and the maneuver-planning running on the vehicles themselves allows a faster reaction.

Both of these planning levels require communication, but decentralized planning without coordination is not dependent on it. This makes it the only planning level that can also be used effectively in mixed traffic when both automated and manually (human-driven) vehicles drive together on roads. Due to the lack of knowledge exchange, however, the behavior of other road users must be approximated, which introduces an additional inaccuracy into the system. The decentralized planning level is therefore robust and independent of communication delays or failures, but cannot achieve the same quality of maneuver planning as decentralized planning with coordination due to the small knowledge base and lack of coordination options. In addition, extra computing effort is required to estimate the behavior of other road users.

With regard to the evaluation metrics in Table 2 it is clear that the different planning levels complement each other. Starting from decentralized planning with coordination, the system can be extended with decentralized planning without coordination for cooperation with manually controlled vehicles. Further, this could also serve as a fallback level in case of communication failure. Further, relevant information can be exchanged via a central backend, e.g., traffic densities in order to cooperate not on the maneuver level, but on the routing level to avoid a jam.

Table 2. Evaluation of the three planning levels for cooperation (circle: full = positive, empty = negative). CP: centralized planning; DP: decentralized planning with coordination; DPWC: decentralized planning without coordination.

	Quality	Communication Needs	Mixed Traffic	Robustness	Computation	Infrastructure	Delay
CP	●	○	○	○	○	○	○
DP	◐	◐	○	◐	●	●	◐
DPWC	○	●	●	●	◐	●	●

5.2. Communication

Connected vehicles form a trending topic mentioned by almost every vehicle manufacture in their vision and current research priorities [27–30]. This can include a connection to the vehicle owner via an app, a connection to an entertainment platform and other cloud services, or a connection to the vehicle service station to transmit maintenance data. However, regarding safety and efficiency, the real-time communication between connected vehicles is the most promising aspect.

In communication, a distinction can be made between direct communication (V2X) and indirect communication via an additional infrastructure (cellular–V2X). For indirect communication, the Long-Term Evolution (LTE) standard, which is widely used in smartphones for mobile data, has established itself. It requires a connection to the LTE radio masts, which cannot be guaranteed especially in rural and highly isolated areas such as tunnels or underground car parks. If the radio mast or the backend were to fail, an entire area would be affected and unable to communicate. Furthermore, the large number of involved nodes in the system increases the potential for manipulation or fault. With the currently available LTE REL-14 (4G), bandwidths of up to 1000 Mbit/s are possible with a latency of 5 ms [31] over an unlimited range. Moreover, with the newest REL-15 to 17 known as 5G, bandwidths of 10 Gbit/s and latency of 1 ms [32] are achievable. However, when changing from one radio mast to the next, which can often occur when driving over the motorway, short delays occur.

The ITS-G5 (IEEE 802.11p) standard [23,33], also known as WAVE, has been explicitly developed for use in communication between vehicles. An ad hoc network will be set up between the vehicles, whereby no infrastructure elements are required, thus there are fewer failure points and attack points, and no running costs arise. The system is therefore more robust and location-independent in comparison to LTE. Bandwidth and latency, however, are strongly dependent on the distance between the vehicles. The maximum range in an open field is approximately 800 m [34], but is reduced to 300 m [35] in autobahn scenarios and less in interconnected cities. The bandwidth and latency lie between 6 and 54 Mbit/s [36] and 1 and 5 ms, respectively [37]. With ITS-G5 only information can be exchanged directly between vehicles and no additional connection to the Internet can be established to receive routing information or traffic jam messages.

Connected vehicles provide interfaces that could be used by adversaries to do damage. Therefore, for connected vehicles and especially their communication, not only safety but also security are important aspects that have to be considered. Examples for such cyber-attacks are false data injection [38], reply attacks [39], and denial-of-service attacks [40]. It is therefore crucial, independent of the communication interface, that such attacks can be detected, blocked, and prevented.

As far as the communication medium is concerned, there are therefore already two strong established standards, with 5G a promising option for the future (Table 3). If a centralized planning level is preferred, c-V2X has to be chosen for the communication with the backend. For decentralized planning with coordination IEEE802.11p is also sufficient. Of course, a combination of both approaches is again possible, whereby routing information is received via c-V2X from a central planning level but maneuver planning is coordinated locally via V2X.

Table 3. Evaluation of the vehicle-to-anything communication (V2X) technologies (circle: full = positive, empty = negative). LTE: Long-Term Evolution.

	Availability	Reliability	Latency	Bandwidth	Range	Local Independence	Costs	Security
ITS-G5	●	●	●	◑	○	●	●	●
LTE 4G	◑	◑	○	◑	●	◑	◑	◑
LTE 5G	○	◑	●	●	●	◑	○	◑

Paul Watzlawick wrote in his first axiom "One cannot not communicate. Activity or inactivity, words or silence all have message value". This is the case even when no dedicated communication medium is involved [41]. Decentralized planning without coordination, where the vehicles "communicate" their intentions through their behavior, uses this. However, since no actual communication via radio is involved, this concept is not discussed here.

Not only the medium is relevant for communication, but also the message protocol and the message type. This choice determines which information is available in which form for maneuver planning and which bandwidth is required for the transmission. Two message formats have already been standardized by the ETSI as Day 1 Messages: CAM and DENM. CAMs contain ego information such as position, speed, vehicle type, etc. and are periodically transmitted at 1–10 Hz via single-hop broadcasting (point-to-multipoint) [42].

DENMs contain information about events that could affect safety or traffic efficiency and include an event type, location, start time and duration. These can be caused by the sending vehicle itself, e.g., an emergency-stop, or they can be detected by sensors, e.g., flash ice. The transmission is event triggered and forwarded over several hops to warn vehicles in a larger environment [43].

Currently, the ETSI is working on the standard for Collective Perception Messages (CPMs) [44], based on the Environmental Perception Message (EPM) [45] as part of the Day 2 Messages. The purpose of the CPM is to share objects detected by the vehicle sensors, thereby augmenting the environmental perception.

These three message types are purely informational and thus allow to improve the perception but do not yet allow a cooperative maneuver coordination. If cooperation on a central or decentralized level with coordination is to be planned, new message standards must be developed and established across manufacturers.

Oliver Sawade et al. developed the Collaborative Maneuver Messages (CMMs) to synchronize several vehicles with a distributed state machine [46]. The CMMs distinguish between three message types: Inform, Request, and Response. While the Inform CMM is used to share information, the other two CMM types allow for starting a negotiation between the vehicles on how to transition to the next state of the distributed state machine.

Bernd Lehmann et al. designed the Maneuver Coordination Message (MCM) [47] that is currently standardized by the ETSI [48] in order to exchange trajectories between vehicles. Different attributes such as "planned" and "desired" allow then to express the need for cooperation, start a negotiation, and finally execute the maneuver.

5.3. Maneuver Planning

The goal of maneuver planning is to generate trajectories, which are then executed by the vehicle. Factors such as safety (no collisions), efficiency, feasibility, traffic rules, and comfort have to be considered. Different classical approaches have already proven themselves feasible and also artificial intelligence methods are advancing in the field [49].

In the case of non-cooperative maneuver planners, vehicles in the environment are regarded as obstacles that cannot be influenced. In order to drive cooperatively, it is therefore necessary to rethink and consider how one's own trajectory affects other road users. Thus, it can be considered how the maneuver can be carried out in a jointly optimized way. One possibility of including the behavior of the surrounding vehicles in the maneuver planning of the ego vehicle is to coordinate the maneuvers of the involved vehicles by V2X communication. In literature, there are already different approaches on how the concept of such a coordination could look like.

Hyldmar et al. implemented cooperative maneuver planning in a miniature fleet by having the vehicles in a lane merge send their desired position for the future via V2X [11]. The receiving vehicles then accelerated or decelerated to allow the sending vehicle to reach the desired position. An experiment with 16 of the miniature cars on a round course showed a more than 30% improvement in throughput through the cooperation.

Sawade et al. developed the Collaborative Maneuver Protocol (CMP) [46], which uses a distributed state machine to synchronize all vehicles of a maneuver in one common state. In each state, there are different roles and tasks for the vehicles and a transition to the next state can only occur if all vehicles agree. The CMP is thus a first step towards cooperative automated vehicles and shows a way in which vehicles can synchronize their plans with each other and collectively decide on the next actions. However, the state machine must be defined in detail for each particular maneuver, e.g., for a cooperative lane merge.

Bernd Lehmann et al. presented a generic concept which creates two trajectories for the vehicle, a plan and a desire, and sends this as an MCM [47]. The planned trajectory conforms to traffic rules and therefore does not lead to collisions with the planned trajectories of other vehicles. The desired trajectory, on the other hand, shows the optimal trajectory for the ego vehicle, which, however, is not traffic rule-compliant as it conflicts with the planned trajectory of another vehicle. This other vehicle then has the possibility to cooperate and adapt its own planned trajectory to enable the desired trajectory of the other vehicle.

If there is no possibility to coordinate one's own behavior with that of other traffic participants, e.g., because not every vehicle is equipped with V2X communication, the behavior of the other road users must be anticipated. In order to implement behavioral planning under these prerequisites, different methodologies are applied in literature.

In the early stage of cooperative driving algorithms, Wei et al. [50,51] modeled behavior planning as a combination of different driving functions covering modules for distance-keeping to the leading vehicle, selection of the desired lane, and a corresponding merge planner. The lane selector acts as a supervisor that determines the best lane to drive in and switches between lane keeping and merging behavior. The lane-keeping module generates a set of constant accelerations, predicts the reaction of the surrounding vehicles, and evaluates the resulting scenarios by a cost function to select the best acceleration strategy. In subsequent work [52], Wei et al. extended their approach by splitting the acceleration in two acceleration segments with equal lengths, and enhanced the behavior prediction of other vehicles by an intention estimation. Since these approaches evaluate all possible strategies, the complexity of the strategies is limited due to the limitation of available computing time. For more complex strategies, search algorithms must be used which do not evaluate all possible behavior plans.

In the field of tree searches, each vehicle has a set of discrete actions from which it can choose. Each action is valid for a specified time step. A maneuver plan is then created from a sequence of actions over the planning horizon. The purpose of the tree search algorithm is to find the best possible combination of actions as a solution of the planning problem. The behavior of surrounding vehicles can also be modelled with a discrete set of actions under the assumption that they make rational decisions. Lenz et al. used a Monte Carlo Tree Search (MCTS) algorithm to solve this search problem [53]. Kurzer et al. [54] extended this approach by integrating macroscopic actions into the MCTS in order to plan over longer time horizons. Furthermore, they used a continuous action space instead of a discrete action set, which is useful in urban scenarios and tight spaces [55].

Another approach is the representation of cooperative behavior planning from a game theoretic perspective. While tree search methods find the best solution according to a given cost function, game theory considers all possible combinations of available strategies of all vehicles (referred to as players or agents). The solution in the form of an equilibrium is the strategy that yields the best reward for the considered player, regardless of which strategy the other players choose. However, the application in the field of cooperative behavior planning [56,57] has been limited to the assessment of two vehicles. Moreover, the scenarios in which the functionality of the approach is demonstrated include only simple scenarios.

Besides classical methods, artificial intelligence is also applied in cooperative maneuver planning. Research focuses on reinforcement learning techniques [58,59], in which the vehicles learn their own behavior based on positive or negative reward. The behavior modeling of the surrounding traffic is done implicitly within the neural network, which can be advantageous because no separate prediction is required, but also has disadvantages because the basis of the decision-making cannot be validated.

6. Problems and Challenges

We have shown that there have already been many developments and achievements in the development of C-ITS, but there are still some challenges to be overcome. For communication, there is disagreement between V2X and C-V2X [60–63], and the international standards are also inadequate for the various message types. It will be necessary to find agreement across manufacturers so that all vehicles can communicate with each other. Only when the equipment rate exceeds a certain threshold will the positive effects of the cooperation, especially with CP and DP, become noticeable.

However, with an increasing number of equipped vehicles and more messages for the coordination of more complex maneuvers, the channel load also increases, resulting in reduced communication range and message failures. Therefore, it will become more and more important to implement optimized message generation rules and congestion control in order to transmit the relevant information in any situation.

In cooperation that is not based on communication, it is important to avoid misunderstandings and to correctly assess the costs and behavior of other road users. With human drivers this is a particular challenge and requires special trust between automated vehicles and humans.

Approaches for cooperative maneuver planners have already been tested in simulation or on a small scale, but the lack of agreement on standards for the messages also makes it unclear what information can be exchanged for maneuver planning purposes.

Various projects like IMAGinE [64], 5G-MOBIX [65], and ICT4CART [66] have already recognized the urge for cooperation and attempted to develop solutions for the challenges mentioned here.

7. Conclusions

Despite traffic rules, there are situations that are not clearly or inefficiently regulated and where cooperation is required to ensure efficient and safe traffic. While human drivers are intuitively able to cooperate and communicate through light signals, gestures, or proactive actions, this is not yet the case with automated vehicles (AVs). Without humans as a cooperative module, the social dilemmas that arise in daily traffic become a major challenge for AVs. While the first AVs can even gain advantages in certain situations as free-riders in the cooperative environment due to the presence of human drivers, with an increasing number of selfish AVs the point will come where the entire road traffic loses efficiency due to lack of cooperation. Therefore, our mission must be to develop not only automated vehicles, but cooperative automated vehicles.

While animals can cooperate with simple and elegant rules, these cannot be applied to our complex road traffic. New methods have to be developed to tackle the challenges of CAV, and here we have provided several characteristics for clustering different cooperative scenarios in order to identify the challenges in a first step. Afterwards we discussed how cooperation could be implemented, and determined that currently no agreement exists on this should be done. While some challenges can be solved at a centralized planning level, other challenges are best tackled with a decentralized approach. In the field of vehicle-to-vehicle communication especially there is a huge disagreement on whether to use ITS-G5 (WLAN) or LTE (cellular) technologies. When discussing cooperative vehicles, however, it is key to realize that cooperation has to start in development. There must be unity among the vehicle manufacturers and there must be open standards that are collaboratively developed in order to foster cooperation in a large scale on our roads. We further showed that there are great synergies between the existing and promising approaches for CAVs that might lead to a combined and unified implementation in the future.

All in all, we highlighted the need for cooperative ITSs, and introduced the characteristics that need to be considered in cooperation and implementation in order to provide an overview and a starting point for further collaborative developments.

Author Contributions: Conceptualization, J.C.M.; methodology, J.C.M.; investigation, J.C.M. and C.K.; writing—original draft preparation, J.C.M. and C.K.; writing—review and editing, J.C.M., C.K., S.E., F.D., and S.K.; supervision, F.D. and S.K.; project administration, F.D. and S.K. All authors have read and agreed to the published version of the manuscript.

Funding: Research supported by MAN Truck & Bus SE and the IMAGinE project (Intelligent Maneuver Automation—cooperative hazard avoidance in real-time). IMAGinE is founded by the German Federal Ministry for Economic Affairs and Energy (BMWi).

Conflicts of Interest: The authors declare no conflict of interest.

References

1. German Law. German Road Traffic Regulations. Available online: https://germanlawarchive.iuscomp.org/?p=1290 (accessed on 14 April 2020).
2. Festag, A. Standards for vehicular communication—From IEEE 802.11p to 5G. Fest-VehicularCom-15. *Elektrotech. Inftech.* **2015**, *132*, 409–416. [CrossRef]
3. Fekete, S.; Vollrath, M.; Huemer, A.K.; Salchow, C. Interaktion im Straßenverkehr: Kooperation und Konflikt. *8. Vdi-Tag. Der Fahr. Im 21. Jhd. Fahr. Fahrunterstützung Und Bedien.* **2015**, *2264*, 325–338.
4. Belbachir, A.; Fallah-Seghrouchni, A.E.; Casals, A.; Pasin, M. Smart Mobility Using Multi-Agent System. *Procedia Comput. Sci.* **2019**, *151*, 447–454. [CrossRef]
5. During, M.; Lemmer, K. Cooperative Maneuver Planning for Cooperative Driving. *IEEE Intell. Transp. Syst. Mag.* **2016**, *8*, 8–22. [CrossRef]
6. Dorri, A.; Kanhere, S.S.; Jurdak, R. Multi-Agent Systems: A Survey. *IEEE Access* **2018**, *6*, 28573–28593. [CrossRef]
7. Khamis, A.M.; Kamel, M.S.; Salichs, M.A. Cooperation: Concepts and General Typology. In Proceedings of the 2006 IEEE International Conference on Systems, Man and Cybernetics, Taipei, Taiwan, 8–11 October 2006; pp. 1499–1505, ISBN 1-4244-0099-6.
8. van Lange, P.A.M.; Joireman, J.; Parks, C.D.; van Dijk, E. The psychology of social dilemmas: A review. *Organ. Behav. Hum. Decis. Process.* **2013**, *120*, 125–141. [CrossRef]
9. Rand, D.G.; Nowak, M.A. Human cooperation. *Trends Cognit. Sci. (Regul. Ed)* **2013**, *17*, 413–425. [CrossRef] [PubMed]
10. Fank, J.; Santen, L.; Knies, C.; Diermeyer, F. "Should We Allow Him to Pass?" Increasing Cooperation Between Truck Drivers Using Anthropomorphism. In *Advances in Human Factors of Transportation*; Stanton, N., Ed.; Springer International Publishing: Cham, Switzerland, 2020; pp. 475–484, ISBN 978-3-030-20502-7.
11. Hyldmar, N.; He, Y.; Prorok, A. A Fleet of Miniature Cars for Experiments in Cooperative Driving. In Proceedings of the 2019 International Conference on Robotics and Automation (ICRA), Montreal, QC, Canada, 20–24 May 2019.
12. Kollock, P. Social Dilemmas: The Anatomy of Cooperation. *Annu. Rev. Sociol.* **1998**, *24*, 183–214. [CrossRef]
13. Raimo, T. On the structural aspects of collective action and free-riding. *Theory Descis.* **1992**, *32*, 165–202.
14. Zimmermann, M.; Schopf, D.; Lütteken, N.; Liu, Z.; Storost, K.; Baumann, M.; Happee, R.; Bengler, K.J. Carrot and stick: A game-theoretic approach to motivate cooperative driving through social interaction. *Transp. Res. Part C Emerg. Technol.* **2018**, *88*, 159–175. [CrossRef]
15. Ploeg, J.; Englund, C.; Nijmeijer, H.; Semsar-Kazerooni, E.; Shladover, S.E.; Voronov, A.; van de Wouw, N. Guest Editorial Introduction to the Special Issue on the 2016 Grand Cooperative Driving Challenge. *IEEE Trans. Intell. Transp. Syst.* **2018**, *19*, 1208–1212. [CrossRef]
16. Stoll, T.; Müller, F.; Baumann, M. When cooperation is needed: The effect of spatial and time distance and criticality on willingness to cooperate. *Cogn. Tech. Work* **2018**, *54*, 163193. [CrossRef]

17. Andrea, B.; Gregor, J.; Alexander, P. EDDI Elektronische Deichsel—Digitale Innovation. 2018. Available online: https://www.deutschebahn.com/resource/blob/4136370/3227eac8b688106dc68e9292f4a173e9/Platooning_EDDI_Projektbericht_10052019_DE-data.pdf (accessed on 14 April 2020).
18. Zhang, L.; Chen, F.; Ma, X.; Pan, X. Fuel Economy in Truck Platooning: A Literature Overview and Directions for Future Research. *J. Adv. Transp.* **2020**, *2020*, 1–10. [CrossRef]
19. Stehbeck, F. Designing and Scheduling Cost-Efficient Tours by Using the Concept of Truck Platooning. *Manag. Sci.* **2019**, *4*, 566–634. [CrossRef]
20. Mohamed, A.B.; Popieul, J.-C.; Chouki, S. Multi-level Cooperation between the Driver and an Automated Driving System during Lane Change Maneuver. In Proceedings of the IEEE Intelligent Vehicles Symposium (IV), Gothenburg, Sweden, 19–22 June 2016.
21. OLG Hamm. Beschluss 4 Ss OWi 629/08. Available online: https://openjur.de/u/31853.html (accessed on 14 April 2020).
22. Knies, C.; Leonhard, H.; Frank, D. Cooperative Maneuver Planning for Highway Traffic Scenarios based on Monte-Carlo Tree Search. In Proceedings of the AAET 2019—Automatisiertes und vernetztes Fahren, Montreal, QC, Canada, 13–17 May 2019.
23. Standards Committee of the IEEE Computer Society. *IEEE Standard for Information Technology—Telecommunications and Information Exchange between Systems—Local and Metropolitan Area Networks—Specific Requirements*; Institute of Electrical and Electronics Engineers: New York, NY, USA, 2010; ISBN 9780738163246.
24. Bjerke, B. LTE-advanced and the evolution of LTE deployments. Bjerke-LTEevolution-11. *IEEE Wirel. Commun.* **2011**, *18*, 4–5. [CrossRef]
25. Reynolds, C.W. Flocks, Herds, and Schools: A Distributed Behavioral Model. *Comput. Graph.* **1987**, *21*, 25–34.
26. Kurzer, K.; Engelhorn, F.; Zöllner, J.M. Accelerating Cooperative Planning for Automated Vehicles with Learned Heuristics and Monte Carlo Tree Search. Available online: https://arxiv.org/pdf/2002.00497.pdf (accessed on 7 February 2020).
27. Volkswagen AG. Strategy TOGETHER 2025+: Shaping Mobility—for Generations to Come. Available online: https://www.volkswagenag.com/en/group/strategy.html (accessed on 7 January 2020).
28. PWC. *Five Trends Transforming the Automotive Industry*. Available online: https://www.pwc.at/de/publikationen/branchen-und-wirtschaftsstudien/eascy-five-trends-transforming-the-automotive-industry_2018.pdf (accessed on 7 January 2020).
29. Daimler AG. CASE—Intuitive Mobility. Available online: https://www.daimler.com/innovation/case-2.html (accessed on 7 January 2020).
30. BMW AG. The BMW Vision iNEXT. Available online: https://www.bmwgroup.com/BMW-Vision-iNEXT (accessed on 7 January 2020).
31. Bill, K. 4G Wireless Technology: When will it happen? What does it offer? In Proceedings of the 2008 IEEE Asian Solid-State Circuits Conference, Fukuoka, Japan, 3–5 November 2008.
32. Mohamed, G.; Georg, J.-M. A Reconfigurable 5G Testbed for V2X and Industry 4.0 Applications. *Elektrotech. Inftech.* **2015**, *132*, 409–416. [CrossRef]
33. ETSI—ITS. Communications Architecture. ETSI-CommunicationArchitecture-10. EN 302 665—V1.1.1—Intelligent Transport Systems (ITS); Vehicular Communications. 2010. Available online: https://www.etsi.org/deliver/etsi_en/302600_302699/302665/01.01.01_60/en_302665v010101p.pdf (accessed on 7 January 2020).
34. Almeida, T.T.C.; de Gomes, L.; Ortiz, F.M.; Junior, J.G.R.; Costa, L.H.M.K. IEEE 802.11p Performance Evaluation: Simulations vs. Real Experiments. In Proceedings of the 21st International Conference on Intelligent Transportation Systems (ITSC), Maui, HI, USA, 4–7 November 2018; pp. 3840–3845.
35. Arena, F.; Pau, G. An Overview of Vehicular Communications. *Future Internet* **2019**, *11*, 27. [CrossRef]
36. Abdeldime, M.S.; Wu, L. The Physical Layer of the IEEE 802.11p WAVE Communication Standard: The Specifications and Challenges. In Proceedings of the World Congress on Engineering and Computer Science, San Francisco, CA, USA, 22–24 October 2014.
37. Gao, S.; Lim, A.; Bevly, D. An empirical study of DSRC V2V performance in truck platooning scenarios. *Digit. Commun. Netw.* **2016**, *2*, 233–244. [CrossRef]
38. Sargolzaei, A.; Yazdani, K.; Abbaspour, A.; Crane III, C.D.; Dixon, W.E. Detection and Mitigation of False Data Injection Attacks in Networked Control Systems. *IEEE Trans. Ind. Inf.* **2020**, *16*, 4281–4292. [CrossRef]

39. Mehdi, H.; Bruno, S.; Emanuele, G. Feasibility and Detection of Replay Attack in Networked Constrained Cyber-Physical Systems. In Proceedings of the 57th Annual Allerton Conference on Communication, Control, and Computing, Monticello, IL, USA, 24–27 September 2019.
40. Biron, Z.A.; Dey, S.; Pisu, P. Resilient control strategy under Denial of Service in connected vehicles. In Proceedings of the 2017 American Control Conference (ACC), Seattle, WA, USA, 24–26 May 2017; pp. 4971–4978, ISBN 978-1-5090-5992-8.
41. Watzlawick, P.; Bavelas, J.B.; Jackson, D.D.; O'Hanlon, B. *Pragmatics of Human Communication. A Study of Interactional Patterns, Pathologies, and Paradoxes*; Pbk. ed.; W.W. Norton, & Co: New York, NY, USA, 2011; ISBN 0393707229.
42. ETSI—ITS. Basic Set of Applications; Part 2: Specification of Cooperative Awareness Basic Service. ETSI-CAM-14. EN 302 637-2—V1.3.1—Intelligent Transport Systems (ITS); Vehicular Communications. 2014. Available online: https://www.etsi.org/deliver/etsi_en/302600_302699/30263702/01.03.01_30/en_30263702v010301v.pdf (accessed on 14 April 2020).
43. ETSI-ITS. Basic Set of Applications; Part 3: Specifications of Decentralized Environmental Notification Basic Service. ETSI-DENM-14. EN 302 637-3—V1.2.—Intelligent Transport Systems (ITS); Vehicular Communications. 2014. Available online: https://www.cisco.com/c/dam/en_us/solutions/industries/docs/trans/its-standards-part-2.pdf (accessed on 14 April 2020).
44. ETSI. Intelligent Transport Systems (ITS). Cooperative Perception Services. Available online: https://portal.etsi.org/webapp/WorkProgram/Report_WorkItem.asp?wki_id=46541 (accessed on 14 April 2020).
45. Gunther, H.-J.; Mennenga, B.; Trauer, O.; Riebl, R.; Wolf, L. Realizing collective perception in a vehicle. In Proceedings of the 2016 IEEE Vehicular Networking Conference (VNC), Columbus, OH, USA, 8–10 December 2016; pp. 1–7, ISBN 978-1-5090-5197-7.
46. Sawade, O.; Schulze, M.; Radusch, I. Robust Communication for Cooperative Driving Maneuvers. *IEEE Intell. Transp. Syst. Mag.* **2018**, *10*, 159–169. [CrossRef]
47. Bernd, L.; Hendrik-Jörn, G.; Lars, W. A Generic Approach towards Maneuver Coordination for Automated Vehicles. In Proceedings of the 2018 IEEE Intelligent Transportation Systems Conference, Maui, HI, USA, 4–7 November 2018; IEEE: Piscataway, NJ, USA, 2018; ISBN 9781728103242.
48. ETSI. Intelligent Transport Systems (ITS); Vehicular Communications; Informative report for the Maneuver Coordination Service. Available online: https://portal.etsi.org/webapp/WorkProgram/Report_WorkItem.asp?wki_id=53991 (accessed on 14 April 2020).
49. Bojarski, M.; Testa, D.D.; Dworakowski, D.; Firner, B.; Flepp, B.; Goyal, P.; Jackel, L.D.; Monfort, M.; Muller, U.; Zhang, J.; et al. End to End Learning for Self-Driving Cars. 2016. Available online: http://arxiv.org/pdf/1604.07316v1 (accessed on 14 April 2020).
50. Wei, J.; Dolan, J.M. A robust autonomous freeway driving algorithm. In Proceedings of the 2009 IEEE Intelligent Vehicles Symposium (IV), Xi'an, China, 3–5 June 2009; pp. 1015–1020, ISBN 978-1-4244-3503-6.
51. Wei, J.; Dolan, J.M.; Litkouhi, B. A prediction- and cost function-based algorithm for robust autonomous freeway driving. In Proceedings of the 2010 IEEE Intelligent Vehicles Symposium (IV), La Jolla, CA, USA, 21–24 January 2010; pp. 512–517, ISBN 978-1-4244-7866-8.
52. Junqing, W.; John, D.; Bakhtiar, L. Autonomous vehicle social behavior for highway entrance ramp management. In Proceedings of the 2013 IEEE Intelligent Vehicles Symposium (IV), Gold Coast City, Australia, 23–26 June 2013.
53. Lenz, D.; Kessler, T.; Knoll, A. Tactical cooperative planning for autonomous highway driving using Monte-Carlo Tree Search. In Proceedings of the 2016 IEEE Intelligent Vehicles Symposium (IV), Gotenburg, Sweden, 19–22 January 2016; pp. 447–455, ISBN 978-1-5090-1821-5.
54. Kurzer, K.; Zhou, C.; Zöllner, J.M. Decentralized Cooperative Planning for Automated Vehicles with Hierarchical Monte Carlo Tree Search. In Proceedings of the 2018 IEEE Intelligent Vehicles Symposium (IV), Changshu, China, 26–30 June 2018; pp. 529–536. [CrossRef]
55. Kurzer, K.; Engelhorn, F.; Zöllner, J.M. Decentralized Cooperative Planning for Automated Vehicles with Continuous Monte Carlo Tree Search. In Proceedings of the 2018 21st International Conference on Intelligent Transportation Systems (ITSC), Maui, HI, USA, 4–7 November 2018; pp. 452–459. [CrossRef]
56. Zhihai, Y.; Jun, W.; Yihuan, Z. A Game-Theoretical Approach to Driving Decision Making in Highway Scenarios. In Proceedings of the 2018 IEEE Intelligent Vehicles Symposium (IV), Changshu, China, 26–30 June 2018; ISBN 9781538644539.

57. Fisac, J.F.; Bronstein, E.; Stefansson, E.; Sadigh, D.; Sastry, S.S.; Dragan, A.D. Hierarchical Game-Theoretic Planning for Autonomous Vehicles. In Proceedings of the International Conference on Robotics and Automation, Montreal, QC, Canada, 20–24 May 2018.
58. Kyushik, M.; Hayoung, K. Deep Q Learning Based High Level Driving Policy Determination. In Proceedings of the 2018 IEEE Intelligent Vehicles Symposium (IV), Changshu, China, 26–30 June 2018; ISBN 9781538644539.
59. Mirchevska, B.; Blum, M.; Louis, L.; Boedecker, J.; Werling, M. Reinforcement Learning for Autonomous Maneuvering in Highway Scenarios. In Proceedings of the 2017 IEEE Intelligent Vehicles Symposium, Los Angeles, CA, USA, 11–14 June 2017.
60. Naik, G.; Choudhury, B.; Park, J.-M. IEEE 802.11bd & 5G NR V2X: Evolution of Radio Access Technologies for V2X Communications. *IEEE Access* **2019**, *7*, 70169–70184. [CrossRef]
61. Vukadinovic, V.; Bakowski, K.; Marsch, P.; Garcia, I.D.; Xu, H.; Sybis, M.; Sroka, P.; Wesolowski, K.; Lister, D.; Thibault, I. 3GPP C-V2X and IEEE 802.11p for Vehicle-to-Vehicle communications in highway platooning scenarios. *Ad Hoc Netw.* **2018**, *74*, 17–29. [CrossRef]
62. Giammarco, C.; Alessandro, B.; Barbara, M.; Masini, A.Z. Performance Comparison between IEEE 802.11p and LTE-V2V In-coverage and Out-of-coverage for Cooperative Awareness. In Proceedings of the 2017 IEEE Vehicular Networking Conference (VNC) 2017, Torino, Italy, 27–29 November 2017.
63. Andrew, T.; Kees, M.; Alessio, F.; Vincent, M. C-ITS: Three observations on LTE-V2X and ETSI-ITS G5. Available online: https://www.nxp.com/docs/en/white-paper/CITSCOMPWP.pdf (accessed on 16 January 2020).
64. IMAGinE—Solutions for Cooperative Driving. Available online: https://imagine-online.de (accessed on 7 January 2020).
65. 5G-MOBIX. Available online: https://www.5g-mobix.com/ (accessed on 16 January 2020).
66. ICT4CART. Available online: https://www.ict4cart.eu/ (accessed on 16 January 2020).

Article

Decision-Making System for Lane Change Using Deep Reinforcement Learning in Connected and Automated Driving

HongIl An and Jae-il Jung *

Department of Electronics and Computer Engineering, Hanyang University, Seoul 133-791, Korea; aviate@hanyang.ac.kr

* Correspondence: jijung@hanyang.ac.kr

Received: 4 April 2019; Accepted: 10 May 2019; Published: 14 May 2019

Abstract: Lane changing systems have consistently received attention in the fields of vehicular communication and autonomous vehicles. In this paper, we propose a lane change system that combines deep reinforcement learning and vehicular communication. A host vehicle, trying to change lanes, receives the state information of the host vehicle and a remote vehicle that are both equipped with vehicular communication devices. A deep deterministic policy gradient learning algorithm in the host vehicle determines the high-level action of the host vehicle from the state information. The proposed system learns straight-line driving and collision avoidance actions without vehicle dynamics knowledge. Finally, we consider the update period for the state information from the host and remote vehicles.

Keywords: lane change; decision-making system; vehicular communication; deep reinforcement learning; collision avoidance; connected and automated vehicle

1. Introduction

Lane change systems have been studied for a long time in the research on autonomous driving [1,2], as well as vehicular communication [3–6]. Lane change systems are continuously drawing attention from academia and industry and several solutions have been proposed to solve this challenging problem. Vehicular communication-based methods can be divided into Long Term Evolution-based [7] or IEEE 802.11p-based method [8]. Moreover, there have been many studies on lane change systems for connected and automated vehicle (CAVs), which combines autonomous driving and vehicular communication [9].

There have been many advances in autonomous vehicle systems in recent years. In particular, end-to-end learning has been proposed, which is a new paradigm that includes perception and decision-making [10]. The traditional method for autonomous navigation is to recognize the environment based on sensor data, generate a future trajectory through decision-making and planning modules, and then follow the given path through control. However, end-to-end learning integrates perception, decision-making, and planning based on machine learning, and produces a control input directly from sensor information [10].

Deep reinforcement learning (DRL) is a combination of deep learning and reinforcement learning. Reinforcement learning involves learning how to map situations to actions to maximize a numerical reward signal [11]. The combination of reinforcement learning and deep learning, which relies on powerful function approximation and representation learning properties, provides a powerful tool [12]. In addition, deep reinforcement learning allows a vehicle agent to learn from its own behavior instead of labeled data [13]. No labeled data are required in an environment in which the vehicle agent will take actions and obtain rewards.

Deep reinforcement learning could be applied to various domains. A typical example is Atari video games. In recent years, a deep Q network (DQN) is a representative example of deep reinforcement learning that shows human-level performance in Atari video games [14]. A DQN has the advantage of learning directly from high-dimension image pixels. It can also be used for indoor navigation [15] and control policies [16] for specific purposes in the robotics field.

Many studies on autonomous driving using deep reinforcement learning have been conducted. Wang et al. [17] suggested a deep reinforcement learning architecture for the on-ramp merging of autonomous vehicles. It features the use of a long short-term memory (LSTM) model for the historical driving information. Wang et al. [18] subsequently proposed automated maneuvers for lane change. The authors considered continuous action space but did not take into account the lateral control issue. Another study was conducted on autonomous driving decision-making [19]. A DQN was used to determine the speed and lane change action of the autonomous vehicle agent. Kaushik et al. [20] developed overtaking maneuvers using deep reinforcement learning. An important feature of this system is the application of curriculum learning. This method, which simulates the way humans and animals learn, enables the effective learning of the vehicle agent. Mukadam et al. [21] suggested decision-making for lane changing in a multi-agent setting. The authors dealt with discrete action space, but it might weaken the feasibility of the solution when applied to real-world problems [18]. In addition, the SUMO simulator was used for evaluation. In this simulator, vehicles run only along the center lane, and the lane change is performed like a teleport. It has a limitation that lateral control cannot be considered as with Reference [18]. Mirchevska et al. [22] proposed a formal verification method to overcome the limits of machine learning in lane change and the authors also defined action space as discrete.

We propose a decision-making system for connected and automated vehicle during lane change. The purpose of our system is to find a policy represented from an actor network, μ, as seen in Figure 1. The actor network, μ, takes the current state of the host vehicle and the remote vehicle from sensors and gives the high-level outputs (throttle, steering wheel). Then the actor network passes the high-level outputs to a controller of the host vehicle. We constructed a simulation environment for autonomous vehicles where an agent can perform trial-and-error simulations.

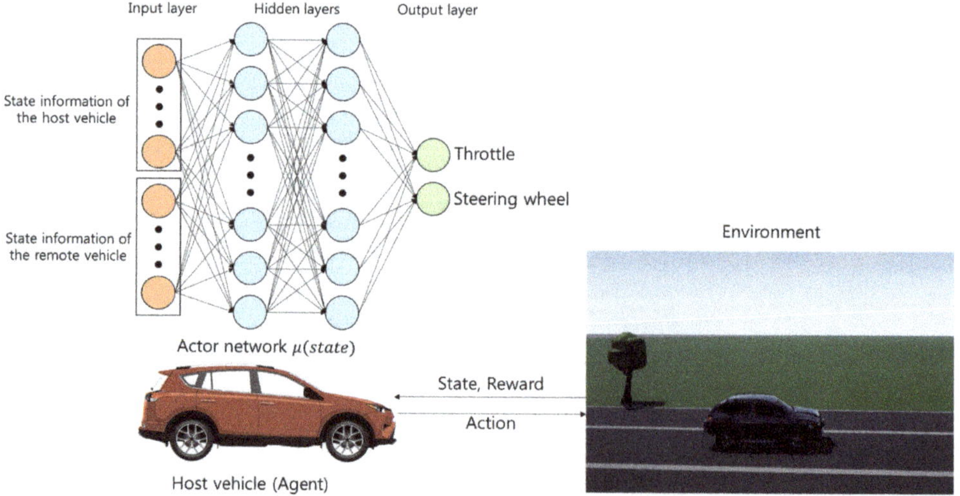

Figure 1. Architecture of proposed decision-making system.

Our contributions are as follows:

- A model that combines deep reinforcement learning with vehicular communication is proposed. We show how to utilize information from a remote vehicle as well as the host vehicle in deep reinforcement learning.
- We handle different state information update periods for the host and remote vehicles. Generally, the state information of the remote vehicle is updated every 100 ms via vehicle communication, whereas the state information of the host vehicle is updated faster. This paper shows the results of agent learning considering these different update periods.
- We handle continuous action space for steering wheel and accelerator to improve the feasibility of the lane change problem. Furthermore, in order that our end-to-end method covers collision-free action, reward function takes the collision reward directly into account.
- We introduce a physically and visually realistic simulator Airsim [23]. Our main focus is to avoid collision between vehicles and realistic longitudinal and lateral control. We have experimented in the simulation environment that can handle both controls.

The remainder of our paper considers the following. Section 2 shows the lane change system architecture and how the agent learns. Section 3 presents the simulation environment for training and evaluation, along with the agent learning results. Finally, the conclusions are presented in Section 4.

2. Decision-Making System for Lane Change

2.1. System Architecture

The architecture of the proposed system consists of the host vehicle (agent) and environment. Just as the agent learns from the environment through interaction in reinforcement learning (RL), the host vehicle interacts with the environment, including the remote vehicle. First, the host vehicle receives the current state of the environment and determines an action that affects the environment. Then, the host vehicle takes the action and the environment gives the host vehicle a reward and the next state corresponding to the action.

More specifically, the host vehicle receives self-state information from the sensors mounted in the agent or state information from the remote vehicle through the vehicular communication device. The state information is provided to the trained actor neural network. The actions needed for a collision-free lane change are generated. After the next time step, the host vehicle will receive a reward and the next set of state information. We designed a reward function that performs a lane change while avoiding collision with the remote vehicle.

2.2. Markov Decision Process

The reinforcement learning problem is defined in the form of a Markov decision process (MDP), which is made up of states, actions, transition dynamics, and rewards. In our system, we assume that the environment is fully observable. A detailed formulation is presented in the next subsection.

2.2.1. States

The host vehicle observes current state, s_t, from the environment at time t. The host vehicle can acquire state information through sensors. A lane change not only relates to vehicle dynamics, but also depends on road geometry [18]. The state space includes speed and heading to handle the longitudinal and lateral dynamics of the host vehicle. The x, y coordinates of the host vehicle are included for road geometry. The information of the remote vehicle is contained for analyzing collision between vehicles. A set of states, S, in our system consists of eight elements. At timestep t, $S = \{s_t^1, s_t^2, s_t^3, s_t^4, s_t^5, s_t^6, s_t^7, s_t^8\}$.

- s_t^1 represents the x coordinate of the host vehicle.
- s_t^2 represents the y coordinate of the host vehicle.
- s_t^3 represents the speed of the host vehicle.

- s_t^4 represents the heading of the host vehicle.
- s_t^5 shows the x coordinate of the remote vehicle.
- s_t^6 shows the y coordinate of the remote vehicle.
- s_t^7 represents the speed of the remote vehicle.
- s_t^8 represents the heading of the remote vehicle.

The values of all elements are converted into values in the range of (0, 1) before being input to the neural network. s_t^1, s_t^2, s_t^3, s_t^4 contain state information related to the host vehicle. Conversely, s_t^5, s_t^6, s_t^7, s_t^8 are associated with the remote vehicle. In the training and evaluation stage, s_t^1, s_t^2, s_t^3, s_t^4 are updated with a period of 0.01 s, and s_t^5, s_t^6, s_t^7, s_t^8 are updated with a period of 0.1 s through vehicular communication.

2.2.2. Actions and Policy

The host vehicle takes actions in the current state. A set of actions, A, consists of 2 elements: $\mathrm{A} = \left\{a_t^1, a_t^2\right\}$.

- a_t^1 represents the throttle of the host vehicle.
- a_t^2 represents the steering wheel of the host vehicle.

The goal of the host vehicle is to learn a policy π that maximizes the cumulative reward [12]. Generally, policy π is a probability function from the states: π: $\mathrm{S} \rightarrow \mathrm{p}(\mathrm{A} = \mathrm{a}|\mathrm{S})$. However, in our system, policy π is a function from states to actions (high-level control input). Given the current state values, the host vehicle can know the throttle and steering wheel values through the policy. In DRL, the policy is represented as a neural network.

2.2.3. Transition Dynamics

The transition dynamics is a function wherein the conditional probability of the next state S_{t+1} is the probability of given state S_t and action A_t. One of the challenges in DRL is that there is no way for the host vehicle (agent) to know the transition dynamics. Therefore, the host vehicle (agent) learns by interacting with the environment. Namely, the host vehicle takes actions using the throttle and steering wheel, which allow it to influence the environment and obtain a reward. The host vehicle accumulates this information and uses it to learn the policy that ultimately maximizes the cumulative reward.

2.2.4. Rewards

The ultimate goal of the agent in DRL is to maximize the cumulative reward rather than the instant reward of the current state. It is important to design a reward function to induce the host vehicle to make the lane changes that we want.

We formulate a lane change task as an episodic task, which means it will end in a specific state [11]. Thus, we define a lane change task as moving into the next lane within a certain time without collision or leaving the road. To meet this goal, a reward function is constructed with three cases at each time step.

1. Collision with the remote vehicle and deviation from the road

$$\mathrm{R}(t) = -3. \tag{1}$$

2. Final time step

    a. Completion of the lane change

$$\mathrm{R}(t) = 1. \tag{2}$$

b. Failure to complete the lane change

$$R(t) = 0. \tag{3}$$

3. Etc.

a. Driving in the next lane

$$R(t) = W_n + W_s \times V_s. \tag{4}$$

b. Driving in the initial lane

$$R(t) = W_i + W_s \times V_s. \tag{5}$$

c. Remainder

$$R(t) = W_s \times V_s. \tag{6}$$

The first case is about penalization. If the host vehicle collides with the remote vehicle in the next lane or leaves the road, the reward is given as above, and the task is immediately terminated. The second case is about the success or failure of the lane change task in the final time step. The lane change is considered successful if the host vehicle is within 0.5 m of the next lane at a certain time step. Otherwise, it is a failure. The last case is about the driving process. This case determines most rewards in one episode. R(t) is the sum of the reward for the corresponding lane and the reward for the corresponding speed. First, the reward related to the lane is similar to the second case. If the host vehicle is within 0.5 m of the center of the lane, it receives the reward. The reward to be given related to the lane is as follows: W_n in the next lane, W_i in the initial lane, and 0 in the remainder. Second, the reward related to the velocity is expressed as the product of speed V_s and weight W_s. The advantage of DRL is that it can achieve the goal of the system without considering the vehicle dynamics of the host vehicle.

2.2.5. Deep Reinforcement Learning

We employed deep reinforcement learning. In our paper, the goal was to find an actor neural network, μ, for collision avoidance during lane change. We applied an actor–critic approach. Therefore, our decision-making system consisted of two neural networks: actor network, μ, and critic network, Q.

A critic $Q(s_t, a_t|\theta^Q)$ is a neural network that estimates the cumulative reward taking state s_t and action a_t with weights θ^Q in time step t. Equation (7) shows the feedforward of this critic network. It is used as a baseline for the actor network. The way is the same as deep-q-network (DQN) [14] in Equation (8) and Equation (9).

$$\text{scalar value} = Q(s_t,\ a_t|\theta^Q). \tag{7}$$

$$y_t = R(t) + \gamma \times Q'(s_{t+1},\ \mu'(s_{t+1}|\theta^{\mu'})|\theta^{Q'}). \tag{8}$$

$$Q(s_t,\ a_t|\theta^Q) \leftarrow y_t - \alpha \times Q(s_t,\ a_t|\theta^Q). \tag{9}$$

where α is the learning rate and y_t is the sum of R(t) and $Q(s_{t+1}, a_{t+1}|\theta^Q)$ with target network technique Q′ and μ′.

An actor network is the central part in the proposed system. An actor $\mu(s_t|\theta^\mu)$ is also a neural network that produce action a_t taking state s_t with weights θ^μ in Equation (10). Then, the actor will provide a high-level action to avoid collision. The actor $\mu(s_t|\theta^\mu)$ is updated through Equation (11) presented in the DDPG algorithm [24].

$$a_t = \mu(s_t|\theta^\mu). \tag{10}$$

$$\nabla_{\theta^\mu}\mathcal{J} \approx \frac{1}{N}\sum_i \nabla_a Q(s, a|\theta^Q)|_{s=s_i, a=\mu(s_i)} \nabla_{\theta^\mu}\mu(s|\theta^\mu)|_{s_i}. \tag{11}$$

where N is minibatch size.

The characteristic of actor–critic method is that the networks used are different for training phase and testing phase. In training phase, the vehicle (agent) learns both actor and critic networks. The critic network is used to learn the actor network. However, in testing phase, the vehicle only uses actor networks in order to produce action. Figure 2 shows the neural networks used at each phase. The left figure relates to the training phase, and the right figure relates to the testing phase.

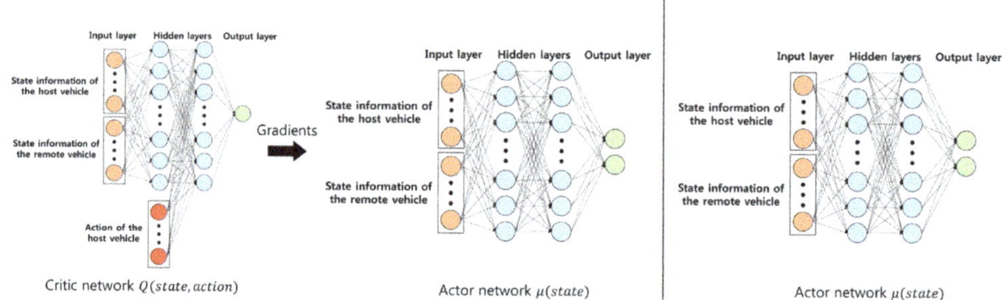

Figure 2. Neural networks in training and testing phase.

2.3. Training Scenario and Algorithm

The host vehicle learns a lane change scenario to avoid a collision on a straight road in Figure 3. The remote vehicle, a black car, is located in the next lane behind the host vehicle in initial state. Both vehicles have same initial speed V_0. The remote vehicle makes a straight run and will run at a faster speed than the initial speed. The host vehicle tries to change lanes and control the speed.

Figure 3. Initial state in lane change scenario for training.

We adopt the DDPG algorithm to allow the host vehicle to learn the lane change [24]. In order to use the DDPG algorithm, several assumptions and modification have been made. We need to use the experience replay memory technique [25] and the separate target network technique to apply the DDPG algorithm to our system. The reason for using the replay memory is to break the temporal correlations between consecutive samples by randomizing samples [12]. A time step was 0.01 s. It was assumed that there was no computation delay for the deep learning during the learning and evaluation. In autonomous driving, the control period is very short (milliseconds). Because of the computational delay during driving, the action was taken in a delayed state that was not the current state. Thus, during the learning and evaluation step, we ignored the computational delay by stopping the driving simulator. The states provided to the DDPG algorithm can be obtained from sensors. The state of the host vehicle is updated at each time step from the sensors of the host vehicle. The host vehicle can obtain the state of the remote vehicle through the vehicular communication device. Unlike the sensors of the host vehicle, it is assumed that the state of the remote vehicle is updated through a basic safety message, which allows it to be obtained every 10 time steps. Algorithm 1 shows the training algorithm with the assumptions and modifications mentioned above for the host vehicle.

Algorithm 1. Modification to Deep Deterministic Policy Gradient (DDPG) Training Algorithm.

Randomly initialize critic network and actor network
Initialize target networks and replay buffer
for episode = 1 **to** E **do**
Receive initial state from sensors in the host vehicle
for step = 0 **to** S-1 **do**
if step % 10 == 0 **then**
Receive state of the remote vehicle from the vehicular communication device
end if
Select action according to the current policy and exploration noise using normal distribution
Simulation pause off
Execute action and observe reward and observe new state
Simulation pause on
Store transition in replay memory
Sample a random minibatch of N transitions from replay memory
Update critic by minimizing the loss
Update the actor policy using the sampled policy gradient
Update the target networks
end for
end for

3. Experiments

3.1. Simulation Setup and Data Collection

Airsim was adopted to train the host vehicle to change lanes [23]. Airsim is an open source platform developed by Microsoft and used to reduce the gap between reality and simulation when developing autonomous vehicles. Airsim offers a variety of conveniences. Airsim is an Unreal Engine-based simulator. The Unreal Engine not only provides excellent visual rendering, but also provides rich functionality for collision-related experiments. When training a host vehicle by trial-and-error, it is possible for the host vehicle to directly experience collision states.

In order to train our neural network, we need to collect data from the environment. By driving the host vehicle and the remote vehicle in Airsim, we are able to collect the data necessary for learning. Airsim supports various sensors for vehicles. All vehicles are equipped with GPS and the Inertial Navigation System (INS). Through these sensors, we can collect desired vehicle status information.

3.2. Training Details and Results

As shown in Figure 3, the host vehicle (red), remote vehicle (black), and straight road were constructed using the Unreal Engine. In the initial state of the training and evaluation, both the host vehicle and remote vehicle had initial speeds of V_0, and the remote vehicle was located at distance d_0 behind the host vehicle in the next lane. After the initial state, the target speed of the remote vehicle was set uniformly in the range of $[V_{min}, V_{max}]$, and the remote vehicle constantly ran at the target speed.

The actor network consisted of two fully connected hidden layers with 64, 64 units. The critic network consisted of two fully connected hidden layers with 64, 64 + 2 (action size: throttle, acceleration) units. Figure 2 shows the architecture of actor network and critic network (left). All hidden layers in the actor network and the critic network used a rectified linear unit (ReLU) activation function. The output layer of the actor network has tanh activation function because of action space range. However, the output layer of the critic network does not have activation function. The weights of the output layer in both the actor network and critic network were initialized from a uniform distribution $[-3 \times 10^{-3}, -3 \times 10^{3}]$. We use the Adam optimizer to update the actor network and the critic network. Table 1 lists parameters related to the road driving scenario and DDPG algorithm.

Table 1. Parameters related to lane change system.

Parameter	Value
Training episodes, E	2000
Width of lanes	3.4 m
Max time step, S	500
Weight corresponding to next lane, W_n	0.01
Weight corresponding to initial lane, W_i	0.001
Weight corresponding to speed, W_s	0.0002
Initial velocity, V_0	11.11 m/s
Initial distance, d_0	10 m
Reply memory size	1,000,000
Minibatch size, N	256
Mean of normal distribution for exploration noise	0
Standard deviation for exploration noise	1
Minimum target speed, V_{min}	16.67 m/s
Maximum target speed, V_{max}	22.22 m/s
Maximum acceleration	4.9 m/s^2
Learning rate for actor	0.001
Learning rate for critic	0.001
Hidden units of actor network	64, 64
Hidden units of critic network	64, 66
Tau for deep deterministic policy gradient (DDPG) algorithm	0.06

Figure 4 shows the cumulative reward (blue) per episode and average cumulative reward (red) in the training step. The cumulative reward is the sum of the rewards the host vehicle obtained from the reward function. The average cumulative reward is the average of only the most recent 100 scores. The lane change success rate was high from 850 to 1050 episodes. The highest average cumulative reward was 3.567 at 1028 episodes.

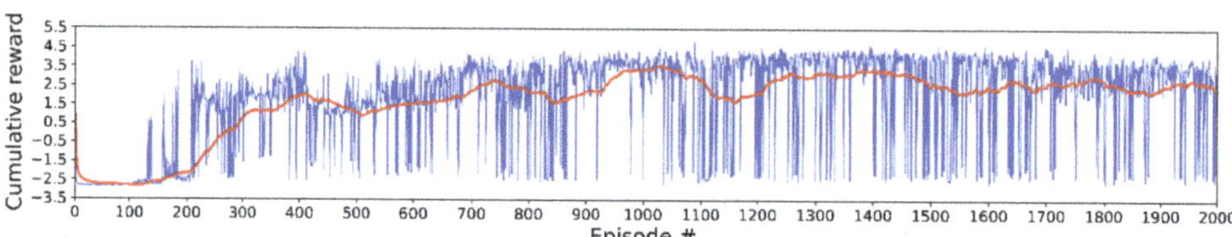

Figure 4. Scores and average scores in total training episodes.

3.3. Evaluation Results

We evaluated the actor network, μ, after 2000 training episodes. We selected the weight to successfully perform the lane change and proceeded to the evaluation. We evaluated the weight of the 1028th episode, which had the highest average score. The lane change was successful, but the host vehicle did not reach a position within 0.5 m from the center of the next lane. We searched heuristically to find a weight that arrived close to the center with a successful lane change before and after the 1028th weight. As a result, the 1030th weight was found, which satisfied both conditions, and the evaluation was made with this weight. The evaluation was conducted in the same environment as the training stage, and a total of 300 episodes were performed. Figure 5 shows the result of the evaluation with the 1030th weight. Figure 5 shows the cumulative reward of the host vehicle. The average cumulative reward was 3.68 and is expressed in a red line. In all of the episodes, the host vehicle successfully performed the lane change without a collision with the remote vehicle.

To evaluate the performance of the host vehicle, three measures were introduced. We analyzed our decision-making system with three metrics. We obtained these measures as well as the cumulative reward for each episode.

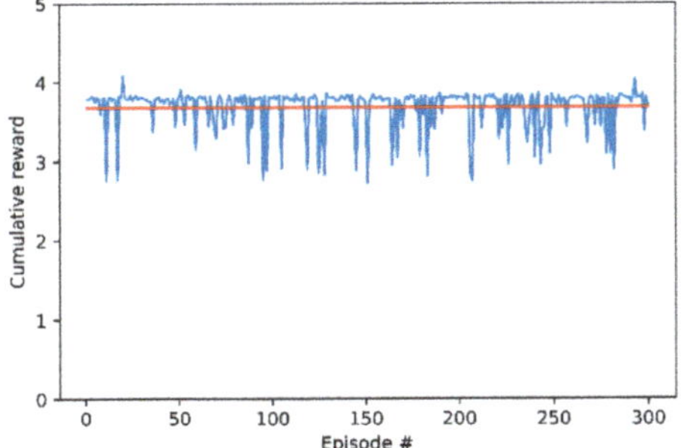

Figure 5. Cumulative reward in evaluation phase.

First, the lane change success rate is the number of lane change successes divided by the total number of lane change trials. We defined the success of the lane changes as arrival in the next lane without collision with the remote vehicle and without leaving the road within a certain time. In Airsim, when a collision occurs between vehicles in simulation, it provides information about the collision. At each time step, the position of the host vehicle can be used to determine if it has left the road or it has arrived in the next lane. For termination of each episode, the result of the lane change was determined. The success rate was 1.0. In other words, the lane change in all episodes was made without collision.

Second, arrival time in next lane is the time when the host vehicle arrives in next lane. For every time step, the position of the host vehicle was checked and if the host vehicle was in next lane at first, the time step was recorded. Figure 6a shows arrival time. This indicates that the host vehicle has made a lane change within the max time step. The time the lane change occurred was before the end of every episode. In a training scenario, the remote vehicle departs from the rear of the host vehicle and passes through the host vehicle. The host vehicle learned that it made a lane change after the remote vehicle passed the host vehicle to avoid collision.

Difference between position x is the gap of position x between the host vehicle and the remote vehicle. The reason for defining this metric is to analyze the lane-change behavior of the host vehicle. When we set the coordinates axes, the x-axis value was designed to be larger in the longitudinal direction of both vehicles. Figure 6b shows the differences between the position x of the host vehicle subtracted from the position x of the remote vehicle. In every episode, the measurements were positive. The remote vehicle preceded the host vehicle at the end of lane change trials. The host vehicle showed a behavior pattern that it tried to change lanes after the remote vehicle had passed.

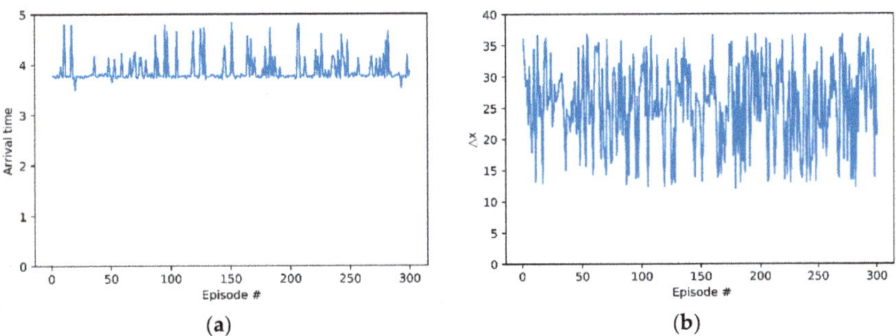

Figure 6. Evaluation results: (**a**) arrival times of the host vehicle in next lane; (**b**) differences between the position x of the host vehicle and the remote vehicle.

Figure 7a presents speed profile of the host vehicle in all episodes. It is not special in the initial time step. The host vehicle travels at various speeds as the lane change is over. Figure 7b shows inter-vehicle distances in all episodes. Figure 7b shows a pattern similar to Figure 7a, Figure 7c,d correspond to the highest reward, and Figure 7e,f depict the lowest reward. We can observe the relationship between speed and inter-vehicle distance in both cases. As inter-vehicle distance between the host vehicle and the remote vehicle decreases, the host vehicle reduces the speed to avoid collision. Conversely, as the inter-vehicle distance increases, the host vehicle does not maintain a reduced speed but increases the speed through acceleration.

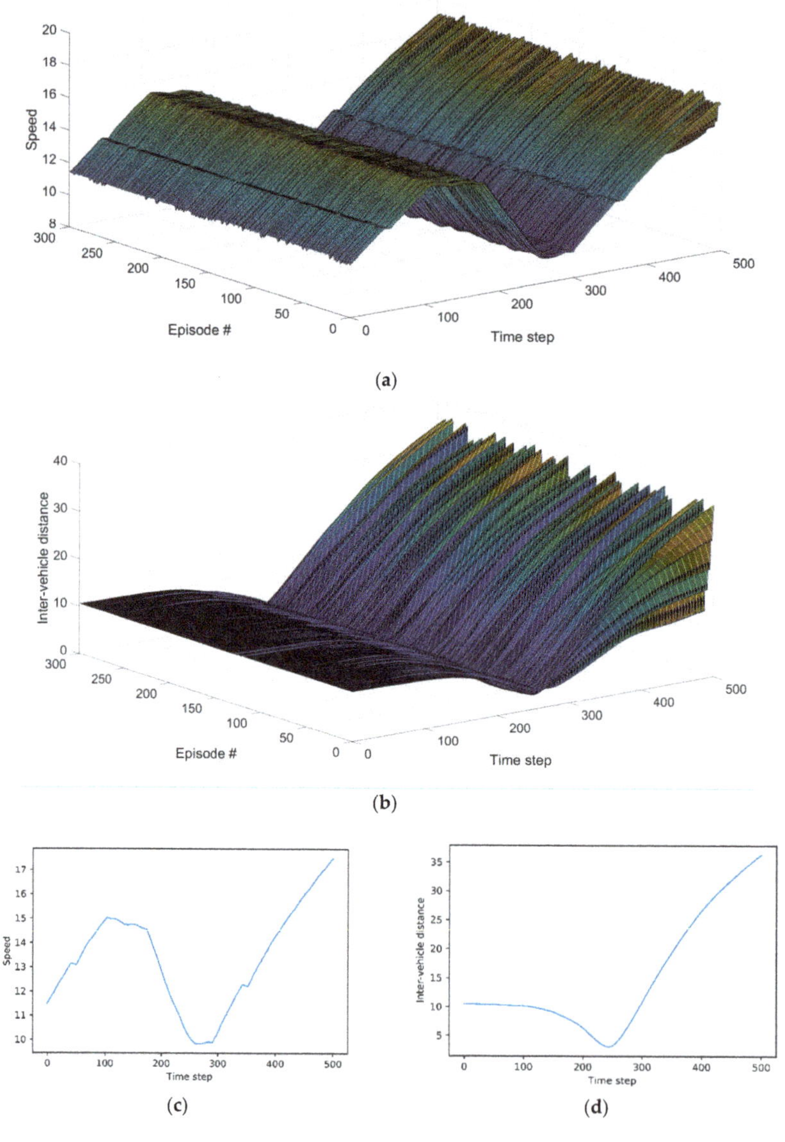

Figure 7. *Cont.*

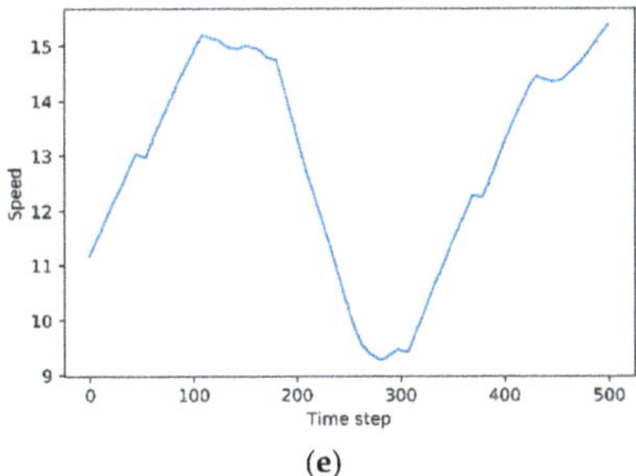

(e)

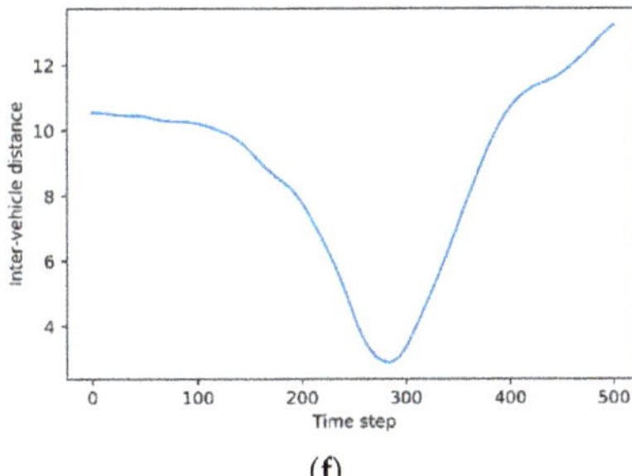

(f)

Figure 7. Evaluation results: (**a**) speed profile of the host vehicle in all episodes; (**b**) inter-vehicle distance in all episodes; (**c**) speed profile of the host vehicle in the highest cumulative reward; (**d**) inter-vehicle distance in the highest cumulative reward; (**e**) speed profile of the host vehicle in the lowest cumulative reward; (**f**) inter-vehicle distance in the lowest cumulative reward.

4. Conclusions

This paper presented a lane change system that uses deep reinforcement learning and vehicular communication. When using both techniques, we proposed the problem that the state information of the host vehicle obtained from installed sensors and the state information of the remote vehicle from the vehicular communication device are updated at different periods. Taking this into account, we modeled the lane change system as a Markov decision process, designed a reward function for collision avoidance, and integrated the host vehicle with the DDPG algorithm. Our evaluation results showed that the host vehicle successfully performed the lane change to avoid collision.

We suggest future research items to improve the proposed system. Our system has limitations in that it only considers a straight road and stable steering when the host vehicle is running. It should be possible to overcome these limitations by considering map information. Also, one of limitations in our system is that only one remote vehicle is considered. An extended model is needed to apply it to real-world problems. At least five remote vehicles need to be considered. For example, the host vehicle, a preceding vehicle and a rear vehicle in the current lane and a preceding vehicle and a rear vehicle in the lane to be moved. In addition, there are studies related with the learning of protocols. In our system, we adopted a pre-defined communication protocol. It is expected to improve safety in a complex lane change environment by introducing a new learning method.

Author Contributions: Project administration, H.A.; supervision, J.-i.J.; writing—original draft, H.A.; writing—review and editing, J.-i.J.

Funding: This work was supported by the National Research Foundation of Korea (NRF) grant, funded by the Korean government (MEST) (NRF-2017R1A2B4012040).

Conflicts of Interest: The authors declare that there are no conflicts of interest regarding the publication of this paper.

References

1. Nilsson, J.; Silvlin, J.; Brannstrom, M.; Coelingh, E.; Fredriksson, J. If, when, and how to perform lane change maneuvers on highways. *IEEE Intell. Transp. Syst.* **2016**, *8*, 68–78. [CrossRef]
2. Cesari, G.; Schildbach, G.; Carvalho, A.; Borrelli, F. Scenario model predictive control for lane change assistance and autonomous driving on highways. *IEEE Intell. Transp. Syst.* **2017**, *9*, 23–35. [CrossRef]
3. Al-Sultan, S.; Al-Doori, M.M.; Al-Bayatti, A.H.; Zedan, H. A comprehensive survey on vehicular ad hoc network. *J. Netw. Comput. Appl.* **2014**, *37*, 380–392. [CrossRef]
4. Ali, G.M.N.; Noor-A-Rahim, M.; Rahman, M.A.; Samantha, S.K.; Chong, P.H.J.; Guan, Y.L. Efficient real-time coding-assisted heterogeneous data access in vehicular networks. *IEEE Internet Things J.* **2018**, *5*, 3499–3512. [CrossRef]

5. Nguyen, H.; Liu, Z.; Jamaludin, D.; Guan, Y. A Semi-Empirical Performance Study of Two-Hop DSRC Message Relaying at Road Intersections. *Information* **2018**, *9*, 147. [CrossRef]
6. Martin-Vega, F.J.; Soret, B.; Aguayo-Torres, M.C.; Kovacs, I.Z.; Gomez, G. Geolocation-based access for vehicular communications: Analysis and optimization via stochastic geometry. *IEEE Trans. Veh. Technol.* **2017**, *67*, 3069–3084. [CrossRef]
7. He, J.; Tang, Z.; Fan, Z.; Zhang, J. Enhanced collision avoidance for distributed LTE vehicle to vehicle broadcast communications. *IEEE Commun. Lett.* **2018**, *22*, 630–633. [CrossRef]
8. Hobert, L.; Festag, A.; Llatser, I.; Altomare, L.; Visintainer, F.; Kovacs, A. Enhancements of V2X communication in support of cooperative autonomous driving. *IEEE Commun. Mag.* **2015**, *53*, 64–70. [CrossRef]
9. Bevly, D.; Cao, X.; Gordon, M.; Ozbilgin, G.; Kari, D.; Nelson, B.; Woodruff, J.; Barth, M.; Murray, C.; Kurt, A. Lane change and merge maneuvers for connected and automated vehicles: A survey. *IEEE Trans. Intell. Veh.* **2016**, *1*, 105–120. [CrossRef]
10. Schwarting, W.; Alonso-Mora, J.; Rus, D. Planning and decision-making for autonomous vehicles. *Annu. Rev. Control Rob. Auton. Syst.* **2018**, *1*, 187–210. [CrossRef]
11. Sutton, R.S.; Barto, A.G. *Reinforcement Learning: An Introduction*, 2nd ed.; MIT Press: Cambridge, MA, USA, 2011.
12. Arulkumaran, K.; Deisenroth, M.P.; Brundage, M.; Bharath, A.A. A brief survey of deep reinforcement learning. *arXiv* **2017**, arXiv:1708.05866. [CrossRef]
13. Wolf, P.; Hubschneider, C.; Weber, M.; Bauer, A.; Härtl, J.; Dürr, F.; Zöllner, J.M. Learning How to Drive in a Real World Simulation with Deep Q-Networks. In Proceedings of the 2017 IEEE Intelligent Vehicles Symposium (IV), Los Angeles, CA, USA, 11–14 June 2017; pp. 244–250.
14. Mnih, V.; Kavukcuoglu, K.; Silver, D.; Rusu, A.A.; Veness, J.; Bellemare, M.G.; Graves, A.; Riedmiller, M.; Fidjeland, A.K.; Ostrovski, G.; et al. Human-level control through deep reinforcement learning. *Nature* **2015**, *518*, 529–533. [CrossRef] [PubMed]
15. Zhu, Y.; Mottaghi, R.; Kolve, E.; Lim, J.J.; Gupta, A.; Fei-Fei, L.; Farhadi, A. Target-Driven Visual Navigation in Indoor Scenes Using Deep Reinforcement Learning. In Proceedings of the 2017 IEEE International Conference on Robotics and Automation (ICRA), Singapore, 29 May–3 June 2017; pp. 3357–3364.
16. Gu, S.; Holly, E.; Lillicrap, T.; Levine, S. Deep Reinforcement Learning for Robotic Manipulation with Asynchronous Off-Policy Updates. In Proceedings of the 2017 IEEE International Conference on Robotics and Automation (ICRA), Singapore, 29 May–3 June 2017; pp. 3389–3396.
17. Wang, P.; Chan, C.-Y. Formulation of Deep Reinforcement Learning Architecture Toward Autonomous Driving for On-Ramp Merge. In Proceedings of the 2017 IEEE 20th International Conference on Intelligent Transportation Systems (ITSC), Yokohama, Japan, 16–19 October 2017; pp. 1–6.
18. Wang, P.; Chan, C.Y.; de La Fortelle, A. A reinforcement learning based approach for automated lane change maneuvers. In Proceedings of the 2018 IEEE Intelligent Vehicles Symposium (IV), Changshu, Suzhou, China, 26 June–1 July 2018; pp. 1379–1384.
19. Hoel, C.-J.; Wolff, K.; Laine, L. Automated Speed and Lane Change Decision Making using Deep Reinforcement Learning. *arXiv* **2018**, arXiv:1803.10056.
20. Kaushik, M.; Prasad, V.; Krishna, K.M.; Ravindran, B. Overtaking Maneuvers in Simulated Highway Driving using Deep Reinforcement Learning. In Proceedings of the 2018 IEEE Intelligent Vehicles Symposium (IV), Changshu, China, 26–30 June 2018; pp. 1885–1890.
21. Mukadam, M.; Cosgun, A.; Nakhaei, A.; Fujimura, K. Tactical decision making for lane changing with deep reinforcement learning. In Proceedings of the NIPS Workshop on Machine Learning for Intelligent Transportation Systems, Long Beach, CA, USA, 9 December 2017.
22. Mirchevska, B.; Pek, C.; Werling, M.; Althoff, M.; Boedecker, J. High-level decision making for safe and reasonable autonomous lane changing using reinforcement learning. In Proceedings of the 2018 21st International Conference on Intelligent Transportation Systems (ITSC), Maui, HI, USA, 4–7 November 2018; pp. 2156–2162.
23. Shah, S.; Dey, D.; Lovett, C.; Kapoor, A. Airsim: High-Fidelity Visual and Physical Simulation for Autonomous Vehicles. In Proceedings of the 11th Conference on Field and Service Robotics, Zürich, Switzerland, 13–15 September 2017; pp. 621–635.

24. Lillicrap, T.P.; Hunt, J.J.; Pritzel, A.; Heess, N.; Erez, T.; Tassa, Y.; Silver, D.; Wierstra, D. Continuous control with deep reinforcement learning. *arXiv* **2015**, arXiv:1509.02971.
25. Lin, L.-J. *Reinforcement Learning for Robots Using Neural Networks;* Carnegie-Mellon University: Pittsburgh, PA, USA, 1993.

Article

Efficiency Analysis and Improvement of an Intelligent Transportation System for the Application in Greenhouse

Tianfan Zhang [1], Weiwen Zhou [2], Fei Meng [3] and Zhe Li [4,*]

1 School of Automation, Northwestern Polytechnical University, Xi'an 710072, China
2 Shanghai Institute of Spaceflight Control Technology, Shanghai 201109, China
3 Department of System Science, University of Shanghai for Science and Technology, Shanghai 200093, China
4 College of Economics and Management, Hubei Engineering University, Xiaogan 432100, China
* Correspondence: lizhe_hbeu@vip.163.com

Received: 30 July 2019; Accepted: 23 August 2019; Published: 28 August 2019

Abstract: In view of the future lack of human resources due to the aging of the population, the automatic, Intelligent Mechatronic Systems (IMSs) and Intelligent Transportation Systems (ITSs) have broad application prospects. However, complex application scenarios and limited open design resources make designing highly efficient ITS systems still a challenging task. In this paper, the optimal load factor solving solution is established. By converting the three user requirements including working distance, time and load into load-related factors, the optimal result can be obtained among system complexity, efficiency and system energy consumption. A specialized visual navigation and motion control system has been proposed to simplify the path planning, navigation and motion control processes and to be accurately calculated in advance, thereby further improving the efficiency of the ITS system. The validity of the efficiency calculation formula and navigation control method proposed in this paper is verified. Under optimal conditions, the actual working mileage is expected to be 99.7%, and the energy consumption is 83.5% of the expected value, which provides sufficient redundancy for the system. In addition, the individual ITS reaches the rated operating efficiency of 95.86%; in other words, one ITS has twice the ability of a single worker. This proves the accuracy and efficiency of the designed ITS system.

Keywords: intelligent transportation systems (ITSs); multiple conditional constraints for ITS; automated guided vehicle; greenhouse environment ; spraying systems

---

1. Introduction

With the aging of the global population and the deepening of urbanization, the contradiction between the loss of agricultural population, rising labor costs and the demand for agricultural production and supply has become an important issue faced by the sustainable development of agricultural production [1]. The development of high-efficiency, high-quality, low-cost intelligent mechatronic systems that can replace human operations is an important means of dealing with the current situation from an engineering perspective [2]. Agricultural labor intensity, high production costs, only harvesting costs can be accounted for more than 25% of total production costs [3,4]. The labor intensity in operations such as pest control is also very high, and directly threatens the health of workers [5]. Automated Guided Vehicle (AGV) as an important part of ITS can effectively reduce the risk of workers in hazardous operations [6], help increase production efficiency and deal with the lack of labor due to the aging of the population [7]. A variety of agricultural robots, including grazing [8], farmingfarming [9], fruit [10] collection, and sorting [11] robots, have been applied.

Compared to industrial robot applications, the development of agricultural robots is limited by the complexity of the work scene and task complexity [12]. The Dutch greenhouses and food factories are widely used and important agricultural production sources [2]. Due to its closed agricultural application scenario, highly standardized, hardened pavement, crop singularity and a high degree of structural work make it an ideal agricultural robot application scenario. However, there is still a lack of more mature applications. Warehousing logistics has similar application conditions to the Dutch greenhouse, although the former environment is much more closed than the latter. However, in the field, especially the development of AGV has important reference [13]. There is no doubt that Amazon's "KIVA" [14,15] and Ali Group's "CaiNiao" [16] ITSs are among the leaders in the field. These robots work in a standardized closed warehouse, which effectively reduces the dependence of human resources, reduce the work intensity of workers, and greatly improves the management efficiency of warehousing and logistics [17].

Even so, the ITSs system for the Dutch greenhouse design is still challenging [13,18]. Although the Dutch greenhouse and storage are semi-closed applications, they have high complexity: (1) firstly, crops rely on natural light rather than constant scattering sources in the warehouse. As time and season change, the lighting conditions in the greenhouse will fluctuate drastically and affect the navigation system. (2) Secondly, the density of crops in the greenhouse is high, the working space is extremely limited, and the irregularity of crop growth makes it possible for the leaves, stems and the like of the crops to hinder the ITS in the established workspace. This requires more redundancy in the design of the ITS. (3) In addition, applications such as pesticide spraying are variable load systems. Moreover, this type of work has strict timeliness requirements, which are determined by the nature of the volatilization and precipitation of pesticides. (4) Finally, due to the patent protection of existing products and research, the technical reference we can obtain is limited, and there are many alternative implementation methods. These factors determine the design of efficient ITSs remains challenging.

In this paper, we propose a high-efficiency ITS design for greenhouse applications, and we can apply this design method to other environments such as warehouses and terminals. Our work makes the following contributions:

- We have provided an optimal load factor solving solution. By converting the three user requirements including working distance, time and load into load-related factors, the optimal result can be obtained among system complexity, efficiency and system energy consumption. The specifications of the main components of the ITS such as drive structure, power and battery components are constrained to a limited range according to the load factor and the work scenario. The specification of constraints helps other researchers determine the design of ITS and the choice of key components.
- We propose a special weighted connected graph structure to support map modeling and navigation applications. Based on this, a specialized visual navigation and motion control system has been proposed. With a fixed control action and boundary-defined Proportion Integral Differential (PID) control algorithm, the repeat path accuracy of the error < 0.02 m can be obtained without relying on third-party positioning. This makes prior path planning, accurate mileage and energy consumption prediction possible. It can help researchers and designers further improve the efficiency of the ITS system.
- We present an efficient example of an ITS system and its detailed design based on the proposed method with optimal calculation results. Test results include working paths and mileage, speed and energy consumption are described in detail under different test conditions, including conditions beyond the rated load. The performance of the system, which is verified in the test environment, is described in detail.

The working environment, requirements, and design goals are described and three known input parameters in detail in Section 3. In Section 4, the overall design framework of the ITS is explored according to the working environment, and then the requirements are converted into three key constraints of load-volume ratio, working distance and work efficiency. In Section 5, a specialized visual navigation and motion control system has been proposed to simplify the path planning, navigation and motion control processes. The ITS prototype designed according to this constraint is elaborated, which is one of the main contributions of this paper. In Section 6, an improved ITS system is implemented as a case, and its detailed design parameters are calculated based on instance greenhouse environment. In addition, then, the validity of the proposed constraints and the efficiency of the ITS prototype are verified by actual verification. Section 8 concludes this paper.

2. Related Works

Agriculture is a labor-intensive industry. With the degeneration of the demographic dividend and the rapid growth of the population, the pressure on agricultural production is increasing. As the most important and labor-intensive operation in agricultural production [5], the pesticides used in pest control have high toxicity and corrosive characteristics [19], which greatly threaten the health of production personnel. Strict and comprehensive protection measures, as well as a lower level of participation of practitioners, further increase the cost of such operations [20]. The United States Environmental Protection Agency (US-EPA) has finalized stronger standards "Agricultural Worker Protection Standard (WPS)" for people who apply for Restricted Use Pesticides (RUPs) [20,21]. These revisions to the Certification of Pesticide Applicators rule will reduce the likelihood of harm from the misapplication of toxic pesticides.

Agricultural automation and intelligent development are one of the effective means of solving these problems [18].The agricultural robot originated from agricultural machinery is a new type of multifunctional agricultural machinery. It is the result of adding intelligent attributes to agricultural machinery [22]. The earliest agricultural robots date back to the autopilot tractors of the 1970s [18]. Up to now, various agricultural robots have emerged, such as shearing robots, milking robots [8], transplanting robots [23–25], grafting robots [23,24], harvesting robots [10,26], and weeding robots [27]. According to the use and operation characteristics, the existing agricultural robots can be roughly divided into livestock management robots, field farming management robots, fruit and vegetable harvesting robots, seedling breeding robots, agricultural product sorting robots, etc.

Due to the complexity of agricultural operations, the complexity of work tasks, and the current level of software and hardware technology development, there are many types of agricultural robots and prototypes, but there are only a handful of cases that can effectively improve the efficiency of production. The more successful cases of agricultural robots are either similar to industrial lines or similar to large agricultural machines such as milking robots, agricultural sorting robots, self-driving tractors, weeding robots, etc. It must be said that, compared with the industrial robot represented by the dexterous mechanical arm [28], the highly intelligent transformers in the traditional concept of human beings [29], there is still a big gap in the development level of agricultural robots. Among the factors such as the agricultural work scene, the complexity of the work tasks and the current development level of hardware and software technology, the complexity of the work scene is the key factor [2,28]. The traditional agricultural production conditions with manual work as the main body do not match the requirements of the robot for the structured working environment, which is an important factor that restricts the agricultural robots from entering the laboratory and entering the farmland. Relative to the industrial structural environment, the special conditions of the sun in the agricultural production environment [12], the disordered crops and the soft and fragile working objects [30] are the objective problems faced by the robots in the agricultural environment [31].

Compared to the open farmland environment, the Dutch greenhouse has the potential to produce under a variety of weather conditions. These closed plant production systems, such as artificially illuminated and highly insulated plant factories, have offered perspectives for urban food production [13]. In terms of economic efficiency, the annual Net Financial Return (NFR) for the Dutch greenhouse is very high [32]. Even so, there are fewer agricultural robot systems available in the greenhouse. Compared with the complex agricultural environment, the industrial application environment is "compendious": the environmental closure and the highly structured operation. This is an important realization condition for the rapid development of industrial robots. Warehousing robots receive special attention among them. There are also a series of intelligent robot systems for greenhouse development, some of which are practical applications, such as Amazon KAVA, Richo M2, River System, Meituan Segway, Postmates, etc. [16,33–45]. Comparison of major ITS systems in recent years is as shown in Table 1.

Table 1. Overview of Intelligent Transportation Systems (ITS) and typical agricultural robot.

Application Field	Name	Size * (L × W × H,m)	Turning Radius (m)	Speed (m/s)	Load (Kg)	Endurance (h)	Open Source
Warehouse	Richo M2 [33]	0.5 × 0.6 × 0.7	0.6	0.5	60	N/A	F
	Kiva [14]	1.0 × 0.6 × 0.4	0.6	1.3	450	N/A	F
	River [15,34]	0.9 × 0.6 × 1.25	0.7	N/A	72	N/A	F
	Butler XL [35]	1.3 × 0.8 × 0.3	1.1	N/A	1300	N/A	F
Delivery	CaiNiao G2 [16]	1.0 × 0.8 × 1.2	0.8	4	50	8	F
	JingDong [36]	1.0 × 0.8 × 0.6	2.2	2	40	6	F
	Yelp EAT24 [46]	1.1 × 0.6 × 1.2	N/A	1.9	12	N/A	F
	Segway [45]	1.2 × 1.0 × 1.2	1.0	11	200	8	T **
	Postmates [37]	0.5 × 0.5 × 0.8	0.5	N/A	25	∼10	F
Agriculture	AGROBOT [38]	4.8 × 6.2 × 3.2	>6	N/A	>100	>10 *	F
	Blueriver [39]	1.2 × 1.0 × 1.0	∼0.8	2.2	N/A	N/A	F
	LettuceBot [40]	4 × 6 × 0.8	∼4.0	N/A	N/A	N/A	F
	RMAX II [41]	3.2 × 3.2 × 0.55	∼3.5	N/A	16	1	F
	BoniRob [47,48]	3.6 ×2.5 × 1.5	∼3.5	N/A	N/A	N/A	T **
Retail	Bossanova	0.3 × 0.3 × 0.5	0.3	0.5	N/A	N/A	F
	ICE RS26 [42]	1.65 × 0.86 × 1.37	>2	1.8	∼130	4	T **
	Navii [43]	0.5 × 0.5 × 1.52	∼0.5	∼0.5	N/A	8∼10	F
	Tally [44]	0.47 × 0.47 × 1.62	∼0.5	∼0.45	N/A	2∼4	F

* These dimensions may be biased due to the lack of published documentation and measurement standards.
** Limited open source.

As shown in Table 1, there are many robot systems that are used in different fields. However, the useful reference for design is very limited. In addition to different application environments and conditions, larger factors may be a limitation of commercial intellectual property. In the above robot system, only the Segway system [45] supports limited open source, and other products or prototype systems do not provide the necessary technical information. Although these systems provide some design references, the role of efficient ITS system design in the Dutch greenhouse application scenario is limited. Therefore, the author's team hopes to analyze the various factors and key conditions in the design one by one based on the project research experience and the basic conditions that need to be given. A normalized conditional formula is given giving detailed design and performance testing, providing a visual reference for other researchers.

3. Problem Description and Objectives

3.1. An Application Case: ITS in a Greenhouse Environment

The application environment comes from an ITS design project of our team. Six Dutch greenhouses with an area of 6000 m^2 are deployed in the research base of the cooperative company. Figure 1 shows the schematic diagram of one of the standard greenhouses. As shown in Figure 1b, we can divide the entire space into three parts: ① operating area, ② working area, and ③ crop growing area. The operating area is mainly used for installation, temporary storage of equipment, and equipment turnover in the working area. Workers are also working in the area. The working area and the crop area are deployed at intervals, and there are $n+1$ crop areas separated by n work areas. The length of each work area is l_n (referred to as height H), the width is d_p and the spacing between work areas is d. The width of the entire working interval is W.

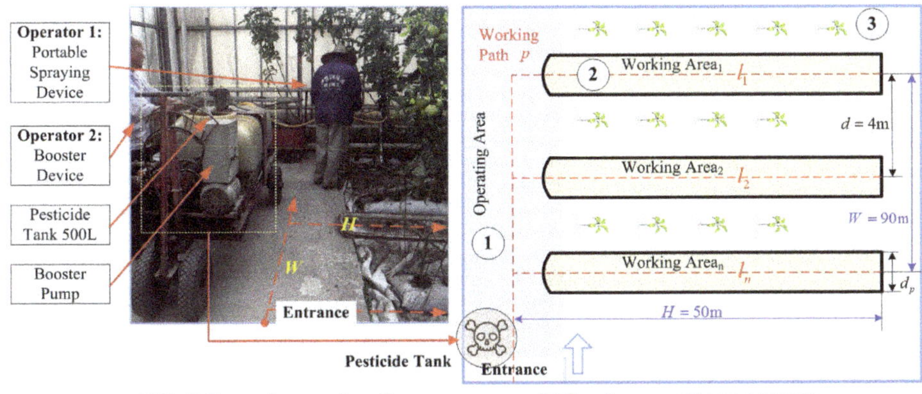

Figure 1. Overview of greenhouse application case.

Taking pest control as an example, the role of pesticide spray control is limited because a large number of insect pests have light-shielding properties and gather behind the foliage. In addition, the role of other control measures is also limited, such as larval boards and other means. This requires workers to use equipment to spray pesticides from the bottom up to remove pests. Corrosive, highly toxic, volatile pesticides are bound to pose a threat to the health of workers, even when protective measures exist. In addition, carrying out the work in such a large range makes the workers have a very high working intensity. This is determined by the characteristics of the pesticide such as volatility and precipitation. The spray work must be completed within the specified time after the pesticide configuration is completed. Otherwise, its medicinal properties will be greatly weakened or even have side effects. Therefore, the use of highly efficient robots to perform similar tasks is not only economical, but also effective in protecting the health of workers.

3.2. Definition of Requirements and Input Conditions

It is challenging to design an ITS that meets user requirements and is optimal especially when there is a limited reference. This requires us to start the analysis from known and user-expected conditions. Firstly, we need to convert the user's requirements into key metrics based on the requirements analysis, as shown in Figure 2. The key metrics are then converted into sub-conditions that affect the ITS design, such as operating speed, load, and output torque, and the boundaries of these conditions are determined. Finally, the optimal values of key conditions are obtained under multiple conditions and their boundary constraints, and the detailed design scheme of ITS is determined.

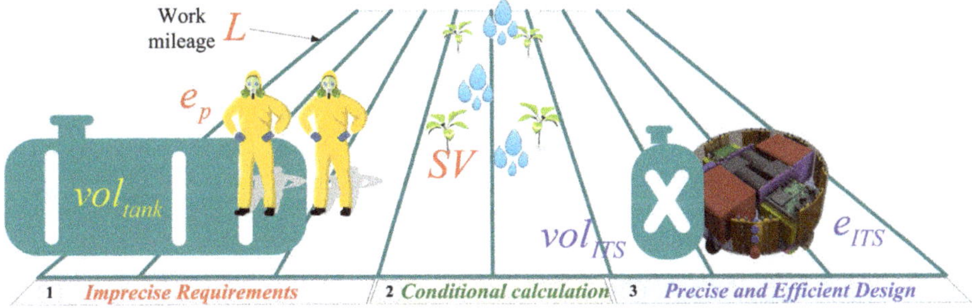

Figure 2. Process between requirements and implementation.

(1) Work efficiency. If the time required for a standard job is T_{norm} and the actual completion time is $T_{practical}$, the productivity e_p can be defined as:

$$e_p = \frac{T_{norm}}{T_{practical}} \times 100\%. \tag{1}$$

This means that the more time is spent, the lower the efficiency. T_{norm} is derived from actual needs and historical work experience, such as $T_{norm} = 5400$ s in this scenario. This is determined by the nature of the liquid and the operating specifications. In addition, in the above conditions, the efficiency of a single worker is $e'_p = 50\%$ because one greenhouse requires two workers to work together. Thus, we can evaluate the efficiency of the designed ITS through e_p.

(2) Work mileage. As shown in Figure 1b, the spraying operation only occurs in the working area, and we refer to such areas as "effective working areas". Obviously, for a given site, the "effective working distance" for each work area is known as $l_1, l_2, \cdots, l_n$. Therefore, the total effective working distance L can be described as the second condition:

$$L = \sum_{i=1}^{n} l_i. \tag{2}$$

If the length of each workspace is the same and the length is H, then $L = n \times H$. The moving distance in the actual job will far exceed this value, whether it is manual or ITS because it needs to work—interval within the interval, switching the working range, and returning the replenishment. In addition, the appropriate ITS system design and path planning system will help reduce these "extra" 'travels'—these will directly affect the actual time spent on the operation $T_{practical}$, which will affect the work efficiency e_p.

(3) Unit workload. Since the total volume of liquid medicine vol_{tank} and the effective working distance L are known, the unit distance dose SV_p can be calculated:

$$SV_p = \frac{vol_{tank}}{L} = \frac{vol_{tank}}{\sum_{i=1}^{n} l_i} = \frac{vol_{tank}}{l_H \times n}. \tag{3}$$

Of course, this value is more appropriately calculated by professional agronomists based on different pests and crops. SV_p will directly interfere with two key issues: (1) determine the specifications of the pumb deployed on the ITS; (2) affect the specification and continuous working time of vol_{ITS}—this will further affect the mission planning of the ITS and the implementation process, and ultimately affect the efficiency of the work.

4. Multiple Conditional Constraints Reasoning

In this section, we will explore in detail and contraction the main conditions affecting efficient MIS design based on known input conditions and expectations. The first problem to be identified is to choose the appropriate overall design for the MIS. An integrated design that differs from existing split operations is more advantageous. As shown in Figure 1b, the hose line connects the separate spray assembly to the pesticide tank assembly in the existing split mode of operation. The main advantage of splite scheme is that the existing syrup supply system can be used directly. In addition, the complexity of the MIS system will be reduced because it does not need to carry a heavy tank and pump system itself. An integrated scheme that differs from existing split operations is more advantageous. However, the solution makes the pipeline system extremely complicated in practical applications. There are two reasons for this: firstly, the ITS needs to enter the working area so that the length of the pipeline will exceed l_H. Secondly, when the ITS switches the working area, the tank also needs to move synchronously to the corresponding operating area, which will increase the complexity of the synchronous control system and the communication system. In comparison, the presticide tank, the sprinkler system, and the ITS system body are brought together in the integrated design. Its main advantage is its compact structure, no external equipment support, and good independence. The key questions here can be expressed using vol_{ITS}, as shown in Figure 3.

ITS volume	Resistance	Output Torque	Motor Power	Motor Speed	Replenishment Count	Practical mileage	Practical Time
vol_{ITS}	F_{res}	F_{out}	P_{moto}	V_{moto}	Pst	L'	$T_{practical}$
Smaller	reducing	reducing	reducing	Fixed	addition	addition	addition
Larger	addition	addition	addition	Fixed	reducing	reducing	reducing

Figure 3. Relationship between main conditions and vol_{ITS}.

It can be seen that, when the fixed V_{moto} is indirectly fixed to the running speed of the AGV, the larger volume is beneficial to reduce the working mileage and working time, which will directly improve the final working efficiency e_p, reducing the distance of reentry due to lesser replenishment count. In addition, the side effect is that it directly increases the system's energy consumption and requires greater output torque—500 L is a big challenge for ITS design. In other words, a viable, efficient ITS system relies on the determination of the optimal vol_{ITS}.

4.1. Load Constraint and Factor

In order to determine the appropriate vol_{ITS}, we will search for favorable clues from known conditions: containers are a huge challenge, but not an incomplete design. However, in the feasible design, we will still find the favorable constraints: First, the width of the working area l_d and the interval spacing will limit the width and length of the MIS and tank, respectively; then, the height and centroid of the tank will follow the volume—increasing and increasing, but the stability of the system will decrease. Finally, greater capacity means greater torque demand and energy consumption pressure.

(1) Load factor. SV_p reflects an indirect relationship between tank and distance, that is, the working distance that the agent in tank can support should be an integer multiple of l_H. It can effectively reduce the number of invalid round trips due to insufficient liquid pesticide. That additional replenishment will generate additional working mileage, which will directly affect the efficiency of ITS. It can be calculated that the amount of medication in a single work area is $vol_{tank}/n = SV_p \times l_H$. In addition, the volume of vol_{veh} should be an integral multiple of it, as a key constraint of an efficient design. We defined m as load factor, and:

$$vol_{ITS} = m \times \frac{vol_{tank}}{n} = m \times SV_p \times l_H, \tag{4}$$

where $m \in [1, n]$ is an integer.

In addition, the number of supplemental options Pst can be calculated when the pump position is fixed:

$$Pst = \lceil \frac{vol_{tank}}{vol_{ITS}} \rceil = \lceil \frac{vol_{tank}}{SV_p \times l_H \times m} \rceil = \lceil \frac{vol_{tank}}{\frac{vol_{tank}}{l_h \times n} \times l_h \times m} \rceil = \lceil \frac{n}{m} \rceil \tag{5}$$

In this way, we simplify the complex capacity problem to the relationship between the two scalars m and n.

4.2. Work Planning and Mileage Calculating

Through Equation (5), we can determine the number of times ITS returns to the pesticide tank, and we can also calculate when it is necessary to return the supply. Therefore, we can divide work planning into two parts: spray and replenishment, as shown in Figure 4. Correspondingly, the calculation of the mileage is also divided into two parts.

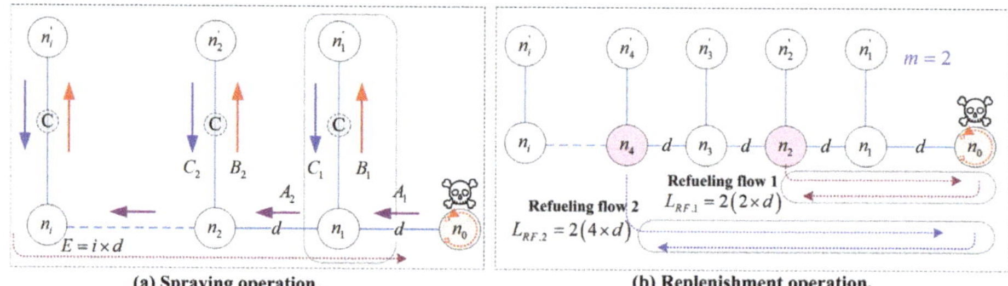

Figure 4. Basic workflow.

(1) Mileage of spray operation.

Each sub spray sequence such as $A_1 \rightarrow B_1 \rightarrow C_1$ consists of three parts: ① entering work area A_1; ② performing spray B_1; ③ returning C_1 ready to enter the next work area. Equation (4) reflects the critical relationship between the operating distances L and vol_{ITS}. In addition, this allows us to ensure that no pesticide shortages occur during the spray step B_1. The spray operation distance $L_{spary.i}$ of each work area can be expressed as:

$$L_{spary.i} = d + l_i \times 2, i \in [1, n]. \tag{6}$$

The total spray operation distance $L_{spary.total}$ can be calculated because the working conditions $l_i = l_H$ are fixed. In addition, we also incorporate the return step E into the process:

$$L_{spary.total} = \sum_{i=1}^{n} L_{spary.i} + L_E = n \times (d + l_H \times 2) + n \times d = 2 \times n(d + l_H). \tag{7}$$

(2) Mileage of refueling operation.

We avoided interrupting the spray operation in the middle of the working area according to Equation (5). At this point, the pesticide exhaustion at node n_i, and the "return-replenishment-reset" process needs to be performed. By defining the serial number of the replenishment as j, we can calculate the line number of the dosing that occurred in i_j via Equation (5):

$$i_j = m \times (j - 1), j \in [1, Pst]. \tag{8}$$

That is, the refueling must occur after one spray is completed. At this point, the ITS will return to the tank to perform the dosing, and its moving distance $L_{spary.i_j}$:

$$L_{supply.i_j} = 2 \times d \times m(j - 1). \tag{9}$$

The total distance of this replenishment operation $L_{supply.total}$ can be calculated:

$$L_{supply.total} = \sum_{i=1}^{Pst} L_{supply.i_j} = \sum_{i=1}^{Pst} 2 \times d \times m(j - 1) = \frac{d \times m \times \lceil \frac{n}{m} \rceil (\lceil \frac{n}{m} \rceil - 1)}{2}. \tag{10}$$

As a key condition, total working mileage L', regarded as an enhancement of the input condition L, can be expressed as:

$$L' = L_{spary.total} + L_{supply.total} = 2 \times n(d + l_H) + \frac{d \times m \times \lceil \frac{n}{m} \rceil (\lceil \frac{n}{m} \rceil - 1)}{2}. \tag{11}$$

4.3. Resistance with Load Capacity

Changes in vol_{ITS} will directly affect the system's resistance F_{res}, and also require output torque $F_{out} > F_{res}$ to drive ITS. Consider the weight WT_{ITS} of the ITS itself and its load WT_{load}, the driving resistance F_{res} can be calculated:

$$F_{out} > F_{res} = (WT_{ITS} + WT_{load})\mu_s e_f g, \tag{12}$$

where the e_f is the empirical coefficient [49], mainly considering the deceleration of motor, transmission part of the loss. In addition, to consider when the work surface water stains will have a reduced friction coefficient, one needs to leave a margin for torque. For example, the efficiency of the primary reducer is about 85~96%. Considering the power loss and redundancy caused by the variable friction coefficient, the effective power is about 70% of the theoretical, that is, $e_f \approx 1.42$. Select the maximum friction coefficient of the asphalt pavement μ_s [50] in order to simplify the calculation. $WT_{load} \approx vol_{ITS}$ when the specific pesticide density is not considered. It can calculate the resistance data under the load and tyre size, according to Equation (12). This is to calculate the motor output torque and even motor drive system selection provided by the basis. Whether a reducer is needed to amplify the output torque depends on the actual size of F_{res}. This will also limit the selection of the motor.

4.4. Speed Condition under Efficiency Constraints

Let us define the average speed of ITS to be $\bar{V}$; then, the actual working time $T_{practical}$ can be calculated by Equation (11):

$$T_{practical} = \frac{L'}{\bar{V}}. \tag{13}$$

Thus, if you want to ensure $e_{ITS} \geq e_p$, ask for $T_{practical} \geq T_{total}$, and we can get:

$$\bar{V} \geq \frac{L'}{T_{total}}. \tag{14}$$

It is more accurate to substitute the total fixed replenishment time $T_{supply} = vol_{tank}/Q$ and the ITS state to determine the loss time T_{loss} into Equation (14). A more accurate average speed can be expressed as $\bar{V}'$:

$$\bar{V}' = \frac{L'}{T_{total} - T_{supply} - T_{loss}}. \tag{15}$$

Then, according to the Equations (12) and (14), the output speed, torque and power of the motor can be pushed back to provide the basis for the selection and further detailed design. The output speed V_{PGR} of the planetary gear reducer (PGR) and the output speed V_{moto} of the motor can be calculated by Equation (16) with different loads according to the rotational speed formula. In addition, we further shrink the selection conditions of the drive system:

$$V_{moto} = V_{PGR} \times Rto_{PGB} = \frac{\bar{V}'}{\pi \times D_{wheel}} \times Rto_{PGB}. \tag{16}$$

4.5. Structural Stability Conditions

A larger tank volume will increase the center of gravity of ITS and affect the stability of the system structure. In actual operation, it is also necessary to consider the ability of ITS to pass through obstacles. In addition, it also needs to consider the system stability under variable load conditions [51]. Calculated in a circular or equivalent circular tank with a radius of r_t, tank height is h_t and vehicle height is h_v. This allows us to calculate the aspect ratio of vehicle diameter and total height, as shown in Figure 5.

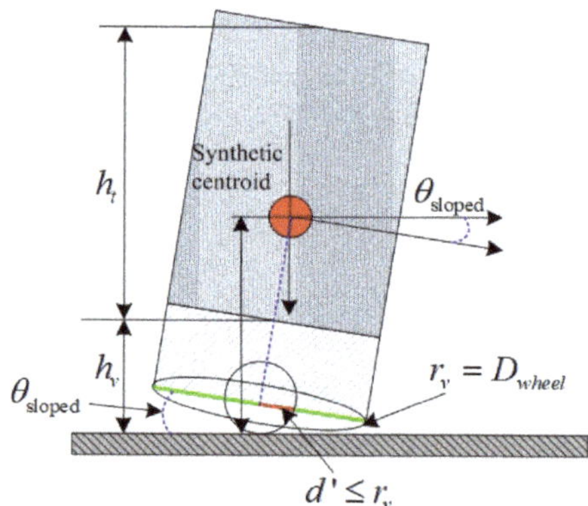

Figure 5. Structural stability conditions.

Excessive height and proportion will affect the stability of the system. Consider the capacity to run the system tilt angle θ_{sloped}. Therefore, the height of the vehicle barycenter can be estimated. At this point, the vertical center of mass must be located within the vehicle support point. If the radius of the AGV is r_v, and the safety height can be calculated taking into account the mass factor h_{mass} :

$$h_{mass} = \frac{r_v \times sin(90^\circ - \theta_{sloped})}{sin\theta_{sloped}}. \tag{17}$$

4.6. Normalization and Optimal Solution of Multiple Conditional Constraints

Although the above conditions can converge the design conditions of the ITS to a large extent, the single is still insufficient to give a specific most preferred result. These conditions are all related to the capacity coefficient m and are divided into two categories: (1) proportional, as vol_{veh}, L, F_{res}; and, (2) inversely, as Pst, $\bar{V}$. Operating speed $\bar{V}$ and output torque F_{res} are the two factors that have the greatest impact on ITS. The speed of work $\bar{V}$ can be replaced by the working mileage L' when the upper limit of the total working time is fixed. In addition, they are all related to the volume of the pesticide tank—further related to the factor m. By mapping them to the same scalar space by normalization, although their units are different, the optimal solution can be calculated. Therefore, taking the factor m as a reference variable, we then employ the Matlab function "mapminmax($parameters, 0, 1$)" to classify the main condition parameters into scalars and compare them.

Through the analysis, we can pass the factors in Sections 3 and 4 through a normalized heuristic method. $\sum_{j=1}^{Pst} L_{spary.i_j}$ in Equation (11) can be simplified using the summation formula of the arithmetic progression $S_n = \frac{n(a_1+a_n)}{2}$, where $a_1, a_2, \cdots, a_n$ is an arithmetic progression. In addition, we can create an optimal evaluation function $f(m)$ related to m because parameters other than m can be obtained through specific examples:

$$\begin{aligned} f(m) &= abs(\text{mapminmax}(L) \Leftrightarrow \text{mapminmax}(F_{res})) \\ &= abs(\text{mapminmax}(2n(d + l_h) + \frac{dm\lceil\frac{n}{m}\rceil(\lceil\frac{n}{m}\rceil - 1)}{2}) \\ &- \text{mapminmax}((WT_{ITS} + \frac{vol_{tank}}{l_h \times n} l_H m)\mu_s e_f G)). \end{aligned} \tag{18}$$

When $f(m) \to 0$, the optimal value of the result and m' can then be calculated.

5. Efficient Visual Navigation and Motion Control System

Before giving the test site and other conditions, we need to briefly introduce the designed ITS navigation system because this will have a direct impact on actual operation and testing. Considering the operating environment in the greenhouse, the design principles of the navigation system are: simple, reliable, accurate and inexpensive. We have proposed and established a special navigation-control scheme which includes a map system, visual navigation system and navigation control system.

5.1. Specialized Undirected Weighted Connected Graph and Fixed Action Control

We propose a special map representation model called: specialized undirected weighted connected graph. As shown in Figure 6a, simplify the expression of the map by adding some restrictions, which is different from the general navigation map expression.

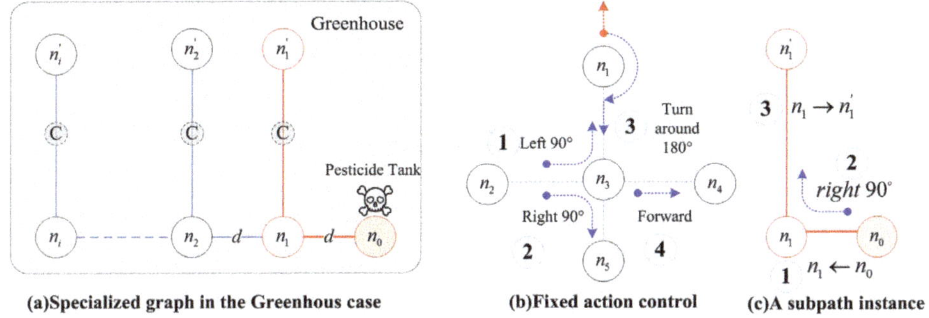

Figure 6. Specialized undirected weighted connected graph and fixed action control.

- All nodes $Node = \{n_1, n_2, \cdots, n_i\}, i \in [1, N]$ in the graph are connected, but loops are not allowed.
- The relative positions between the nodes are limited to four types of up, down, left, and right. For example, for a node N_i, the node adjacent to it is only allowed to appear in its $0°, 90°, 180°$ and $270°$ in four positions. This also limits the direction of ITS movement to four, which can significantly reduce the complexity of graph and control algorithms..
- The association between the nodes is given a weight of w_j according to the distance between the nodes. For example, the weight between nodes n_0 and n_1 is $w_1 = d$, and the weight between nodes n_1 and n_1' is $w_1' = H$. This weight value helps to control the distance heuristics in the algorithm.

This design can significantly reduce the cost of map modeling and path planning algorithms. Since the starting point and the target node are given, only one optimal/shortest path is generated. For example, the shortest path between n_0 and n_1' can be expressed as $n_0 \rightarrow n_1 \rightarrow n_1'$. Nodes can be expressed only by one edge, and there is no need to equidistantly set a large number of auxiliary nodes (a small number of auxiliary nodes that do not participate in the calculation help the system's own state check, avoiding the possibility of falling into the algorithm when moving long distances. In the problem of defects).

Based on this specialized graph navigation map, a "fixed action control" scheme was proposed as shown in Figure 6b. There are only four sports programs here: ① Turn Left 90°; ② Turn Right 90°; ③ Turn Around 180° and ④ Forward. These four types of actions are further divided into two types of events to match the event processing mechanisms: forward and turn. Thus, whenever the vision system recognizes a node identifier, the algorithm will prioritize whether to ignore the point to continue straight or perform a "special turn". Take the $n_2 \rightarrow n_3 \rightarrow n_1 \rightarrow n_3$ process as an example:

First, AGV will perform a forward action until the center of the ITS p overlaps with N_3; then, the "left90" action will be triggered, i.e., the AGV will turn left and its azimuth is 90° to point to n_1. Then, go straight to the n_1 node and trigger the "trun180" action, i.e., ITS rotates 180° in place and points to n_3. Finally, go straight to the n_3 node to complete the task.

This control method is obviously not optimal, such as moving the farthest distance and consuming more time when cornering. However, this simple control method does not require complex algorithm support.

5.2. Vision-Based Navigation System

Each node *Node* or n_i uses a "Quick Response Code (QR Code)" to mark it [17]. Thus, the ITS knows its current relative position, thereby constructing a complete path navigation system and supporting the concrete navigation algorithm. In addition, in a "QR Code Annotated Map" (QR-CAM) visual navigation system based on a specialized graph [52,53], as shown in Figure 7a, guide line connections are used between nodes to provide effective assistance in visual guidance and control.

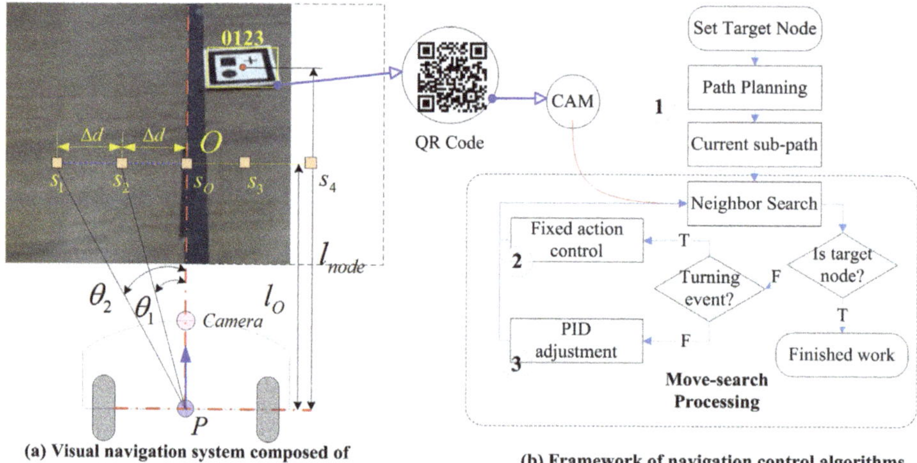

Figure 7. High-precision navigation scheme based on guide line and QR code assisted.

(1) We can determine the actual distance between the center point of the ITS, P, and the QR Code, using the "Pixel-Distance" formula established by camera calibration. ITS will trigger the "turn" event and perform the turn immediately after the l_{node} distance. In addition, when the event is triggered, the cumulative state of the system is reset—the accumulated error is also set to zero, which increases the reliability of the navigation and control algorithms.

(2) Set 5 evaluation points $S = \{s_1, s_2, s_o, s_3, s_4\}$ on both sides of the image center. The distance between the evaluation points is Δd. Thus, when the center of the ITS is near the guide line, its deflection angle can be calculated as $\theta_1 = \arctan(\Delta d / l_O)$ and $\theta_2 = \arctan(2 \times \Delta d / l_O)$. This allows a quick evaluation of the positional relationship between the ITS and the guide line, thus providing a p-value reference for the PID control (if the settings are appropriate, in an ideal environment, only proportional adjustment can be used to complete the effective control of the ITS).

(3) The specific control process is divided into three parts, as shown in Figure 7b. In addition, this process can be called "Setting-Find-Task Execution". At first, the staff sets the task target and work path through the terminal equipment such as the wired or wireless touch panel on the vehicle [54]. In addition, then, the vehicle will then complete the movement between the nodes in units of sub-paths and eventually reach the target node. For example, the n_0 to n_1' path can be split into $n_o \rightarrow n_1$ "forward", "turn right 90°", $n_1 \rightarrow n_1'$ "forward" as shown in Figure 6c. That is, all tasks can be

planned and calculated in advance, just like Programmable Logic Controller (PLC), and each task can be completed in steps—this can greatly reduce the computational burden caused by dynamic programming and real-time control. The motion control only has two simple control processes: ② "turn" and ③ "go forward". This ensures that the vehicle can move precisely along the guide line, inspired by θ_1 and $theta_2$, based on the PID control algorithm. The deviation can be controlled at 0.5% or ±2 m, even if there are machining and assembly errors.

(4) Repeat the "move-search" process according to the planned route until all tasks have been completed or reached a "termination point". (Note: The user can manually set the termination node or when the ITS needs to return to the service station when it detects a failure [55].)

6. Improved ITS System Selection and Implementation

6.1. Qualification Calculation Based on Instance

The example parameters of the greenhouse environment can be given according to the greenhouse structure shown in Figure 1 and the user requirements and input conditions in Section 3. Table 2 lists instances of the main parameters.

Table 2. Instance of parameters in a greenhouse environment.

Params Name	Symbols	Quantity	Params Name	Symbols	Quantity
greenhouse width	W	90 m	pesticide tank	vol_{tank}	500 L
greenhouse height	H	50 m	Pump flow	Q	1.17 L/s
total operating time	T_{total}	1.5 h	working area count	n	22
area path length	$l_H = H$	50 m	operationg area width	d_o	2.5 m
space path length	$l_W = W$	90 m	area path width	d_p	0.6 m
space path spacing	d	4 m	spray volume	SV_p	0.46 L/m
worker efficiency	e'_p	50%	required efficiency	$e_p = 2 \times e'_p$	100%
frictional coefficient (asphalt pavement)	μ_s	0.6 [50]	gravitation-acceleration	g	9.8 m/s
vehicle self-weight	WT_{ITS}	25 Kg	load weight	WT_{load}	$m\times$ 22.7 Kg
wheel size	D_{wheel}	3~16 inches	experience factor	e_f	1.42

The above example parameters are brought into the formula one by one. Bring the parameters listed in Table 2 into the formula one by one, derived according to the theory. In addition, finally, determine the optimized ITS selection design.

1. First of all, the quality factor $m \in [1, 22]$ can be determined according to the working area count $n = 22$.
2. Then, the total working distance can be calculated according to the factor m that $L' \in [2376, 3300]$.
3. In addition, the total effective working time limit can be calculated: $T' = T_{total} - T_{supply} - T_{loss} \in [4640, 4955]$ s, where the total replenishment time is $T_{supply} = vol_{tank}/Q \approx 430$ s. The loss time is $T_{loss} = 15 \times Pst \in [15, 330]$ s and the time for the device status to confirm the fixed loss is about 15 s per replenishment.
4. The average speed of the system can be determined by mileage and time, which is $\bar{V}' \in [0.51, 0.71]$ m/s.

When the three factors of workload, distance, and time are known, the resistance of the ITS system, F_{res}, and the output torque of the motor can be calculated. As shown in Figure 8 by a two-coordinate graph ('plotyy'), here is a comparison of the resistance and the required output torque of three different radius drive wheels with different loads.

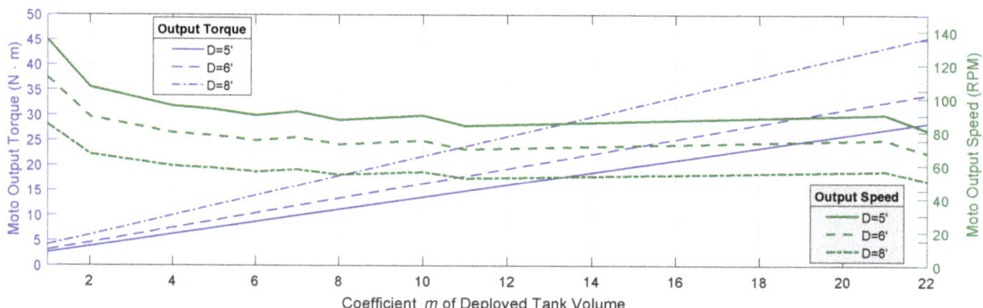

Figure 8. Output torque and speed comparison according to coefficient m and wheel diameter D.

It can be seen that as the load factor m increases: (1) The average output speed of the motor are 55, 72 and 88 rpm with different D_{wheel} or D. Respectively, minor changes when $m > 3$. (2) However, the output torque is increasing rapidly, $Fres \in (13.7 \sim 21.9)$ N · m when $m = 10$. When $m = 22$, it grows to 28.4∼45.5 N · m, which is over the limit of ordinary DC motors and reducers.

Because the general motor is characterized by low torque and low torque, it is difficult to meet the drive of large loads. Therefore, it is necessary to use PGR to increase the output torque. The range of conditions for output torque can be determined by Figure 8 and the motor selection code. "**Limitation conditions (3)**" $F_{out} = 10$ N · m when $m > 10$ to offset the resistance according to Equation (12). In addition, the unit N of F_{res} or F_{out} has been converted to N · m according to the power wheel radius.

The output torque is related to the driving wheel diameter, the motor torque and the deceleration ratio of the reducer. The ITS can only move when $F_{res} \leq F_{out}/D_{wheel}$ (especially static state). In addition, the $V_{PGR} \in [50 \sim 140]$ rpm and $V_{moto} \in [300 \sim 1400]$ rpm based on drive wheel diameter 5–8 inchs.

According to the analysis of Section 4.6, the key conditions of F_{res} and L are combined with factor m, and the results are as shown in Figure 9.

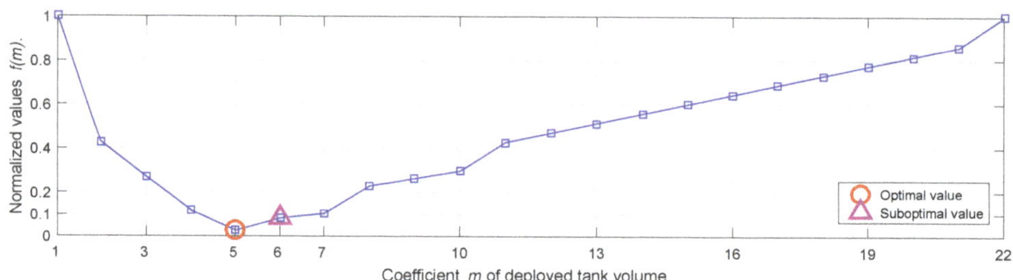

Figure 9. The solution of the optimal value $f(m)$ is based on the load factor m.

Substituting Equation (18) into Equation (4) according to the given condition, we can see that $m' = 5$.

Larger volume capacity provides more redundancy and shorter distance, such as nozzle loss during liquid spray. It should be noted that the excessive load can lead to a significant increase in the probability of ITS system failure and difficulty in diagnosing [56]. Compared to $m' = 5$, when $m' = 6$, the center shifts by only 0.08 m and the height of tank is about 0.65 m, which is still within the design tolerance. In addition, the working distance is reduced by 2.2%.

Based on the obtained optimal coefficient m', the structural stability of the system can be verified. The working area $d_p = 0.6$ m and the bottom radius of the container is $0.4 \leq r_t \leq 0.5$ m. In addition, $r_v \in [0.5, 0.6]$ m. Bring these parameters to Equation (17) to calculate $h_{mass} = 1.45$–2.25 m. In addition, considering that the liquid in the container may be shaken , the safety height should be set at least 80% of the limit value at design time. Therefore, the safety height $h'_{mass} = h_{mass} \times 80\%$ is 1.45–1.8 m.

Because at this time the bottom area of tank is about 0.196 m^2, it can thus be limited to the car cabinet volume within 200 L, that is, $m < 9$. Although this result is larger than the optimal value ($m = 5$) we obtained in the normalization condition calculation, since the height of the vehicle is easy to estimate, the calculation condition can be contracted earlier by this method.

6.2. An ITS Instance for Greenhouse Spraying Application

We designed and implemented a complete ITS system based on the above analysis and given the greenhouse application scenario [55]. Figure 10a shows the schematic diagram of a full-featured prototype. Figure 10b shows a 3D view of the ITS. More information on subsequent versions can be found in 2017—the 15th China International Agricultural Trade Fair (http://www.xinhuanet.com/politics/2017-09/22/c_129710586.htm) and the 24th China Yangling Agricultural High-tech Achievements Expo (http://www.chinanews.com/cj/shipin/cns-d/2017/11-05/news739705.shtml).

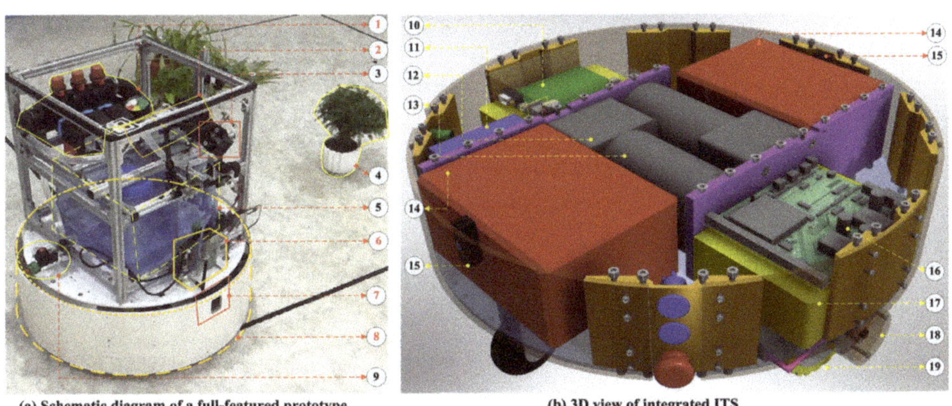

(a) Schematic diagram of a full-featured prototype (b) 3D view of integrated ITS

Figure 10. An ITS instance for greenhouse spraying application.

We can see the composition of the system for pesticide spraying applications by Figure 10a. It consists of nine parts, including a full-featured version of the entire ITS base system, and the final version is the result of this prototype's improvement and optimization. It can be divided into two parts from the perspective of use:

(1) Pesticide spraying subsection. ⑤ Two 15 lwater tanks form the medicine box and the main load; ② a self-priming booster pump (E-CHEN EC-RV-03L) is used to draw the liquid from the chamber and provide sufficient pressure to the ⑨ nozzle. ① A solenoid valve system monitors the system pressure while controlling each nozzle. In addition, a variety of filtration systems are available for different liquids to prevent clogging of the above components.

(2) ITS and control system. In the prototype system, ⑧ the power system and ECU parts of the ITS are integrated into the chassis. The main processer ⑦ (NVIDIA TK1) and the ③ first-view visual sensor (wireless image transmission system based on GoPro Hero 5) are external. Although the ⑦ image sensor (Logitech C920, CH: Switzerland) for visual navigation is installed in the ITS chassis, its data are still processed by ③ [52].

Among them, ③ is used to identify information of each object in the environment, such as the plant or pest as ④ and is also used for remote control of the ITS by the user from the first perspective. ⑦ is mainly responsible for the identification of the node QR-Code and the guide line, and provides upper decision information for path planning and navigation. Eventually, this decision information will be sent to the ECU in the ITS for specific implementation. More details of the actual operation can be seen in the "video abstract".

The ITS (version "BigPan-III") shown in Figure 10b is an upgraded version of the prototype. All the necessary components, including the power supply, drive system and controller, are integrated together, and the high degrees of independence and autonomy are the advantages of this transitional version. Nine major modules are divided into two categories:

(1) Control system. Compared to the prototype, the external intelligent processing device (6) is integrated into the ITS (10) after it has been redesigned and miniaturized. The image sensor (7) is replaced with a smaller camera module (45 (L)*15W (W)*15(H) mm) so that it can be mounted to the (15) position. In addition, the front and rear dual cameras allow ITS to achieve the same level of navigation in an environment that cannot be turned. The ECU (16) is updated to accommodate high temperature and high humidity environments. It integrates four currents, four voltages, three-axis accelerometers and three-axis angle sensors. This information will be combined with the rotary encoder (18) to perform attitude fusion calculations inside the Electronic Control Unit (ECU) (core chip is Freescale MC9S12X, Austin, TX, USA) to obtain the current speed of the ITS (relative) information such as position and power consumption. This information will all be uploaded to the intelligent controller (10) for decision, and the generated issued instructions will be sent to the (11) actuator to drive the actual movement of the ITS.

(2) Power and drive system. Two motors (13) (5PC120GU-24 rated power 120 W, 1800 rpm) are installed in a symmetrical manner, with the 5-inch drive wheel to form a differential drive system. The output section is equipped with a gear unit (12) (5GU10K supports 180 RPM output speed). This provides an output torque of >8.5 N·m at an output speed of 120 rpm (0.75 m/s. A battery (14) (48 V 12 Ah) allows the system to operate for at least one hour at maximum power consumption, while a second backup battery doubles the system's endurance to meet demanding mission's requirements.).

The specifications of an experimental ITS are shown in Table 3.

Table 3. Specifications of an experimental agricultural vehicle.

Components	Specification	Components	Specification
Moto power	$P_{moto} = 120\ \text{W} * 2$	ECU	Freescale MC9S12X $\bar{P}_{ECU} = 5$ W
Moto output torque	$8.5\ \text{N} \cdot \text{m}$	Pump	$P_{pume} = 120$ W (rated)
Total ITS weight	25 Kg	Main Controller	NVIDIA Tegra K1/X2 $\bar{P}'_{ECU} = 10$ W
Rated load	150 Kg 250 Kg MAX	Main battery	48 V 12 Ah $\times$ 1*
Wheel diameter	5 inch	Size (include tank)	0.52(L) $\times$ 0.52(W) $\times$ 0.75(H) m

Therefore, the total power consumption of the system is approximately $P_{MAX} = 2 \times P_{moto} + P_{ECU} + P_{pump} = 410$ W. The total energy demand approximately 615 $\text{W} \cdot \text{h}$. In practice, not all of the time is the maximum power—for example, the ECU and control computer average power consumption of about 15 W. While the pump system works less than half its working time, it can consider its average power consumption as 60 W. According to full load operation, the total power consumption of the system is about $\bar{P}_{total} = 315$ W and the total energy consumption is 472.5 $\text{W} \cdot \text{h}$. Therefore, the optional lithium battery specification is 36 V· 16 Ah or 48 V · 12Ah. Except for the battery, the electricity consumption is lower than 36V, which still meets the safety requirements (GB/t 3805–2008) [57]. Considering the effect of actual loss, battery life and temperature on the battery, the battery with larger capacity should be selected in the actual selection. We can give the overall architecture of the ITS system when the two key components of the motor and battery are determined.

7. System Efficiency Testing and Result Analysis

7.1. Test Environment Settings

We have selected a complete cabin as shown in Figure 1b for system testing. The main parameters are shown in Table 2. It is then converted to a digitized map of the XY markers to facilitate the description of the test results, as shown in Figure 11.

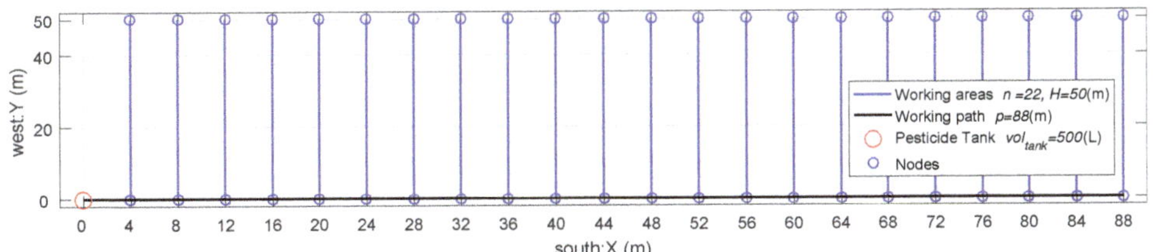

Figure 11. Instance of the greenhouse environment.

In order to more clearly show the movement of the AGV under different loads, i.e., different load factor m. We define $replenishment - work$ as the task group WS. Thus, when $m = 6$, WS_1 can be described according to Section 5:

$$WS_1 = \mathbf{Fill}(n_0) \rightarrow (n_1 \rightarrow n_1' \rightarrow n_1) \rightarrow (n_2 \rightarrow n_2' \rightarrow n_2) \cdots (n_6 \rightarrow n_6' \rightarrow n_6) \rightarrow \mathbf{Fill}(n_0), \tag{19}$$

where a pesticide filling job **Fill()** needs to be performed before each working group starts or end in n_0. For a 70 L/min replenishment pump, it takes about 430 s to complete the filling of $vol_{tank} = 500$ L tank. In addition to $m = 22$, the AGV will return to the medicine box several times for replenishment. Since the total dose is the same, the total dosing time is still 430 s, but this time will be allocated to each replenishment process. Thus, the working distance of each group is $L_{WS_i}, i \in [1, Pst]$ and the total distance L' can be calculated:

$$L_{WS_i} = \begin{cases} 2 \times m \times l_H + 2 \times m \times d, i = 1, \\ 2 \times m \times d \times i + WS_{i-1}, i > 1. \end{cases} \tag{20}$$

The motion control method proposed in this paper divides the motion of ITS into two types: linear and fixed turn. Therefore, through Section 6, we know that the spraying operation and the straight travel speed $\bar{V}_2 \approx 0.67$ m/s. In addition, within the rated load range, the acceleration both are fixed at 0.2 m/s^2.

7.2. DAQ System

As shown in Figure 12 and Table 4, we built a DAQ system [55] to obtain the main operating parameters of the ITS runtime in order to accurately evaluate its work efficiency. An ADI AD7606 DAQ model (8-channel DAS with 16-bit) is integrated into the ECU for the acquisition and conversion of numerical data. Four of them are used for voltage detection, and the other four are used with Allegro ACS758LCB-050B linear hall current sensors (Worcester, MA, USA) for current detection. Two AVAGO HEDS-9140 encoders (San Jose, CA, USA) monitor the operating speed of the two drive wheels. An MPU-6050 triaxial accelerometer sensor is used to obtain ITS acceleration information for Inertial Measurement Unit (IMU) assisted positioning and measurement.

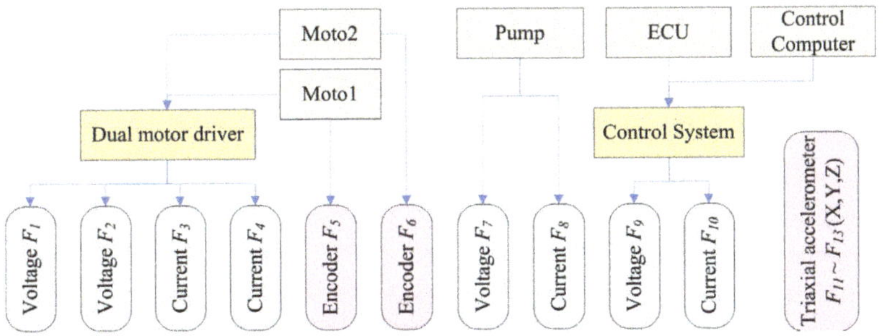

Figure 12. Multi-channel sensor signal acquisition.

Table 4. Sensor parameter setting.

Part	Target	Sensor	Feature	Values	Rate
(1)	Moto 1	Current	F_1	50 mA	
(2)	Moto 1	Voltage	F_2	50 mV	
(3)	Moto 2	Current	F_3	50 mA	
(4)	Moto 2	Voltage	F_4	50 mV	
(5)	Moto 1	Encoder	F_5	1024 PPR	
(6)	Moto 2	Encoder	F_6	1024 PPR	10 Hz
(7)	Pump	Voltage	F_7	50 mV	
(8)	Pump	Current	F_8	50 mA	
(9)	Control System	Voltage	F_9	50 mV	
(10)	Control System	Current	F_10	50 mA	
(11)–(13)	Triaxial accelerometer	XYZ	$F_{11} \sim F_{13}$	16,384 LSB/g	

7.3. System Test and Result Comparison

The test mainly includes five items: working path, total mileage, speed, power and total power consumption. Based on load factor $m = 1 \to 8$, eight sets of effective experiments were carried out here. It should be noted that the rated load of the given ITS instance is $m = 6$, i.e., 138 Kg. The system's power performance will gradually decrease when the rated load is exceeded, such as the system load at $m = 9$. At the limit, the acceleration is only 0.03 m/s, and the maximum speed is only 0.1 m/s. This is much lower than the efficiency users need. Eight sets of valid experiments were carried out, and the corresponding load factor was set to $m = 1, 2, 3, 4, 5, 6, 7, 8$.

(1) The results of the work path are as shown in Figure 13.

The task group represented by the different color trajectories shows the impact of different load factor m on the task flow. Obviously, a larger m makes the task less fragmented.

(2) The results of the center speed of the ITS are as in Figure 14. It can be seen that, due to the existence of the "fixed action control" described in Section 6.1, the speed curve of ITS is very similar to that of PLC, showing a high degree of regularity.

(3) The results of the work mileage comparison are as shown in Figure 15:

It can be seen that, when the load factor m changes, it will cause changes in the WS process, the working path, and the work history, and exhibit similar changes. In addition, the change in total working distance is consistent with that shown in Figure 9, where the total working distance is the shortest when $m = 6$.

A summary of the above main test results are shown in Table 5.

Table 5. Test results and comparison.

Load Factor (m)	Mileage (m)			Consumption	Discharge	Time (s)	Efficiency (%)
	Ideal	Actual	Mileage Error (%)	(Wh)	Rate (%)	Actual	Actual
1	4224	4214.7	99.78	532.8	92.5	8644	62.47
2	3256	3247.6	99.74	524.2	91.0	6775	79.70
3	2880	3040.1	105.56	483.8	84	6353	85.00
4	2856	2968.1	103.94	496.5	86.2	6191	87.22
5	2616	2768.5	105.83	490.2	85.1	5831	92.61
6	2664	2656.5	99.72	481.1	83.5	5633	95.86
7	2544	2704.8	106.32	484.4	84.1	5701	94.72
8 *	2568	2712	105.61	535.7	93	5749	93.93

* Exceeding rated load.

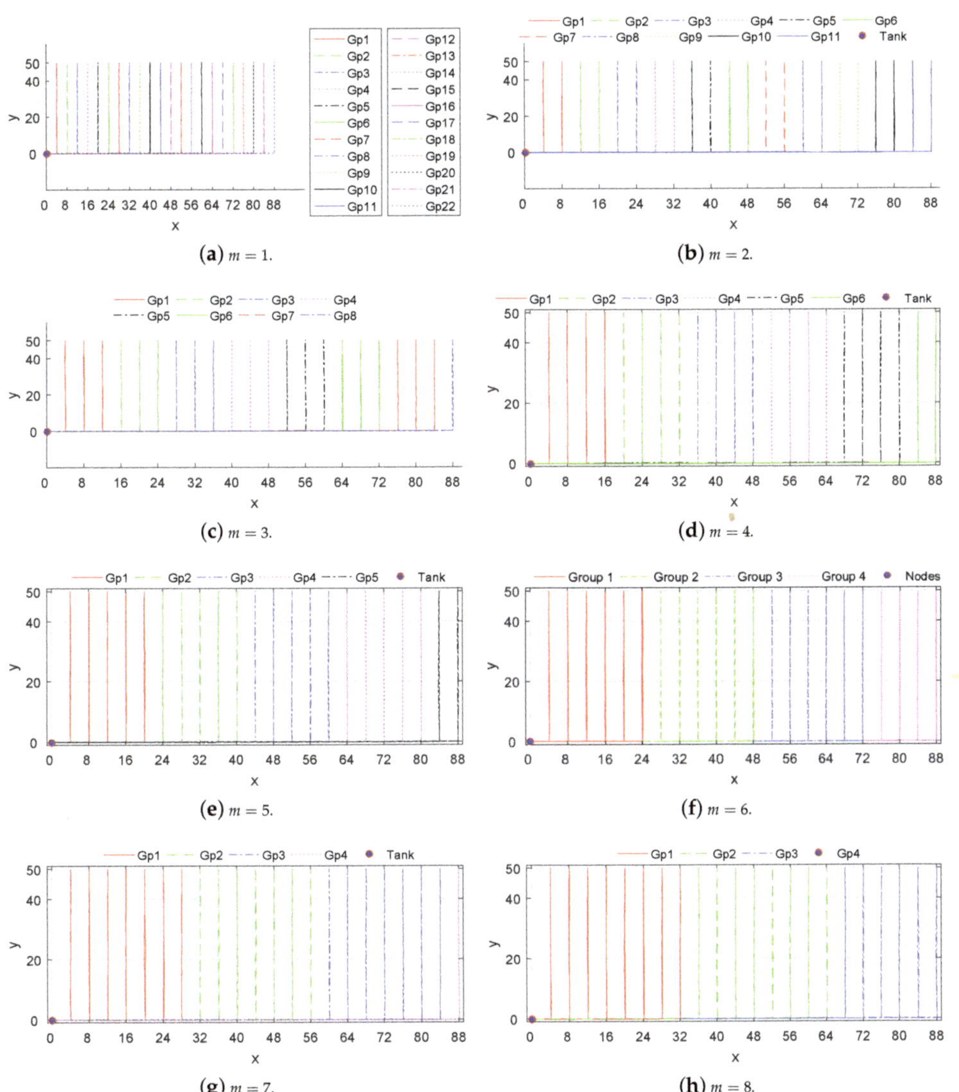

(a) $m = 1$. **(b)** $m = 2$. **(c)** $m = 3$. **(d)** $m = 4$. **(e)** $m = 5$. **(f)** $m = 6$. **(g)** $m = 7$. **(h)** $m = 8$.

Figure 13. Results of the work path.

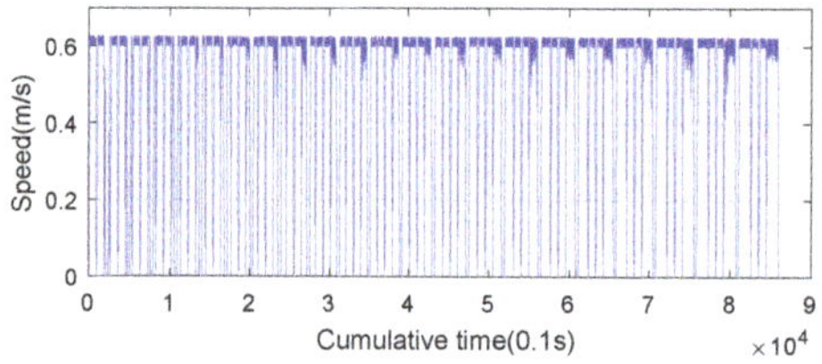

(**a**) $m = 1$.

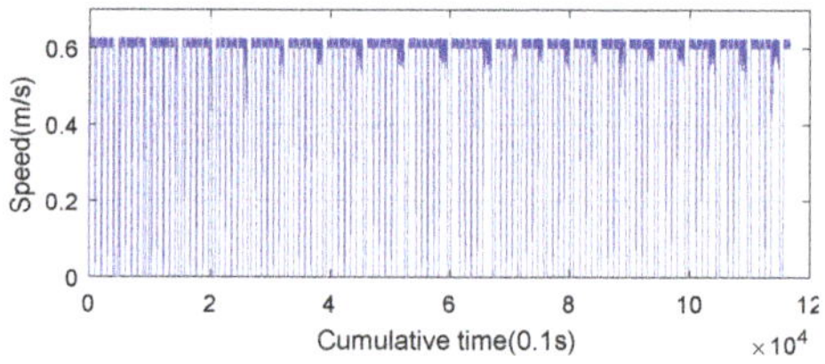

(**b**) $m = 2$.

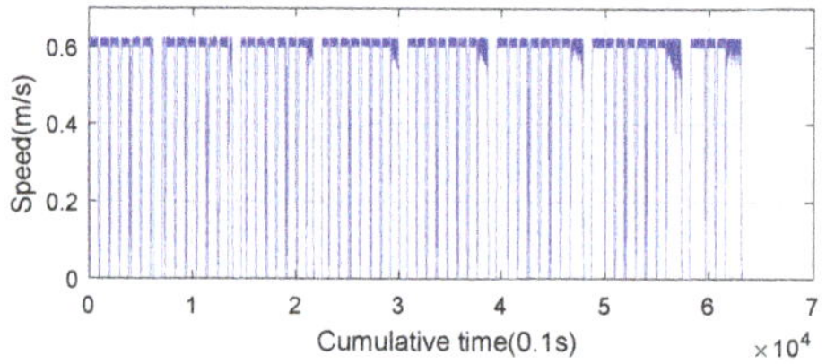

(**c**) $m = 3$.

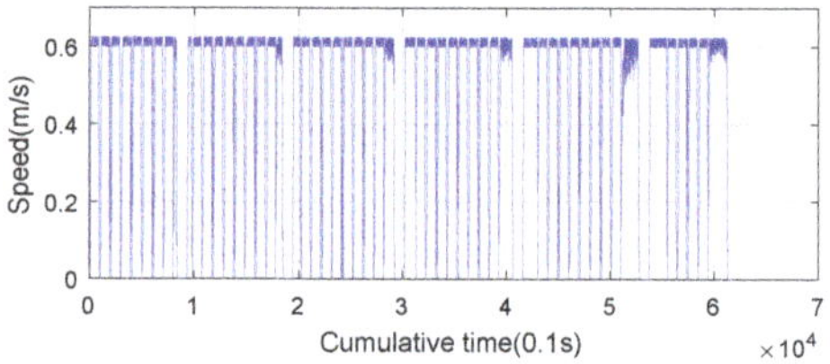

(**d**) $m = 4$.

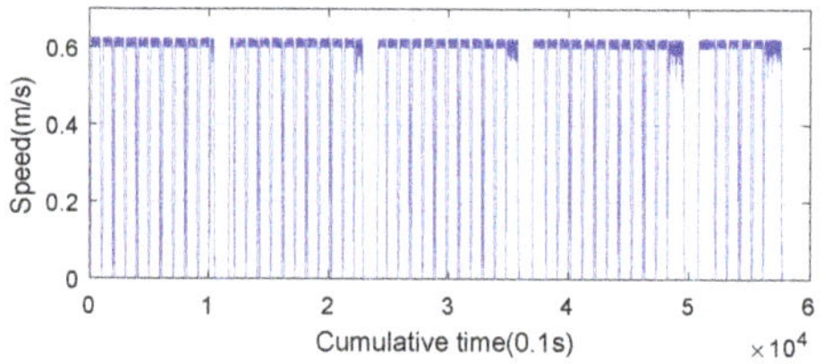

(**e**) $m = 5$.

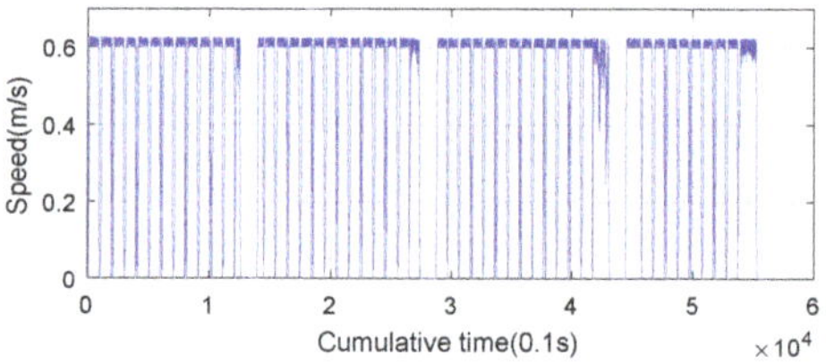

(**f**) $m = 6$.

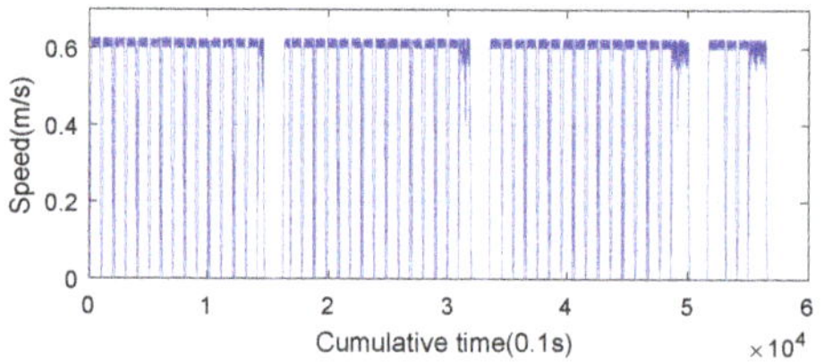

(**g**) $m = 7$.

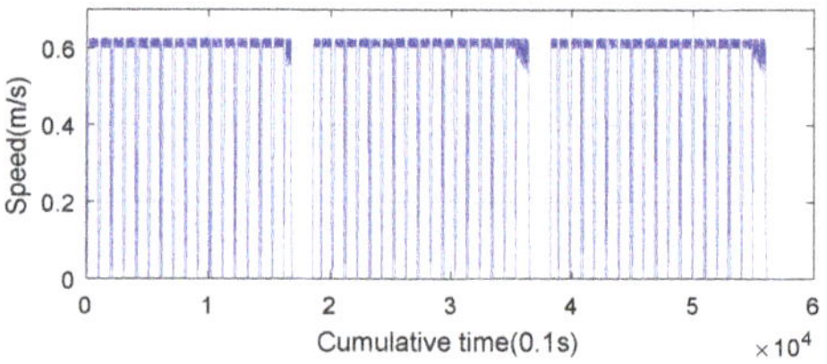

(**h**) $m = 8$.

Figure 14. Comparison of working speed under different load factor m.

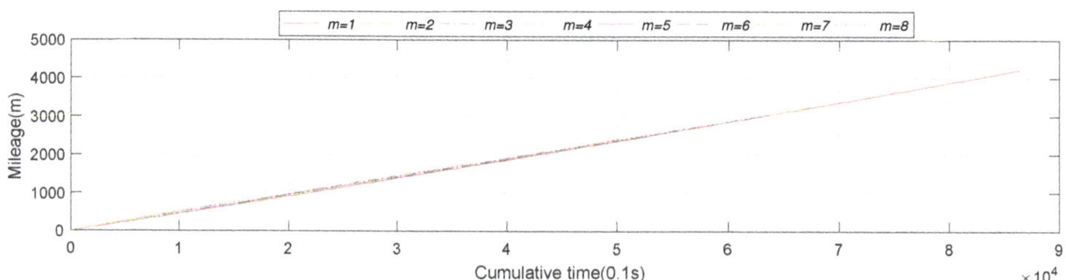

Figure 15. Total mileage comparison at different m.

We can draw the following conclusions:

- The trend of the above results is consistent with that shown in Figure 9, indicating that the optimal Equation (18) expression is valid for the design of the ITS system.
- When $m = 5$, the main indicators of the system are optimal, and the $m = 6$ sub-optimal result is only 3% different. This provides some redundancy for system design.
- The average error between the actual working mileage and the ideal calculated value is only 2%, which proves the accuracy of the navigation and control system designed in this paper. When $m = 6$, the actual working distance is 99.7% of the ideal value. Due to the presence of sensor errors and the response speed of the PID controller, the AGV is unlikely to reach exactly above the target node. It is possible for the AGV to stop during the previous sample-execution cycle or to stop during the next cycle. Thus, the AGV has a $Deltad < 0.1$ m error from the target node, and the accumulation of the error will cause a positive and negative deviation between the total mileage of the AGV and the ideal value.
- In the above test of m, although the predicted value is exceeded by about 5%, the total system energy consumption is lower than the battery power supply capacity. Existing power supply systems are feasible, especially if a backup battery is enabled.
- In terms of efficiency, $e_{ITS} = 95.86\%$ when $m = 6$ is equivalent to two workers due to the maximum output speed limit, in other cases in which system efficiency is lower than e_p, but still higher than the efficiency of a single worker e'_p. This proves the high efficiency of the ITS system designed in this paper.

The test results demonstrate the effectiveness of the design method presented in this paper. The proposed ITS system demonstrates features such as efficiency, accuracy, and energy control for the application in greenhouse.

8. Conclusions

In this work, a high-efficiency ITS system for greenhouse applications based on our proposed ITS efficiency formula was designed and validated. Designers can significantly reduce the range of critical components such as drive systems, energy supply systems, ITS mechanical frames, etc., by shrinking complex requirements to three main conditions. This provides guidance for the rapid design of high efficiency ITS systems required for composite applications. First, we present a real greenhouse application scenario. Then, according to the scenario, the designing process of the efficiency formula is elaborated. Finally, the efficiency of the designed ITS system is proved by experiments. In the future, the authors will further study the efficient ITS design issues in complex application environments (not limited to greenhouse applications); in addition, multi-ITS collaboration is an important research direction to further improve work efficiency.

Author Contributions: Project administration, F.M. and T.Z.; data curation, Z.L. and W.Z.; funding acquisition, F.M. and W.Z.; methodology, T.Z. and W.Z.; resources, F.M.; software, Z.L.; validation, T.Z. and W.Z.; writing, original draft, T.Z.; writing, review and editing, Z.L. and F.M.

Funding: This research was partially supported by the Natural Science Foundation of Hubei Provincial (2018CFB314), the Science and Technology Commission of Shanghai Municipality (16391902702), and "Shuguang Program" supported by the Shanghai Education Development Foundation and Shanghai Municipal Education Commission (No.15SG43)—scientific Research Project of College of Technology, Hubei Engineering University(2016Hgxky01).

Conflicts of Interest: The authors declare no conflict of interest.

Abbreviations

The following abbreviations are used in this manuscript:

AC	Alternating Current Power
ADC	Analog-to-Digital Converter
AGV	Automated Guided Vehicle
AI	Artificial Intelligence
AWPS	Agricultural Worker Protection Standard
CAN	Controller Area Network
CPU	Central Processing Unit
CUDA	Compute Unified Device Architecture
DAQ	Data Acquisition
DC	Direct Current Power
ECU	Electronic Control Unit
IMS	Intelligent Mechatronic Systems
IMU	Inertial Measurement Unit
ITS	Intelligent Transportation Systems
MMC-AGV	Multiple Conditional Constraints for AGV
MOSFET	Metal-Oxide-Semiconductor Field-Effect Transistor
NFR	Net Financial Return
PGR	Planetary Gearboxes Reducer
PLC	Planetary Gearboxes Reducer
QR	Quick Response
RUPs	Restricted Use Pesticides
PWM	Pulse Width Modulation
US-EPA	United States Environmental Protection Agency

References

1. Heller, P.S. The Challenge of an Aged and Shrinking Population: Lessons to be Srawn from Japan's Experience. *J. Econ. Ageing* **2016**, *8*, 85–93. [CrossRef]
2. Wang, R.; Sun, B. Development Status and Expectation of Agricultural Robot. *Bull. Chin. Acad. Sci.* **2015**, *30*, 803–809.
3. Font, D.; Palleja, T.; Tresanchez, M.; Runcan, D.;Moreno, J.M.;Martinez, D.; Teixido, M.; Palacin, J. A Proposal for Automatic Fruit Harvesting by Combining a Low Cost Stereovision Camera and a Robotic Arm. *Sensors* **2014**, *14*, 11557–11579. [CrossRef] [PubMed]
4. Feng, Q.; Cheng, W.; Zhou, J.; Wang, X. Design of Structured-light Vision System for Tomato Harvesting Robot. *Int. J. Agric. Biol. Eng.* **2014**, *7*, 19–26.
5. Gil, E.; Arno, J.; Llorens, J.; Llop, J.; Rosellpolo, J.R.; Gallart, M.; Escola, A. Advanced Technologies for the Improvement of Spray Application Techniques in Spanish Viticulture: An Overview. *Sensors* **2014**, *14*, 691–708. [CrossRef] [PubMed]
6. Silva, T.; Dias, L.M.; Nunes, M.L.; Pereira, G.; Sampaio, P.; Oliveira, J.A.; Martins, P.J. Simulation and economic analysis of an AGV system as a mean of transport of warehouse waste in an automotive OEM. In Proceedings of the 19th International IEEE Conference on Intelligent Transportation Systems (ITSC) 2016, Rio de Janeiro, Brazil, 1–4 November 2016; pp. 241–246.
7. Rampini, A.A.; Viswanathan, S. Financial Intermediary Capital. *Rev. Econ. Stud.* **2019**, *86*, 413–455. [CrossRef]
8. Drach, U.; Halachmi, I.; Pnini, T.; Izhaki, I.; Degani, A. Automatic Herding Reduces Labour and Increases Milking Frequency in Robotic Milking. *Biosyst. Eng.* **2017**, *155*, 134–141. [CrossRef]
9. Cancar, L.; Sanz, D.; Hernandez, D.D.; Cerro, J.D.; Barrientos, A. Precision Humidity and Temperature Measuring in Farming Using Newer Ground Mobile Robots. *Robot* **2014**, 443–456.
10. Zujevs, A.; Osadcuks, V.; Ahrendt, P. Trends in Robotic Sensor Technologies for Fruit Harvesting: 2010–2015. *Procedia Comput. Sci.* **2015**, *77*, 227–233. [CrossRef]

11. Struik, P.C.; Kuyper, T.W. Sustainable Intensification in Agriculture: The Richer Shade of Green. A review. *Agron. Sustain. Dev.* **2017**, *37*, 39. [CrossRef]
12. Emmi, L.; Gonzalezdesoto, M.; Pajares, G.; Gonzalezdesantos, P. Integrating Sensory/Actuation Systems in Agricultural Vehicles. *Sensors* **2014**, *14*, 4014–4049. [CrossRef] [PubMed]
13. Graamans, L.; Baeza, E.J.; Den, D.A.V.; Tsafaras, I.; Stanghellini, C. Plant Factories Versus Greenhouses: Comparison of Resource Use Efficiency. *Agric. Syst.* **2018**, *160*, 31–43. [CrossRef]
14. Li, J.; Liu, H. Design Optimization of Amazon Robotics. *Autom. Control Intell. Syst.* **2016**, *4*, 48–52. [CrossRef]
15. Amazon Robotics. Amazon's Kiva Robotics. Technical Report, Mind Commerce. 2016. Available online: https://robots.ieee.org/robots/kiva/ (accessed on 10 August 2016).
16. Caiiao. The "Cainiao-G" Is Specially Designed to Solve the Last Mile Delivery at the End of the Express Delivery Industry. Technical Report, Mind Commerce. 2016. Available online: https://www.cainiao.com/markets/cnwww/lab-xiao-g (accessed on 1 September 2016).
17. Wurman, P.R.; Dandrea, R.; Mountz, M. Coordinating Hundreds of Cooperative, Autonomous Vehicles in Warehouses. *AI Mag.* **2008**, *29*, 9–19.
18. Krasniqi, X.; Hajrizi, E. Use of IoT Technology to Drive the Automotive Industry from Connected to Full Autonomous Vehicles. *IFAC-PapersOnLine* **2016**, *49*, 269–274. [CrossRef]
19. Tago, D.; Andersson, H.; Treich, N. Pesticides and Health: A Review of Evidence on Health Effects, Valuation of Risks, and Benefit-Cost Analysis. *Adv. Health Econ. Health Serv. Res.* **2014**, *24*, 203–295. [PubMed]
20. Toumi, K.; Joly, L.; Vleminckx, C.; Schiffers, B. Exposure of Workers to Pesticide Residues During Re-entry Activities: A review. *Hum. Ecol. Risk Assess.* **2019**, 1–23. [CrossRef]
21. EPC. Revised Certification Standards for Pesticide Applicators. Technical Report, Mind Commerce. 2017. Available online: https://www.epa.gov/pesticide-worker-safety/revised-certification-standards-pesticide-applicators (accessed on 4 January 2017).
22. Kondo, N. Automation on Fruit and Vegetable Grading System and FoodTraceability. *Trends Food Sci. Technol.* **2010**, *21*, 145–152. [CrossRef]
23. Lee, J.M.; Kubota, C.; Tsao, S.J.; Bie, Z.; Echevarria, P.H.; Morra, L.; Oda, M. Current Status of Vegetable Grafting: Diffusion, Grafting Techniques, Automation. *Sci. Hortic.* **2010**, *127*, 93–105. [CrossRef]
24. Rouphael, Y.; Kyriacou, M.C.; Colla, G. Vegetable Grafting: A Toolbox for Securing Yield Stability under Multiple Stress Conditions. *Front. Plant Sci.* **2018**, *8*, 2255. [CrossRef]
25. Aravind, K.R.; Raja, P.; Perezruiz, M. Task-based agricultural mobile robots in arable farming: A review. *Span. J. Agric. Res.* **2017**, *15*, 1–16. [CrossRef]
26. Kishore, N.; Avinash, V.; Chennakeshwar.; Pranaka, P.; Reddy, V.K.; Srinivaslu, S.; Ravindran, M. Multi-Purpose Agricultural Robot. *Int. J. Adv. Res. Ideas Innov. Technol.* **2017**, *3*, 1696–1701.
27. Xiong, Y.; Ge, Y.; Liang, Y.; Blackmore, S. Development of a prototype robot and fast path-planning algorithm for static laser weeding. *Comput. Electron. Agric.* **2017**, *142*, 494–503. [CrossRef]
28. Roblagomez, S.; Becerra, V.M.; Llata, J.R.; Gonzalezsarabia, E.; Torreferrero, C.; Perezoria, J. Working Together: A Review on Safe Human-Robot Collaboration in Industrial Environments. *IEEE Access* **2017**, *5*, 26754–26773. [CrossRef]
29. Massa, D.; Callegari, M.; Cristalli, C. Manual Guidance for Industrial Robot Programming. *Ind. Robot- Int. J.* **2015**, *42*, 457–465. [CrossRef]
30. Garciasantillan, I.; Guerrero, J.M.; Montalvo, M.; Pajares, G. Curved and Straight Crop Row Detection by Accumulation of Green Pixels From Images in Maize Fields. *Precis. Agric.* **2018**, *19*, 18–41. [CrossRef]
31. Christian, F.; Johannes, M.; Christian, F. Discrimination of plants and weed by multi-sensor fusion on an agricultural robot. In Proceedings of the International Conference of Agricultural Engineering, Zurich, Switzerland, 6–10 July 2014; pp. 1–8.
32. Vanthoor, B.H.E.; Stigter, J.D.; Van, H.E.J.; Stanghellini, C.; De Visser, P.H.B.; Hemming, S. A Methodology for Model-based Greenhouse Design: Part 5, Greenhouse Design Optimisation for Southern-Spanish and Dutch Conditions. *Biosyst. Eng.* **2012**, *111*, 350–368. [CrossRef]
33. Richo. Richo M2 Industry AGV. Technical Report, Mind Commerce. 2015. Available online: https://industry.ricoh.com/agv/ (accessed on 1 March 2015).
34. RiverSystem. River AGV. Technical Report, Mind Commerce. 2016. Available online: https://6river.com/ (accessed on 29 March 2016).

35. Christina, L. GreyOrange to Unveil New Butler XL for End-to-End Supply Chain Automation in Larger Warehouses. Technical Report, Mind Commerce. 2018. Available online: https://www.greyorange.com/press/GreyOrange-to-unveil-new-Butler-XL-for-end-to-end-supply-chain-automation-in-larger-warehouses (accessed on 27 February 2018).
36. Finance, D.T. "Cainiao" VS "Jingdong": Which is the Logistics Robot? Technical Report, Mind Commerce. 2016. Available online: https://www.yicai.com/news/5106442.html (accessed on 19 September 2016).
37. Arielle, P.G. Postmates Robotics. Technical Report, Mind Commerce. 2018. Available online: https://www.wired.com/story/postmates-delivery-robot-serve/ (accessed on 13 December 2018).
38. Durmus, H.; Gunes, E.O.; Kirci, M.; Ustundag, B.B. The Design of General Purpose Autonomous Agricultural Mobile-Robot: "AGROBOT". In Proceedings of the 4th International Conference on Agro-Geoinformatics (Agro-Geoinformatics 2015), Istanbul, Turkey, 20–24 June 2015; pp. 49–53.
39. Stacey, H. Blue River Gets $3.1M for a Weed-Whacking Robot. Technical Report, Mind Commerce. 2018. Available online: https://gigaom.com/2012/09/10/blue-river-gets-3-1m-for-a-weed-whacking-robot/ (accessed on 10 September 2012).
40. Vegetable Growers News. LettuceBot Wins 2017 AE50 Award. Technical Report, Mind Commerce. 2018. Available online: https://vegetablegrowersnews.com/news/lettucebot-wins-2017-ae50-award// (accessed on 3 January 2017).
41. Yamaha. The Yamaha RMAX is the High Performance Standard in Remotely-Piloted Helicopters Designed for Agriculture Purposes. Technical Report, Mind Commerce. 2016. Available online: https://www.yamahamotorsports.com/motorsports/pages/precision-agriculture-rmax (accessed on 5 March 2016).
42. Braincorp. Commercial Floor Scrubbers Powered by BrainOS. Technical Report, Mind Commerce. 2018. Available online: https://www.braincorp.com/intelligent-floor-scrubbers (accessed on 12 October 2018).
43. LoweBot. Robots in Retail. Technical Report, Mind Commerce. 2019. Available online: https://www.fellowrobots.com/navii-2/ (accessed on 11 May 2019).
44. Simbe. Say Hello to Tally. Technical Report, Mind Commerce. 2019. Available online: https://www.simberobotics.com/platform/tally/ (accessed on 20 March 2019).
45. Dude. Out-of-Sale Distribution Scenes Shake the Future of Unmanned Driving Technical Report, Mind Commerce. 2018. Available online: https://www.leiphone.com/news/201806/ZD8u4PMNC3vZfjS1.html (accessed on 27 June 2018).
46. Steve, C. Marble Delivery Robots Working for Yelp Eat24 in San Francisco. Technical Report, Mind Commerce. 2017. Available online: https://www.roboticsbusinessreview.com/rbr/marble_delivery_robots_yelp_eat24/ (accessed on 12 April 2017).
47. Peter, B.; Ulrich, W.; Michael, D.; Amos, A. Navigation System of the Autonomous Agricultural Robot "BoniRob". In Proceedings of the IEEE/RSJ International Conference on Intelligent Robots and Systems (IROS 2012), Vilamoura, Algarve, Portugal, 7–12 October 2012; pp. 1–7.
48. Simbe. Workshop on Agricultural Robotics: Enabling Safe, Efficient, Affordable Robots for Food Production. Technical Report, Mind Commerce. 2012. Available online: https://www.cs.cmu.edu/~mbergerm/agrobotics2012/ (accessed on 11 October 2012).
49. Finley, W.R.; Hodowanec, M.M.; Holter, W.G. An Analytical Approach to Solving Motor Vibration Problems. In Proceedings of the IEEE Industry Applications Society 46th Annual Petroleum and Chemical Technical Conference, San Diego, CA, USA, 13–15 September 1999; pp. 217–232.
50. Chen, L.; Luo, Y.M.; Bian, M.;Qin, Z.;Luo, J.;Li, K. Estimation of Tire-road Friction Coefficient based on Frequency Domain Data Fusion. *Mech. Syst. Signal Process.* **2017**, *85*, 177–192. [CrossRef]
51. Boada, B.L.; Boada, M.J.L.; Zhang, H. Sensor Fusion Based on a Dual Kalman Filter for Estimation of Road Irregularities and Vehicle Mass Under Static and Dynamic Conditions. *IEEE/ASME Trans. Mechatronics* **2019**, *24*, 1075–1086. [CrossRef]
52. Milioto, A.; Lottes, P.; Stachniss, C. Real-Time Semantic Segmentation of Crop and Weed for Precision Agriculture Robots Leveraging Background Knowledge in CNNs. In Proceedings of the IEEE International Conference on Robotics and Automation (ICRA) 2018, Brisbane, QLD, Australia, 23–27 May 2018; pp. 2229–2235.
53. Li, Z.; Zhang, H.; Mu, D.; Guo, L. Random Time Delay Effect on Out-of-Sequence Measurements. *IEEE Access* **2016**, *4*, 7509–7518. [CrossRef]

54. Zhang, T.; Li, Z.; Cheng, Z.;Jing, X. Design and performance verification of an optimized multi-agent system. *Eur. J. Electr. Eng.* **2019**, *21*, 99–105. [CrossRef]
55. Zhang, T.; Li, Z.; Deng, Z.; Hu, B. Hybrid Data Fusion DBN for Intelligent Fault Diagnosis of Vehicle Reducers. *Sensors* **2019**, *19*, 2504–2527. [CrossRef] [PubMed]
56. Zhang, H.; Wang, J. Active Steering Actuator Fault Detection for an Automatically-Steered Electric Ground Vehicle. *IEEE Trans. Veh. Technol.* **2017**, *66*, 3685–3702. [CrossRef]
57. China Standardization Administration. Extra-low voltage(ELV)—Limit values(GB/T 3805-2008). Technical Report, Mind Commerce. 2008. Available online: http://www.gb688.cn/bzgk/gb/newGbInfo?hcno=2A3598C5E4A0EB6BDBD6D1BD52681A6A (accessed on 22 January 2008).

MDPI
St. Alban-Anlage 66
4052 Basel
Switzerland
Tel. +41 61 683 77 34
Fax +41 61 302 89 18
www.mdpi.com

Electronics Editorial Office
E-mail: electronics@mdpi.com
www.mdpi.com/journal/electronics

www.ingramcontent.com/pod-product-compliance
Lightning Source LLC
LaVergne TN
LVHW070143170726
843515LV00007B/1853
* 9 7 8 3 0 3 6 5 0 5 0 6 0 *